Grassland Management for Sustainable Agroecosystems

Grassland Management for Sustainable Agroecosystems

Special Issue Editors
Abad Chabbi
Gianni Bellocchi

MDPI • Basel • Beijing • Wuhan • Barcelona • Belgrade

Special Issue Editors
Abad Chabbi
Institut national de recherche pour l'agriculture,
l'alimentation et l'environnement (INRAE)
France

Gianni Bellocchi
Institut national de recherche pour l'agriculture,
l'alimentation et l'environnement (INRAE)
France

Editorial Office
MDPI
St. Alban-Anlage 66
4052 Basel, Switzerland

This is a reprint of articles from the Special Issue published online in the open access journal *Agronomy* (ISSN 2073-4395) from 2018 to 2020 (available at: https://www.mdpi.com/journal/agronomy/special_issues/grassland_management_sustainable_agroecosystems).

For citation purposes, cite each article independently as indicated on the article page online and as indicated below:

LastName, A.A.; LastName, B.B.; LastName, C.C. Article Title. *Journal Name* **Year**, *Article Number*, Page Range.

ISBN 978-3-03928-222-7 (Pbk)
ISBN 978-3-03928-223-4 (PDF)

Cover image courtesy of Célia Pouget.
Institut national de recherche pour l'agriculture, l'alimentation et l'environnement (INRAE)

© 2020 by the authors. Articles in this book are Open Access and distributed under the Creative Commons Attribution (CC BY) license, which allows users to download, copy and build upon published articles, as long as the author and publisher are properly credited, which ensures maximum dissemination and a wider impact of our publications.

The book as a whole is distributed by MDPI under the terms and conditions of the Creative Commons license CC BY-NC-ND.

Contents

About the Special Issue Editors vii

Gianni Bellocchi and Abad Chabbi
Grassland Management for Sustainable Agroecosystems
Reprinted from: *Agronomy* **2020**, *10*, 78, doi:10.3390/agronomy10010078 1

Jean L. Steiner, Patrick J. Starks, James P.S. Neel, Brian Northup, Kenneth E. Turner, Prasanna Gowda, Sam Coleman and Michael Brown
Managing Tallgrass Prairies for Productivity and Ecological Function: A Long-Term Grazing Experiment in the Southern Great Plains, USA
Reprinted from: *Agronomy* **2019**, *9*, 699, doi:10.3390/agronomy9110699 6

Xavier Úbeda, Meritxell Alcañiz, Gonzalo Borges, Luis Outeiro and Marcos Francos
Soil Quality of Abandoned Agricultural Terraces Managed with Prescribed Fires and Livestock in the Municipality of Capafonts, Catalonia, Spain (2000–2017)
Reprinted from: *Agronomy* **2019**, *9*, 340, doi:10.3390/agronomy9060340 28

Simone Ravetto Enri, Alessandra Gorlier, Ginevra Nota, Marco Pittarello, Giampiero Lombardi and Michele Lonati
Distance from Night Penning Areas as an Effective Proxy to Estimate Site Use Intensity by Grazing Sheep in the Alps
Reprinted from: *Agronomy* **2019**, *9*, 333, doi:10.3390/agronomy9060333 40

Brian K. Northup, Patrick J. Starks and Kenneth E. Turner
Soil Macronutrient Responses in Diverse Landscapes of Southern Tallgrass to Two Stocking Methods
Reprinted from: *Agronomy* **2019**, *9*, 329, doi:10.3390/agronomy9060329 47

Brian K. Northup, Patrick J. Starks and Kenneth E. Turner
Stocking Methods and Soil Macronutrient Distributions in Southern Tallgrass Paddocks: Are There Linkages?
Reprinted from: *Agronomy* **2019**, *9*, 281, doi:10.3390/agronomy9060281 65

Donato Andueza, Fabienne Picard, Philippe Pradel and Katerina Theodoridou
Feed Value of Barn-Dried Hays from Permanent Grassland: A Comparison with Fresh Forage
Reprinted from: *Agronomy* **2019**, *9*, 273, doi:10.3390/agronomy9060273 81

Shengfang Ma, Yuting Zhou, Prasanna H. Gowda, Liangfu Chen, Patrick J. Starks, Jean L. Steiner and James P. S. Neel
Evaluating the Impacts of Continuous and Rotational Grazing on Tallgrass Prairie Landscape Using High-Spatial-Resolution Imagery
Reprinted from: *Agronomy* **2019**, *9*, 238, doi:10.3390/agronomy9050238 92

Patrick J. Starks, Jean L. Steiner, James P. S. Neel, Kenneth E. Turner, Brian K. Northup, Prasanna H. Gowda and Michael A. Brown
Assessment of the Standardized Precipitation and Evaporation Index (SPEI) as a Potential Management Tool for Grasslands
Reprinted from: *Agronomy* **2019**, *9*, 235, doi:10.3390/agronomy9050235 107

Yuting Zhou, Prasanna H. Gowda, Pradeep Wagle, Shengfang Ma, James P. S. Neel, Vijaya G. Kakani and Jean L. Steiner
Climate Effects on Tallgrass Prairie Responses to Continuous and Rotational Grazing
Reprinted from: *Agronomy* **2019**, *9*, 219, doi:10.3390/agronomy9050219 . . . . **123**

Alan J. Franzluebbers, Patrick J. Starks and Jean L. Steiner
Conservation of Soil Organic Carbon and Nitrogen Fractions in a Tallgrass Prairie in Oklahoma
Reprinted from: *Agronomy* **2019**, *9*, 204, doi:10.3390/agronomy9040204 . . . . **138**

Patricia Poblete-Grant, Philippe Biron, Thierry Bariac, Paula Cartes, María de La Luz Mora and Cornelia Rumpel
Synergistic and Antagonistic Effects of Poultry Manure and Phosphate Rock on Soil P Availability, Ryegrass Production, and P Uptake
Reprinted from: *Agronomy* **2019**, *9*, 191, doi:10.3390/agronomy9040191 . . . . **149**

Nicolas Puche, Nimai Senapati, Christophe R. Flechard, Katia Klumpp, Miko U.F. Kirschbaum and Abad Chabbi
Modeling Carbon and Water Fluxes of Managed Grasslands: Comparing Flux Variability and Net Carbon Budgets between Grazed and Mowed Systems
Reprinted from: *Agronomy* **2019**, *9*, 183, doi:10.3390/agronomy9040183 . . . . **162**

Gabriel Y. K. Moinet, Andrew J. Midwood, John E. Hunt, Cornelia Rumpel, Peter Millard and Abad Chabbi
Grassland Management Influences the Response of Soil Respiration to Drought
Reprinted from: *Agronomy* **2019**, *9*, 124, doi:10.3390/agronomy9030124 . . . . **193**

Guadalupe Tiscornia, Martín Jaurena and Walter Baethgen
Drivers, Process, and Consequences of Native Grassland Degradation: Insights from a Literature Review and a Survey in Río de la Plata Grasslands
Reprinted from: *Agronomy* **2019**, *9*, 239, doi:10.3390/agronomy9050239 . . . . **206**

Pradeep Wagle and Prasanna H. Gowda
Tallgrass Prairie Responses to Management Practices and Disturbances: A Review
Reprinted from: *Agronomy* **2018**, *8*, 300, doi:10.3390/agronomy8120300 . . . . **227**

About the Special Issue Editors

Abad Chabbi is a plant ecologist and soil biogeochemist. He has worked at the Louisiana State University, USA; the Faculty of Environmental Science in Cottbus, Germany; the University of Pierre & Marie, Curie (UPMC), France; and the INRAE (Institut national de recherche pour l'agriculture, l'alimentation et l'environnement) Research Council, where he has been leading the National Observatory for Environmental Research-Agro-Ecosystems, Biogechemical Cycles and Biodiversity since 2009. Abad Chabbi is currently Director of Research at INRAE, and his research centers on the linkage between soil carbon sequestration, nutrient availability, and stoichiometry in the plant–soil system and their relations in land use management and climate change.

Gianni Bellocchi holds a PhD in agronomy and environment from the Sant'Anna School of Advanced Studies (Pisa, Italy). He is currently Senior Scientist at INRAE (Institut national de recherche pour l'agriculture, l'alimentation et l'environnement), Grassland Ecosystem Research Unit. With expertise in agro-climatic and hydrological modeling, his research and publication interests include carbon–nitrogen cycles and biodiversity in grassland ecosystems, with an interest in sequestration processes and gas emission balances using simulation models.

Editorial

Grassland Management for Sustainable Agroecosystems

Gianni Bellocchi [1] and Abad Chabbi [2],*

1. UCA, INRAE, VetAgro Sup, UREP, 63000 Clermont-Ferrand, France; gianni.bellocchi@inrae.fr
2. INRAE, URP3F, 86600 Lusignan, France
* Correspondence: abad.chabbi@inrae.fr; Tel.: +33-549-556-178

Received: 11 December 2019; Accepted: 29 December 2019; Published: 6 January 2020

Abstract: Knowledge on sustainable grassland management is available in the large body of literature. However, it is unclear where to look for it, and what is really relevant to the many interrelated challenges of sustainable grassland management. This special issue illustrates options to fill some of those gaps. This editorial introduces the Special Issue entitled "Grassland Management for Sustainable Agroecosystems". Two review articles deal with (i) concepts for monitoring grassland degradation (by Tiscornia et al. *Agronomy* **2019**, *9*, 239) and (ii) impacts of alternative management practices and disturbances (by Wagle and Gowda et al. *Agronomy* **2018**, *8*, 300). One paper (by Steiner et al. *Agronomy* **2019**, *9*, 699). summarized a series of papers of the special issue. Other topics covered include four main aspects: (I) Landscape features (Ravetto Enri et al. *Agronomy* **2019**, *9*, 333), two papers by Northup et al. *Agronomy* **2019**, *9*, 329, Northup et al. *Agronomy* **2019**, *9*, 281, and Ma et al. *Agronomy* **2019**, *9*, 238; (II) climate (Zhou et al. *Agronomy* **2019**, *9*, 219, Starks et al. *Agronomy* **2019**, *9*, 235, and Moinet et al. *Agronomy* **2019**, *9*, 124); (III) soil fertility (Franzluebbers et al. *Agronomy* **2019**, *9*, 204, Poblete-Grant et al. *Agronomy* **2019**, *9*, 191); and (IV) one on modeling (Puche et al. *Agronomy* **2019**, *9*, 183). Two additional papers are from Andueza et al. *Agronomy* **2019**, *9*, 273 (on the feed value of barn-dried hay) and Úbeda et al. *Agronomy* **2019**, *9*, 340 (on the role of prescribed burns).

Keywords: continuous and rotational grazing; phosphorous availability and uptake; potential management; site use intensity; soil macronutrient responses and distributions; soil organic carbon and nitrogen; soil respiration

---

1. Introduction

Grasslands should no longer be considered only for global food supply contributing to ruminant milk and meat productions, but also, and above all, as a source of production of ecosystem services that contribute to the sustainability of agriculture. The intensification of agricultural production has been accompanied by a specialization of production systems, which has led to a spatial separation of agriculture and livestock and an excessive standardization of territories [1]. This has led to unacceptable environmental impacts on society. Grasslands, like forests, play a key role in this regard (i) by strongly coupling the cycles of C, N, and P, thus limiting emissions to the hydrosphere and atmospheres, and (ii) in acting positively on the dynamics of biodiversity. However, the intensification of grasslands aimed at maximizing their production function tends to decouple the C and N cycles through the animal and therefore minimize their environmental function [2]. A compromise must therefore be sought through reasoned intensification. In this regard, the importance of studies on grasslands has intensified during the last years, as awareness has grown that societies and individuals gain important benefits from these ecosystems [3]. The location of grassland vegetation in the agricultural landscape contributes greatly to soil erosion control, and this is often coupled to other services relating to water supply and regulation, carbon sequestration, and soil fertility [4–6]. These roles are widely recognized by actors in the agricultural world and by public policies. On the recognition that species-rich grasslands are a source of multiple ecosystem services, several contributions have provided the ground for innovative management practices that can increase fodder production, while, at the same time, enhancing other services like pollination, biological control, climate regulation, and soil conservation [7–9].

The renewed interest in grasslands, observed since the turn of the 21st century, is precisely driven by a greater recognition of the importance of grassland and livestock grazing systems in relation to soil conservation, biodiversity promotion, farming communities' stabilization, and the provision of a wealth of natural ecosystem services [10]. The special issue is an excellent overview of the management actions taken on different types of grasslands worldwide, with particular care on the Great Plains of North America. As known, this area is one of the most hit zones of the world because native grasslands have mostly disappeared in the past century due to agricultural expansion. The remaining prairie ecosystems are important for livestock grazing and provide additional benefits including habitat for a diversity of avian, terrestrial, and aquatic species, and the regulation of hydrological and carbon cycles.

2. Special Issue Overview

The special issue of Agronomy entitled "Grassland Management for Sustainable Agroecosystems" publishes 15 articles that provide insights into challenges of, and possible solutions to, sustainable grassland management. It contains two review articles dealing with: (1) Concepts for monitoring grassland degradation and (2) the impacts of alternative management practices and disturbances, including restoration. We add that while describing the initial results from a major grazing experiment in the Southern Great Plains of the USA, Steiner et al. ([11]. introduced and summarized a series of papers in this special issue. This is probably the article that most readers who are unfamiliar with this field of research should begin with.

The review by Tiscornia et al. [12] provides the background and context to the study of grassland degradation, clarifying the related drivers, processes, and consequences. This is important because grassland degradation has multi-casual drivers. The possibility of a common interpretation of the degradation issue over different scales is certainly fascinating. In fact, this paper is both a literature and an expert analysis on how the grassland degradation is defined and studied across the world, and more particularly, for the pasture ecosystems of the Region of Río de La Plata (across the central-eastern part of Argentina, most of Uruguay, and southern Brazil). The importance of natural grasslands to food security, the provision of ecosystem services, and the economy of many developing countries such as Uruguay, Argentina, and Brazil, implies that their degradation is a major political, economic, and environmental issue. This importance is reinforced by the experts' perception that the effects of livestock production intensification and climate change can increase degradation. The review paper has the merit of proposing a novel conceptual model, where multiple drivers (e.g., rainfall variability and overgrazing) operate simultaneously and interact in different site-specific conditions like soil types, bioclimatic zones, and history of grazing. To advance in the knowledge of grassland degradation process, the authors represented such interactive effects in a conceptual state and transition models of the main processes affected (e.g., decreased primary production by a reduced leaf area index, reduced plant species diversity by an increment of interspecific competitive exclusion for light) and the consequences of management practices.

The review [13] of Wagle and Gowda compiled several information about how multiple aspects of tallgrass prairie respond to a variety of treatments applied in experimental settings, with some of those actually being common management practices. In fact, tallgrass prairie grasslands can range from unmanaged low productive systems to highly managed, high-productive systems. Notably, whether nitrogen (N) fertilization is a widely used management practice to improve the productivity of managed prairie grasslands for grazing or hay harvests, a judicious use of N fertilization is recommended to minimize undesirable and invasive species and to maintain healthy native prairie stands. There are increasing efforts from government and private organizations to restore diverse prairies in agricultural sites, but restored prairies show a tendency of community composition to shift towards C4 grass dominance and lower soil quality compared to native remnants. Prescribed fire and grazing together can create highly productive, diverse, and heterogeneous grassland systems because of the modulation of the fire effect on plant diversity by reducing the dominance of C4 grasses and increasing the diversity and richness of species.

The role of prescribed burns was also addressed in the study conducted by Úbeda et al. [14] in Catalonia (Spain). When alternating with livestock (goat) grazing, it may be an acceptable way to promote abandoned terraces into pasture and reduce the risk of fire.

The other papers addressed three broad topics: Landscape features, climate, and soil fertility. We add the contribution of Andueza et al. (France) [15], who highlighted a higher feed value (for digestibility and intake) of barn-dried hay (hay is cut, wilted in the field, and then further dried in a burn) obtained from permanent grasslands than fresh forage.

2.1. Landscape Features

Sustainable grassland management involves highly complex challenges. There is a need to take into account all framework conditions, including the soil resources, water supply, and the climatic conditions, among others. A study conducted in the Italian Alps (Ravetto Enri et al.) [16] provided new knowledge to support sheep grazing management, as it highlighted that environmental and management factors, such as distance from night penning areas, distance from water, and slope, can be reliable proxies of the pasture use intensity by grazing sheep. This is particularly relevant, considering that Alpine pastures are characterized by a high spatial heterogeneity due to changes in topography and vegetation.

Landscape features included within paddocks, including organization of water and other features, require higher attention because they influence the (non-uniform) distribution of plant-available macronutrients fluxes beyond the stocking method adopted. The potential to define grazing effects on macronutrient distribution was addressed in two contributions by Northup et al. [17,18] in central Oklahoma (USA), while Ma et al. [19] highlighted the need of high spatial and temporal resolution images to monitor the tallgrass prairie landscapes (notably shrub encroachment, for which grazing management effects are hardly detectable).

2.2. Climate Change

Numerous indicators confirm the reality of climate change, which should affect several grasslands and forage systems not only through the average evolution of climatic variables (temperature, precipitation, and CO_2 concentration), but also by their interactions and increasing their variability. Considering the inertia of the climate system, grassland management adaptation is now inevitable whatever additional efforts we manage to deploy to reduce greenhouse gas emissions. Contributions from American grassland systems indicating that adaptive grazing management (adjustment in stocking rates and season of use to adapt to changing climatic conditions) instead of a fixed management system might be better for farmers to cooperate with changing climatic conditions (Zhou et al.) [20]. Thanks to the study of Starks et al. [21] in Konza Prairie, Kansas (USA), the standardized precipitation and evaporation index (SPEI) is offered as a suitable predictive tool of aboveground grassland or forage mass. Yet not conclusive about which of mowing or grazing enhance drought resilience, Moinet et al. [22] found in a French study that different management practices and decisions strongly contribute to determine the response of soil carbon dynamics to changes in soil water content. In other words, grassland management practices can change the relationship between soil respiration and its water content.

2.3. Soil Fertility and Modeling

Grazing lands typically have greater soil organic C and N contents than other agricultural land uses. An important aspect of regaining full functionality of grasslands focuses on how animals are stocked and allowed to graze available forage. Grazing livestock is an important regulator of how C and N in grasslands are partitioned in the ecosystem, but this may not be discernible in the short-medium time (Franzluebbers et al.) [23]. Though further testing in long-term field experiments is required, Poblete-Grant et al. [24]. evoked the use of poultry manure compost as a strategy to replace inorganic fertilizers. It is known that current and past management practices interact with soil and environmental factors in determining pasture and animal productivity, and soil organic carbon stocks. These multiple, complex, and interacting factors can realistically only be assessed through modeling approaches, as the number of possible interactions exceeds the number that could feasibly be studied experimentally. At the same time, models need to be tested and verified against real-world observations to build confidence in their ability to accurately describe the patterns we are interested in, in particular, the effect of different management options in enhancing soil organic carbon storage without compromising pasture productivity. The contribution from Puche et al. [25] is a modeling work on mown and grazed paddocks in France, equipped with eddy-covariance (EC) measuring systems. The model used, CenW, confirms the difficulty of simulating grazing systems where EC devices may not completely detect C losses.

3. Conclusions

Grasslands produce forage to support livestock farming. Grassland ecosystems are also a reservoir of biological diversity and soil carbon, and the lever of multiple ecosystem services supporting the life and the economy of various communities and territories. This special issue gathered research investigations that relied on the grassland management, with various objectives and methodologies, showing to which extent grassland ecosystems are the object of multiple researches from large communities of scientists. Furthermore, this special issue also demonstrated ways to address the complexity of sustainable grassland management and its current and future challenges with respect to climate change and landscape features.

Author Contributions: A.C. and G.B. wrote this editorial for the introduction of the Special Issue, entitled "Grassland Management for Sustainable Agroecosystems", of Agronomy, and edited the Special Issue. All authors have read and agreed to the published version of the manuscript.

Funding: This research received no external funding.

Acknowledgments: We thank all authors who submitted their valuable papers to the Special Issue, entitled "Grassland Management for Sustainable Agroecosystems", of Agronomy.

Conflicts of Interest: The authors declare no conflict of interest.

References

1. Lemaire, G.; Gastal, F.; Franzluebbers, A.; Chabbi, A. Grassland-cropping rotations: An avenue for agricultural diversification to reconcile high production with environmental quality. *Environ. Manag.* **2015**, *65*, 1065–1077. [CrossRef] [PubMed]
2. Rumpel, C.; Chabbi, A. Plant-soil interactions control CNP coupling and decoupling processes in Agroecosystems with perennial vegetation. In *Agro-Ecosystem Diversity: Reconciling Contemporary Agriculture and Environment Quality*; Lemaire, G., De Faccio Carvalho, P.C., Kronberg, G., Recous, S., Eds.; Elsevier Academic Press: Cambridge, MA, USA, 2019; pp. 3–13.
3. Bengtsson, J.; Bullock, J.M.; Egoh, B.; Everson, C.; Everson, T.; O'Connor, T.; O'Farrell, P.J.; Smith, H.G.; Lindborg, R. Grasslands—More important for ecosystem services than you might think. *Ecosphere* **2019**, *10*, e02582. [CrossRef]
4. Souchère, V.; King, C.; Dubreuil, N.; Lecomte-Morel, V.; Le Bissonnais, Y.; Chalat, M. Grassland and crop trends: Role of the European Union Common Agricultural Policy and consequences for runoff and soil erosion. *Environ. Sci. Policy* **2003**, *6*, 7–16. [CrossRef]
5. Pilgrim, E.S.; Macleod, C.J.A.; Blackwell, M.S.A.; Bol, R.; Hogan, D.V.; Chadwick, D.R.; Cardenas, L.; Misselbrook, T.H.; Haygarth, P.M.; Brazier, R.E.; et al. Interactions among agricultural production and other ecosystem services delivered from European temperate grassland systems. *Adv. Agron.* **2010**, *109*, 117–154.

6. Hou, R.; Yu, R.; Wu, J. Relationship between paired ecosystem services in the grassland and agro-pastoral transitional zone of China using the constraint line method. *Agric. Ecosyst. Environ.* **2017**, *240*, 171–181.
7. Holland, J.M.; Douma, J.C.; Crowley, L.; James, L.; Kor, L.; Stevenson, D.R.; Smith, B.M. Semi-natural habitats support biological control, pollination and soil conservation in Europe. A review. *Agron. Sustain. Dev.* **2017**, *37*, 31. [CrossRef]
8. Wehn, S.; Hovstad, K.A.; Johansen, L. The relationships between biodiversity and ecosystem services and the effects of grazing cessation in semi-natural grasslands. *Web Ecol.* **2018**, *18*, 55–65. [CrossRef]
9. Johansen, L.; Taugourdeau, S.; Hovstad, K.A.; Wehn, S. Ceased grazing management changes the ecosystem services of semi-natural grasslands. *Ecosyst. People* **2019**, *15*, 192–203. [CrossRef]
10. Sollenberger, L.E.; Kohmann, M.M.; Dubeux, J.C.B.; Silveira, M.L. Grassland management affects delivery of regulating and supporting ecosystem services. *Crop Sci.* **2019**, *59*, 441–459. [CrossRef]
11. Steiner, J.L.; Starks, P.J.; Neel, J.P.; Northup, B.; Turner, K.E.; Gowda, P.; Coleman, S.; Brown, M. Managing Tallgrass Prairies for Productivity and Ecological Function: A Long-Term Grazing Experiment in the Southern Great Plains, USA. *Agronomy* **2019**, *9*, 699. [CrossRef]
12. Tiscornia, G.; Jaurena, M.; Baethgen, W. Drivers, process, and consequences of native grassland degradation: Insights from a literature review and a survey in Río de la Plata grasslands. *Agronomy* **2019**, *9*, 239. [CrossRef]
13. Wagle, P.; Gowda, P.H. Tallgrass prairie responses to management practices and disturbances: A review. *Agronomy* **2018**, *8*, 300. [CrossRef]
14. Úbeda, X.; Alcañiz, M.; Borges, G.; Outeiro, L.; Francos, M. Soil Quality of abandoned agricultural terraces managed with prescribed fires and livestock in the municipality of Capafonts, Catalonia, Spain (2000–2017). *Agronomy* **2019**, *9*, 340. [CrossRef]
15. Andueza, D.; Picard, F.; Pradel, P.; Theodoridou, K. Feed value of barn dried hays from permanent grassland: A comparison with fresh forage. *Agronomy* **2019**, *9*, 273. [CrossRef]
16. Ravetto Enri, S.; Gorlier, A.; Nota, G.; Pittarello, M.; Lombardi, G.; Lonati, M. Distance from Night Penning Areas as an Effective Proxy to Estimate Site Use Intensity by Grazing Sheep in the Alps. *Agronomy* **2019**, *9*, 333. [CrossRef]
17. Northup, B.K.; Starks, P.J.; Turner, K.E. Soil macronutrient responses in diverse landscapes of southern tallgrass to two stocking methods. *Agronomy* **2019**, *9*, 329. [CrossRef]
18. Northup, B.K.; Starks, P.J.; Turner, K.E. Stocking methods and soil macronutrient distributions in Southern tallgrass paddocks: Are There Linkages? *Agronomy* **2019**, *9*, 281. [CrossRef]
19. Ma, S.F.; Zhou, Y.; Gowda, P.H.; Chen, L.; Starks, P.; Steiner, J.L.; Neel, J.S.N. Evaluating the impacts of continuous and rotational grazing on tallgrass prairie landscape using high-spatial-resolution imagery. *Agronomy* **2019**, *9*, 238. [CrossRef]
20. Zhou, Y.T.; Gowda, P.H.; Wagle, P.; Ma, S.F.; Neel, J.S.N.; Kakani, V.G.; Steiner, J.L. Climate effects on tallgrass prairie responses to continuous and rotational grazing. *Agronomy* **2019**, *9*, 219. [CrossRef]
21. Starks, P.J.; Steiner, J.L.; Neel, J.S.N.; Turner, K.E.; Northup, B.K.; Gowda, P.H.; Brown, M.A. Assessment of the standardized precipitation and evaporation index (SPEI) as a potential management tool for grasslands. *Agronomy* **2019**, *9*, 235. [CrossRef]
22. Moinet, G.Y.K.; Midwood, A.J.; Hunt, J.E.; Rumpel, C.; Millard, P.; Chabbi, A. Grassland management influences the response of soil respiration to drought. *Agronomy* **2019**, *9*, 124. [CrossRef]
23. Franzluebbers, A.J.; Starks, P.J.; Steiner, J.L. Conservation of Soil Organic Carbon and Nitrogen Fractions in a Tallgrass Prairie in Oklahoma. *Agronomy* **2019**, *9*, 204. [CrossRef]
24. Poblete-Grant, P.; Biron, P.; Bariac, T.; Cartes, P.; de La Luz Mora, M.; Rumpel, C. Synergistic and antagonistic effects of poultry manure and phosphate rock on soil P availability, ryegrass production, and P Uptake. *Agronomy* **2019**, *9*, 191. [CrossRef]
25. Puche, N.; Senapati, N.; Flechard, C.R.; Klumpp, K.; Kirschbaum, M.U.F.; Chabbi, A. Modeling carbon and water fluxes of managed grasslands: Comparing flux variability and net carbon budgets between grazed and mowed systems. *Agronomy* **2019**, *9*, 183. [CrossRef]

© 2020 by the authors. Licensee MDPI, Basel, Switzerland. This article is an open access article distributed under the terms and conditions of the Creative Commons Attribution (CC BY) license (http://creativecommons.org/licenses/by/4.0/).

Article

Managing Tallgrass Prairies for Productivity and Ecological Function: A Long-Term Grazing Experiment in the Southern Great Plains, USA

Jean L. Steiner [1], Patrick J. Starks [2,*], James P.S. Neel [2], Brian Northup [2], Kenneth E. Turner [2], Prasanna Gowda [3], Sam Coleman [4] and Michael Brown [2]

1. United States Department of Agriculture-Agricultural Research Service, Kansas State University, Manhattan, KS 66502, USA; jlsteiner@ksu.edu
2. United States Department of Agriculture-Agricultural Research Service, El Reno, OK 73036, USA; Jim.neel@usda.gov (J.P.S.N.); Brian.northup@usda.gov (B.N.); Ken.turner@usda.gov (K.E.T.); michaelbrown@atlinkwifi.com (M.B.)
3. United States Department of Agriculture-Agricultural Research Service, Stoneville, MS 38776, USA; Prasanna.gowda@usda.gov
4. United States Department of Agriculture-Agricultural Research Service, Azle, TX 76020, USA; colespec@gmail.com
* Correspondence: Patrick.starks@usda.gov; Tel.: +1-405-262-5291

Received: 3 July 2019; Accepted: 28 October 2019; Published: 30 October 2019

Abstract: The Great Plains of the USA is one of largest expanses of prairie ecosystems in the world. Prairies have been extensively converted to other land uses. The remaining prairie ecosystems are important for livestock grazing and provide benefits including habitat for avian, terrestrial, and aquatic species, carbon regulation, and hydrologic function. While producers, land management agencies, and some researchers have promoted livestock management using rotational stocking for increased production efficiency and enhanced ecosystem function, scientific literature has not provided a consensus on whether rotational stocking results in increased plant biomass or animal productivity. To address this research need, we established long-term grazing research using an adaptive management framework to encompass a wide range of production and ecological interactions on native grassland pastures. This paper describes objectives, design, and implementation of the long-term study to evaluate productivity and ecological effects of beef cow–calf management and production under continuous system (CS) or rotational system (RS) on native tallgrass prairie. Findings from 2009 to 2015 indicate that plant biomass and animal productivity were similar in the two grazing management systems. There were some indicators that forage nutritive value of standing biomass and soil nutrient content were enhanced in the RS system compared with the CS, yet individual calf body weight (BW) at weaning was greater in the CS. This prepares us to engage with producers to help determine the focus for the next phase of the research.

Keywords: long-term agroecosystem research network; LTAR; rotational grazing

---

1. Introduction

Efficiency of forage use in livestock production systems is important for global food security. Pastures, rangeland, and forest lands suited to grazing ruminant livestock constitute 35% of the U.S. land area [1]. Pastures and rangelands are a dominant land use and support ruminant production in every climatic regime on every continent [2]. In 2000, about 35 million km^2 were in permanent pasture, representing 30% of the world's land area and over 70% of the agricultural area [3]. Animal meat and dairy products are a critical component of human food systems, including in some of the least food-secure regions of the world. It has been estimated [4] that 1 billion people depend on livestock for income and food security, including 70% of the world's 880 million rural-poor populations. It has been suggested [5] that as protein and micronutrient concentrations of grain crops decline with increasing atmospheric carbon dioxide (CO_2) concentrations, an additional 122 million people could struggle with protein-deficient diets, 175 million people would experience zinc-deficient diets, and 1.4 million women of child-bearing age and children under 5 could face a 20% increase in occurrence of anemia. These estimates are based on the 2050 projections for human population and atmospheric CO_2 concentration. Food security challenges and biodiversity conservation must be addressed as a linked process [6] based on increasing productivity while enhancing biodiversity and other ecosystem services from existing agricultural lands, particularly in the face of changing climate. [7,8]. Understanding interactions between climate variability and agricultural systems is essential to ensure a sustainable food supply while maintaining agricultural soil and water resources for the next generation [9,10].

Pasture systems around the world are heterogeneous, ranging from those dominated by native vegetation, such as in the central portion of North America, Argentina, and parts of Asia, to those that are dominated by domesticated pasture, such as in Europe, New Zealand, eastern North America, and parts of Asia [11]. The native vegetation can range from relatively intact ecosystems to vegetation that may have regrown on abandoned cropland. Grasslands face numerous challenges, some long-standing and others of growing importance. As human populations have expanded, lands that were formerly grassland have been converted to urban and cropland or fragmented into smaller holdings. In general, remaining grasslands are located on drier, steeper, or less fertile areas in every region of the world.

Prairies, particularly tallgrass prairies, have been extensively converted to other land uses. However, the Great Plains of the USA remains one of largest expanses of prairie ecosystems in the world [12–14]. These remaining prairie ecosystems are important for livestock grazing and provide numerous benefits of diverse habitat for avian, terrestrial, and aquatic species [15–17], carbon regulation [18,19], and hydrologic function [20,21]. The Northern Great Plains has exhibited cooling and wetting over recent decades, while the Southern Great Plains (SGP) has exhibited increased aridity [19]. Challenges to these critical ecosystems in the SGP include encroachment by woody species [22–25], increased aridity and drought [19,26], climate change [27], and fragmentation of the landscape [28]. Additional resource concerns include low-productivity land and unstable stream networks. To address complex, interactive challenges such as these, an interdisciplinary systems research approach is needed [29].

Synthesis of a large body of literature indicated stocking rate to be the key management decision [30]. Low to moderate stocking rates provided the optimal vegetation required for livestock and wildlife productivity and the best economic returns. High-intensity, rotational stocking was advanced in the 1960s to improve forage harvest efficiency, increase livestock-carrying capacity, and boost economic returns [31]. Adoption of conservation cropping and adaptive multi-paddock grazing management was proposed to reduce the carbon footprint of agriculture in North America and promote a wide range of ecosystem services from agricultural lands [32]. However, a review of number of grazing studies found no plant or animal production benefit to rotational stocking compared with continuous stocking [33]. Subsequently, [34] recommended a research framework that would incorporate human elements such as goal setting, experiential knowledge, and decision making along with ecological variables and an adaptive framework for managing rangelands systems. Published studies of grazing systems indicated that grazed grasslands accumulated soil organic carbon (SOC) at a faster rate than undefoliated grasslands and that low to moderate livestock stocking rates favored SOC accumulation compared with high stocking rates [7]. Additionally it was found that more digestible forages defoliated at optimal maturity may decrease methane emissions by grazing ruminants. However, the effects of methods of applying grazing pressure on the C balance of grasslands have been mixed, particularly as stocking methods have regularly been shown to have little effect on productivity of grasslands at moderate stocking rates [35–41].

Adaptive management of grazing systems may incorporate periods of intensive grazing in a rotational framework, but with a focus on goal setting, monitoring, and adaptive response to conditions in the field. Some ranchers in the SGP have reported improved rangeland health and increased productivity and/or profitability after adopting holistic, rotational, or intensively managed grazing systems, but relatively little research focuses on these types of systems. To address this research need, we established long-term grazing research within an adaptive management framework to encompass a wide range of production and ecological interactions on native grassland pastures, specifically to evaluate productivity and ecological effects of beef cow–calf management and production under continuous and rotational stocking systems on native tallgrass prairie. The purpose of this paper is to describe the objectives, design, and implementation of the long-term study and present productivity indicators from 2009 to 2015. Additional papers in this issue present findings related to plant and soil responses to grazing management systems.

2. Stakeholder Engagement

The need for research was identified in collaboration with ranchers from the Oklahoma Grazing Lands Conservation Association. In spring of 2008, a workshop was convened at the United States Department of Agriculture (USDA) Agricultural Research Service (ARS), Grazinglands Research Laboratory, El Reno, OK, USA to identify producer concerns and priorities and identify research needed to develop the information requested. The producers were interested in research on continuous versus rotational stocking management, and, in particular, the impacts of high-stocking-rate, short-duration rotations. They recommended that the number of animals available be concentrated into herds that are managed to balance grazing periods with forage recovery periods to keep the forage in the rotational system physiologically "young". For the continuous stocking system, the producers identified winter management as an important factor impacting the next growing season. The producers recommended that fall- and winter-dormant forage be rationed (limited grazed) but grazed down prior to onset of spring growth to create litter and open the canopy to sunlight. The goal was to use adaptive management to optimize productivity and ecosystem performance for each management system. Responses that producers and researchers identified were grazing system effects on plant vegetative community and soil biology, as well as overall beef production and profitability. Approximately 250–300 ha of land was identified to support the experiment. After discussion, the research group determined that continuous stocking (CS) and rotational stocking (RS) management treatments would be established, with an embedded experiment that included exclosures for ungrazed (UG) land treatments and for high-stocking-density, short-duration (HDSD) grazing treatments. Results from the HDSD treatments are not reported in this paper but are reported by Northup et al. [42,43], and those from the UG plots are reported by Starks et al. [44].

3. Research Questions

In 2013, this experiment was incorporated into the Long-Term Agroecosystems Research (LTAR) Common Experiment [45,46]. The LTAR Common Experiment was established at 18 sites across the U.S., with each site implementing research with two treatments, Business as Usual (BAU, in this case CS) or Aspirational (ASP, in this case adaptive, RS). Research teams at each site and across sites focused on sustainability from the perspective of production, environment, and social or cultural responses. Table 1 summarizes research questions, based on input from the 2008 producer workshop, articulated into the LTAR framework of linked ecosystem services and metrics used to quantify the system effects on sustainability.

Table 1. Research questions and metrics to address ecosystem services of alternate grazing systems.

Sustainability Pillars	Ecosystem Services	Research Questions	Metrics
Production	Primary production Secondary production	How does grazing management system affect gross primary production? Is animal performance impacted by management system? How do management system and climate interact in controlling productivity?	Biomass production Forage nutritive value Breeding efficiency Weaning weights per unit area Drought indices
Environment	Climate regulation Soil nutrient cycling Soil biodiversity Plant biodiversity	How does management system affect plant and soil biodiversity, soil carbon and nutrients?	Soil nutrient budgets: C, N, macronutrients Soil organic matter (Phospholipid Fatty Acid analysis—future)

Grazed tallgrass prairies also provide considerable social and cultural services, which are generally realized at landscape, watershed, and regional scales. While no research questions were developed around these ecosystem services in this field-scale experiment, the research site has provided education benefits through graduate and postdoctoral research projects. Additionally, on a regional scale, such land uses provide supporting services such as wildlife habitat and watershed function and extensive recreational services including hunting and birding.

4. Site Description

The USDA Agricultural Research Service, Grazinglands Research Laboratory is located in El Reno, Oklahoma (35°33′29″ N, 98°1′50″ W, 414 m above mean sea level). The laboratory is located on 2700 ha of land, of which approximately 1200 ha are tallgrass prairie, most of which have been grazed at low stocking rates (animal unit (AU) ha^{-1}) over the past several decades. In addition, a prescription of spring burns had been applied on a 3–5 year interval to limit invasion by woody species. However, after implementation of the study in 2009, the paddocks included in the study have experienced no burns of any kind.

4.1. Climate

The SGP has pronounced seasonal, year-to-year, and multi-year variations in air temperature and precipitation [47,48], which are key drivers of ecological processes and agricultural production [49]. Average annual minimum air temperature over the study region is about 9.2 °C. Typically, July and August are the warmest months of the year, while the coolest months coincide with the driest months of December, January, and February (mean −2.3 °C). Average monthly temperature peaks in July at 28 °C. Average annual precipitation over Central Oklahoma, USA is about 870 mm. The "wet" season is generally the months of April through June (spring). On average, the spring months receive approximately 326 mm of rainfall or 38% of the annual precipitation. The "dry" season occurs in the winter months of December, January, and February. These months contribute about 13% to the annual precipitation, with an average seasonal precipitation of 111 mm. Long-term monthly mean precipitation and temperature for 1994–2018 (period of record at the site) are shown in Figure 1.

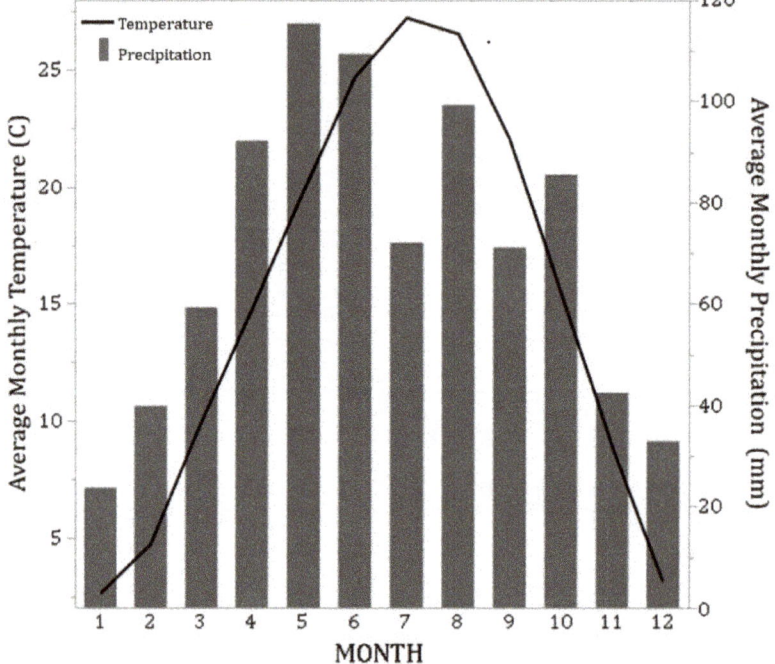

Figure 1. Monthly mean temperature precipitation for the El Reno site of the Oklahoma Mesonet, 1994–2018. Month 1 = January ... Month 12 = December. Data Source: Oklahoma Climate Survey, http://climate.ok.gov.

The average monthly soil volumetric water content (θv) for the 0–100 cm layer indicates the strong intra- and interannual variability in the moisture regime (Appendix A Figure A1, methods in Appendix A.2). The general pattern is that the soil recharges with moisture in autumn, winter, and spring during the cool months and is depleted during the hot summer growing season. The 2011 and 2012 time periods experienced the driest summer conditions (June–August), with the difference being that 2012 entered the summer growing season with high soil water storage due to a wet autumn and winter and spring. Values for the 5, 10, 20, 50, and 100 cm layers are shown in Figures A2–A6.

The SGP experienced a persistent pluvial period lasting most of the last quarter of the 20th century [47], with a return to drier conditions in the early 21st century. Since 1990, extreme precipitation events have increased [50]. An increased frequency of dipole events–wet years following drought years was reported by [51]. The dipole events in the SGP can be extremely damaging as intense precipitation acts on depleted soil and vegetative surfaces, resulting in runoff, erosion, and flooding risks. Additionally, a high frequency of severe storms [52] with hail, extreme winds, and tornados can produce catastrophic losses to agricultural crops, livestock, and agricultural infrastructure. The variable and changing climate makes long-term research essential in order to determine interactive impacts of climate and management on a broad range of production and ecosystem responses.

4.2. Soils and Vegetation

The site is located in the Central Great Plains ecoregion [53], with dominant grass species being big bluestem (*Andropogon gerardii*), little bluestem (*Schizachyrium scoparium*), and Indiangrass (*Sorghastrum nutans*). The soils included the Norge silt loam (fine-silty, mixed, active, thermic Udic Paleustolls), Pond Creek silt loam (fine-silt, mixed, superactive, thermic Pachic Argiustolls), Kirkland–Pawhuska complex (fine, mixed, superactive, thermic Udertic Paleustolls), and Bethany silt loam (fine, mixed, superactive, thermic Pachic Paleustolls).

Analysis of particle sizes of soils in the study area (see methods below) showed the texture of the 0–30 cm layer was statistically similar in all field replicates and that there were no significant differences in pH or electrical conductivity (EC) (Table A1). The sand and clay fractions each accounted for 21% to 26% of the soil in each field replicate, and pH and EC were about 6 in all replicates.

4.3. Layout and Experimental Design

We established two replicates of two grazing systems on native tallgrass prairie, each with a dedicated herd of beef cattle (Figure 2). Two of the herds are managed using CS on assigned paddocks, and two herds are managed using RS management assigned to two groups of 10 sub-paddocks under an adaptive rotational grazing framework. Land area and initial animal information for each replicate in 2009 are given in Table 2. Cow numbers for each herd were based on an animal cow/calf unit (500 kg cow body weight (BW)) and a targeted stocking rate of ~3 ha per cow/calf unit. Bodyweight of cows was measured at the time of weaning in 2009 to determine stocking rate for each replicate. Calves remain with the cows from March through weaning in October. Bulls are present with each cow herd during the June 1 to July 31 breeding season. The 10 sub-paddocks in the RS treatments ranged from 4.3 to 9.5 ha in RSa and from 5.3 to 12.9 ha in RSb.

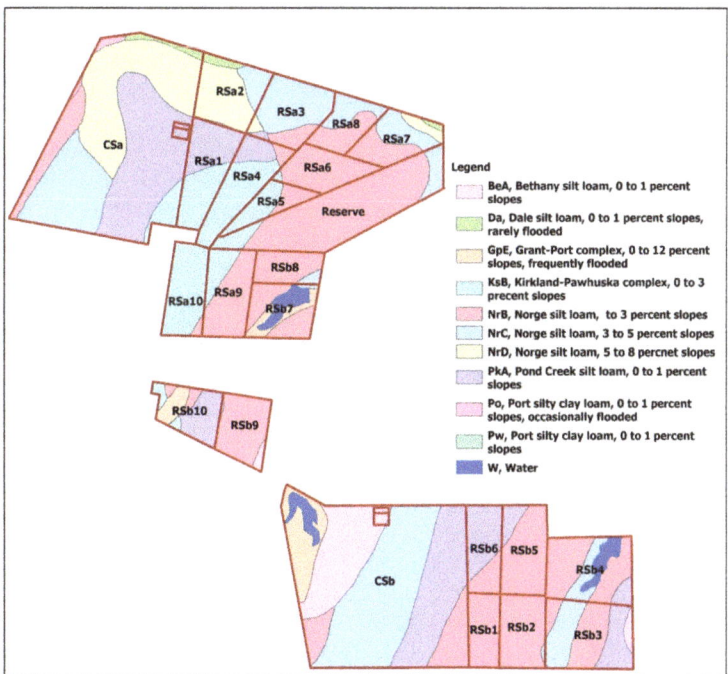

Figure 2. Map of research site at the USDA-ARS Grazinglands Research Laboratory, showing continuous (CSa, CSb) and rotational (RSa, RSb) locations on the USDA Natural Resources Conservation Service soils map. White boxes indicate exclosures for embedded high-density, short-duration grazing treatments described in Appendix A.

Table 2. Paddock size, cow numbers, average cow bodyweight, cow animal unit (AU), and stocking rate (500 kg per AU) for 2009.

Treatment & replicate	Area	Cows	Cow Bodyweight	Stocking Rate
	ha	#	Kg	Cow AU ha^{-1}
Continuous—CSa	58.6	20	614	0.42
Continuous—CSb	62.7	21	613	0.41
Rotational—RSa	78.1	25	584	0.37
Rotational—RSb	82.8	25	584	0.35

5. Materials and Methods

5.1. Soil Parameters

In 2009, 40 soil sampling sites were located within each field replicate using a stratified sampling design, where the strata were composed of the soil mapping units in the area (downloaded from the USDA Natural Resources Conservation Service (NRCS) Geospatial Data Gateway—https://datagateway.nrcs.usda.gov/GDGOrder.aspx). The number of soil samples for a given strata within a field replicate was based upon an area-weighted average of the soil mapping units. Thus, if 50%, 30%, and 20% of a field replicate was underlain by soil mapping units X, Y, and Z, then map unit X would be assigned 20 sampling sites, whereas Y would be assigned 12 and Z eight. Sampling points within a given strata were randomly located and their positions recorded by a hand-held GPS device. Soil sampling in subsequent years was guided by these GPS coordinates.

Soil cores were extracted using a Giddings (Windsor, CO, USA) probe equipped with a 30 cm long barrel, having an inside diameter of 4 cm (nominally). Soil cores down to 60 cm were collected in 2009, from which bulk density (ρ_b) and particle fractions of sand, silt, and clay were determined for the 0–6, 6–12, 12–20, 20–30, and 30–60 cm intervals at all sampling locations. Bulk density was determined from the oven-dry soil weights and sampling volume, and particle size fractions were determined using the hydrometer method [54]. In 2012 and 2017, ρ_b was measured in the 0–15 and 15–30 cm intervals. Thus for 2009, the 0–6 and 6–12 cm layer ρ_b values were averaged to represent the 0–15 cm layer, and the 12–20 and 23–30 data were averaged to represent the 15–30 cm layer.

Particle size fractions, EC, and pH were determined only from the 2009 soil core data. The EC and pH were measured using a VWR Scientific conductivity meter (Control Company, Friendship, TX, USA). Additional methods for soil measurement and analysis are given in references [42,43,55].

5.2. Forage Biomass and Nutritive Parameters

Two and five exclosures measuring 1.07 m × 6.4 m were randomly located within the CS and RS replicates, respectively, for the purpose of measuring end-of-year biomass. The vegetative biomass in these exclosures was clipped to a height of ≈1 cm above the soil surface using a sickle bar mower and then weighed fresh. A grab sample was collected for determination of dry (forced-air oven, 65 C) biomass on a kg ha^{-1} basis for each paddock.

In addition to the end-of-year biomass samples from exclosures, forage biomass and nutritive values were measured during the season. The CS paddocks were sampled from four to 11 times, depending upon year. Both CS replicates were sampled at least once a month from June through August. The biomass in the RS sub-paddocks was sampled prior to the introduction of cattle (three to four times per year, depending upon year). Samples were collected by clipping all standing vegetation within a 0.5 m^2 sampling frame to within 1 cm of the soil surface and determining dry mass as indicated above. Four random forage subsamples per replicate were collected for each sample date.

Aliquots of oven-dried forage subsamples were ground, packed into sample cuvettes, and scanned on a benchtop near-infrared spectrometer (NIRS; Foss-NIR Systems, Silver Spring, MD, USA). The NIRS scans were analyzed in WinISI 1.5 (FOSS North America, Eden Prairie, MN, USA) for the purpose of identifying a representative subset of forage samples for wet chemistry determinations of total nitrogen

(N), acid detergent fiber (ADF), neutral detergent fiber (NDF), and in vitro true digestibility (IVTD). Analysis for ADF, NDF, and IVTD followed the procedures outlined by Ankom Technology (Macedon, NY, USA; https://www.ankom.com/analytical-methods-support/fiber-analyzer-a2000). Total N was determined through combustion analysis (Elementar Americas, Inc., Mt. Laurel, NJ, USA). The laboratory-measured values were then used in conjunction with the NIRS scans to develop and validate calibration equations to predict total N, ADF, NDF, and IVTD. Once satisfactorily calibrated and validated, the calibration equations were applied to the NIRS scans of the remaining forage samples to predict total N, ADF, NDF, and IVTD. Additional methods and analysis for plant data are detailed in reference [44].

5.3. Animal Parameters

The cows utilized for experimental purposes were Angus-cross cows with Brahman influence. These cows were bred to terminal sires to maximize heterosis (i.e., sire breed was not represented within the cow). Cows were bred to calve in the spring, and generally calved between mid-March and mid-May. Calves were typically weaned from the cows in October. At weaning, individual cow and calf weights were taken, as well as cow body condition score (BCS). The BCS score was based on a 1–9 scoring system, with a score of 5–6 being desirable, 1–4 being too thin, and 7–9 being overly fat. During the course of the year, each herd had ad libitium access to fresh water and a commercially available mineral. During the fall through spring seasons, all dry cows were supplemented with a 40% crude protein cattle cube to meet production-stage nutritional requirements. Cows were supplemented with a 20% crude protein, high-energy cube to meet production-stage nutritional requirements during early spring, after calving. Within the RS treatment, herds were moved to the next paddock in the rotation (sub-paddocks 1 through 10) when available forage biomass was reduced to approximately 50% of that available upon start of grazing, determined by visual examination of pastures by trained personnel. Cattle were not supplemented with conserved forage (hay) except during periods of snow or ice cover (at this research location, the snow or ice cover does not occur frequently nor persist for long periods). In those instances, cattle within each system were offered native hay at the rate of approximately 30 kg dry matter (DM) per head per day until snow or ice cover had cleared.

All applicable national and/or institutional guidelines for the care and use of cattle were followed. Cattle were cared for in accordance with the standards of the Guide for the Care and Use of Agricultural Animals in Agricultural Research and Teaching [56]. Care of cows and calves grazing pastures followed approved standard operating procedures (SOPs) for beef cattle care and management. Short-term confinement of cows and calves in small paddocks (HDSD treatments) was conducted under approved protocol #GRL-2016-10-19-1. Rumen fluid collected from rumen-cannulated beef steers and used to determine forage digestibility (IVTD) in the laboratory was conducted under approved protocol #GRL-2018-6-12-1. All animal care and use SOPs and protocols were reviewed and approved by the Institutional Animal Care and Use Committee, Grazinglands Research Laboratory, El Reno, OK, USA.

5.4. Statistical Analysis

Means of ρ_b, percent sand and clay, pH, EC, and biomass means were compared using Tukey's Honestly Significant Difference (HSD) test. All analyses were performed in JMP 13 Pro (SAS Institute, Cary, NC, USA), and the level of significance was set at $\alpha = 0.05$. Descriptive statistics and linear least-squares regression of means ($\alpha = 0.5$) of plant biomass, herbage nutritive value, and soil properties were conducted in JMP Pro 14 (SAS Institute, Cary, NC, USA). Animal performance data were analyzed by ANOVA in SAS (SAS Institute, Cary, NC, USA) that included year as a random effect and grazing system (GS) and rep as fixed effects. Year × GS × rep was the error term.

6. Results

6.1. Temperature and Precipitation

Box plots of temperature and precipitation during the 2009–2015 experimental years (Figure 3) show that the experimental years were highly variable but representative of the long-term climate in the region (Figure 1). Precipitation was especially variable during the months of May and July over the 6 yr period. Average annual precipitation was below normal of 871 mm for the first four years of the study (557 to 795 mm), above normal in 2013 (1157 mm), below normal in 2014 (610 mm), and above normal in 2015 (1273 mm).

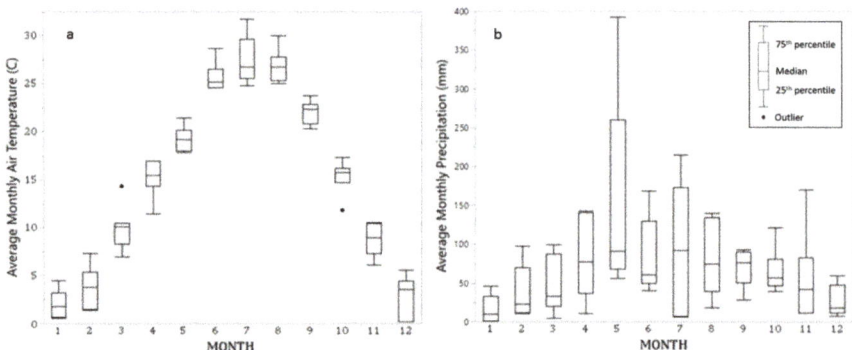

Figure 3. Box plots of monthly (**a**) temperature and (**b**) precipitation for 2009–2015, El Reno, Oklahoma. Month 1 = January ... Month 12 = December. Data Source: Oklahoma Climate Survey, http://climate.ok.gov

6.2. Soil parameters

Soil Bulk Density

Averaged across measurements collected in 2009, 2012, and 2017, the ρ_b of the 0–15 cm layer was 1.12 g cm^{-3} under CS management compared with 1.16 g cm^{-3} in RS management. While RS was higher density than CS ($p < 0.05$), the difference was minimal. There was no treatment effect on ρ_b in the 15–30 cm layer across the three years. However, both treatment and year effects were significant, with ρ_b being higher in 2012 than in 2009 and 2017 (data not shown). Appendix A Table A2 shows that the initial samples from 2009 had slightly higher ρ_b in the 0–15 cm layer in the RS replications than in the CS replicates, but that there was no difference between treatments in the surface layer in other years or at the 15–30 cm layer in any year. Samples from 2009 would not be expected to show treatment differences, since the treatments were initiated in 2008, but represent variability at the site. Changes in soil characteristics after 2009 could be related to treatment; however no differences were found in 2012 or 2017.

6.3. Forage

6.3.1. Above-Ground Biomass

Biomass production was highly variable over the study and was highest in 2013 and lowest in 2011 and 2014 (Table 3). The year 2010 also showed a high level of biomass production, but was statistically lower ($p < 0.05$) than that of 2013, while statistically greater than biomass production during the remaining years. Biomass production in 2010, 2009, 2012, 2014, and 2011 was 78%, 58%, 50%, 39%, and 25%, respectively, of amounts measured in 2013. No statistically significant differences between treatments were observed in end of season, which was expected as the samples were from ungrazed exclosures. There was also no difference by treatment in biomass collected through the growing seasons (data not shown).

Table 3. Avearge annual forage biomass production by year, pooled across treatments.

Year	Biomass [1] kg ha^{-1}
2009	4563 C
2010	6084 B
2011	2015 D
2012	3992 C
2013	7813 A
2014	3127 CD

[1] Site-specific means in the same column not connected by the same letter are statistically different ($\alpha = 0.05$) based on Tukey's Honestly Significant Difference (HSD).

6.3.2. Monthly Mean and Variability of Forage Nutritive Values

Monthly means of nutritive value during 2009 through 2014 illustrate the annual pattern of forage quality and level of variability (Figure 4). On average, N concentrations peaked in the May–June timeframe ($\approx 1.0\%$) and sharply decline thereafter, reaching minimum values by November ($\approx 0.6\%$). From November through March, total N remained low, but began to increase in April (Figure 4a). The ADF (Figure 4b) and NDF (Figure 4c) were highest in January through April and in November and December (≈ 50 and 81%, respectively). Lowest values of ADF and NDF occurred in June and July (≈ 41 and 74%, respectively). Typically, digestibility (IVTD, Figure 4d) was highest in June and July ($\approx 60\%$) and was generally above 50% in May through August. As indicated by the error bars in Figure 4, interannual variability was considerable for all forage nutritive variables shown, especially in the months April through July.

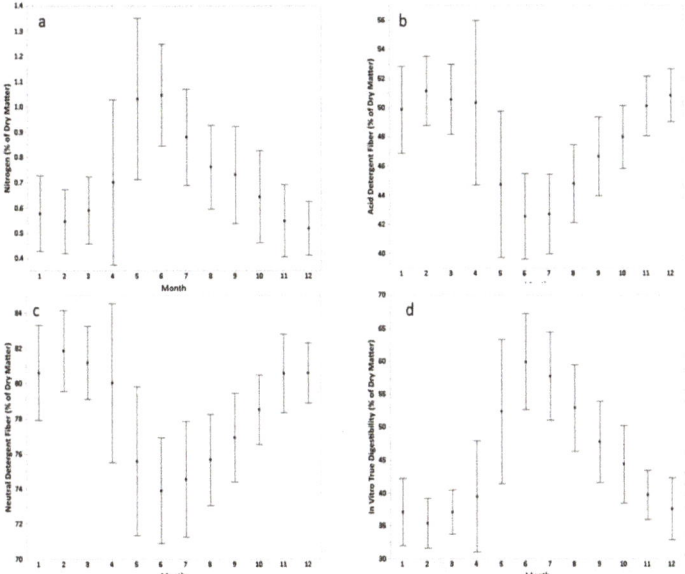

Figure 4. Monthly mean concentrations of (**a**) total nitrogen, (**b**) acid detergent fiber, (**c**) neutral detergent fiber, and (**d**) in vitro true dry matter digestibility of standing biomass. Error bars are ± 1 standard deviation. Month 1 = January ... Month 12 = December.

6.3.3. Forage Nutritive Values during Summer Grazing Season

Mean concentrations of total N (Table 4) were highest in 2009, but statistically similar to that of 2010, and lowest in 2011. Mean ADF concentration was highest in 2011 but statistically similar to that observed in 2009 and 2014. ADF was lowest in 2010, which was similar to that measured in 2013. Mean NDF was highest in 2011 (significantly different than all other years) and lowest in 2013. Mean IVTD was highest in 2010 and lowest in 2011. Statistical differences were observed for some forage nutritive values at the treatment level. While small, when they differed, total N concentration was higher in RS than in CS in 2011, 2013, and 2015; IVTD was higher in RS than in CS in 2012–2015; and ADF and NDF were lower in RS than in CS in 2013 and 2015.

Table 4. Comparison of treatment (CS = continuous stocking, RS = rotational stocking) means of concentrations (% of dry matter) of total nitrogen (N), acid detergent fiber (ADF), neutral detergent fiber (NDF), and in vitro true digestibility (IVTD) by year of study.

	N		ADF		NDF		IVTD	
Year	CS	RS	CS	RS	CS	RS	CS	RS
2009	1.0	1.0	46.0	45.3	76.3	76.1	54.5	56.8
2010	0.95	0.93	42.5	42.9	75.3	76.4	59.2	59.0
2011	0.60 [b]	0.65 [a]	48.1	47.3	79.6	78.9	41.2	41.7
2012	0.72	0.75	47.0	46.4	77.1	76.9	42.1 [b]	45.4 [a]
2013	0.74 [b]	0.87 [a]	47.0 [a]	45.0 [b]	77.2 [a]	75.6 [b]	45.6 [b]	52.1 [a]
2014	0.63	0.65	48.8	49.7	79.9	79.6	42.2 [b]	46.0 [a]
2015	0.72 [b]	0.82 [a]	49.2 [a]	46.5 [b]	78.5 [a]	76.3 [b]	39.4 [b]	47.2 [a]
Mean	0.74 [b]	0.80 [a]	47.0 [a]	46.3 [b]	77.8 [a]	77.0 [b]	45.8 [b]	49.5 [a]

Mean values within the same row and forage nutritive variable with unlike superscripts differ ($p < 0.05$).

6.4. Animal Parameters

6.4.1. Stocking Rate

Stocking rate by replicate and year are shown in Figure 5. Analyzed across the experimental period, the CS and RS stocking rates did not differ ($p = 0.05$). As discussed below, stocking rates in CS replicates were reduced in the latter years of the study.

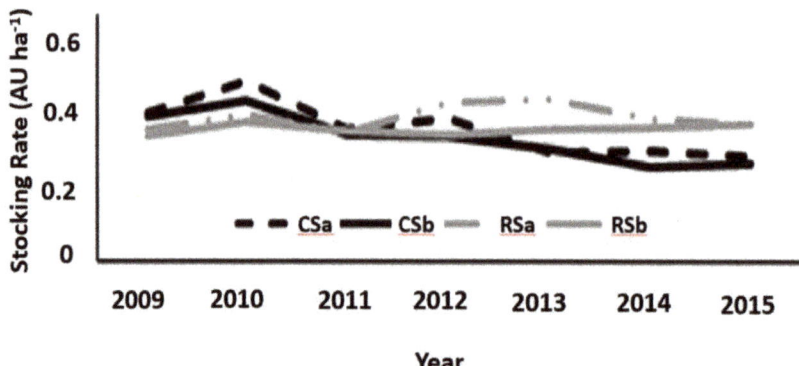

Figure 5. Stocking rate for each replicate from 2009 to 2015.

6.4.2. Animal Performance

Cow–calf performance measures are presented by management and year, since no management X year interactions were observed (Table 5). Years 2010 and 2011 were not utilized for statistical analysis. In 2010, data collection was compromised by faulty livestock scales. In spring 2011, a tornado caused catastrophic damage to our animal research facilities, including pasture areas, which hindered the research and collection of data. All animal measurements were taken in the fall, at the time of calf weaning. Fall cow BCS and calf weaning weight (WW) were higher in the CS treatment compared with RS, which could be indicators of either herbage nutritive value or intake (Coleman et al. [57]). Cow BCS varied between years, with 2014 being greater than 2015 ($p < 0.05$), and both of those years being greater than previous years. Calf WW also impacted by year ($p < 0.01$), with 2013 being the highest and 2015 being the lowest. When slight differences in stocking rate and calf weaning percentage were considered, no differences were observed in WW per cow AU or total calf kg WW ha^{-1} due to grazing system.

Table 5. Effect of grazing system (GS) and year on cow body weight (BW), body condition score (BCS), calf weaning weight (WW), WW per cow AU, and WW per ha.

Measurement	Continuous Stocking		Rotational Stocking		p value
	Mean	SE	Mean	SE	
Cow BW, kg	640	6.0	630	4.9	0.2122
Cow BCS	5.8 a	0.06	5.6 b	0.05	0.0021
Calf WW, kg	235 a	7.6	215 b	6.2	0.0477
WW ha^{-1}, kg	59	2.1	61	1.7	0.6342

Measurement [1]	Year										p value
	2009	SE	2012	SE	2013	SE	2014	SE	2015	SE	
Cow BW, kg	638	7.8	640	8.7	637	9.2	623	8.8	637	9.0	0.6538
Cow BCS	5.5 c	0.07	5.5 c	0.08	5.6 bc	0.09	6.1 a	0.08	5.8 b	0.09	<0.0001
Calf WW, kg	232	9.9	219	11.0	243	11.7	225	11.1	208	11.4	0.2730
WW Cow-AU^{-1}, kg	183	8.1	170	9.0	195	9.6	181	9.1	168	9.4	0.2689
WW ha^{-1}, kg	74 a	2.7	59 b	3.0	57 b	3.1	58 b	3.0	53 b	3.1	<0.0001

[1] Within individual measurement and across years, means with unlike superscripts differ $p < 0.05$.

7. Discussion

The 2011–2012 drought resulted in poor plant productivity and vigor, particularly on the "A" replicates; increased brushy vegetation was observed in the CSa replicate (reference [58], this issue). Based on the observation of forage availability following drought periods in 2011 and 2012, stocking rate was reduced slightly on the CS grazing treatments in 2013. However, the 2009–2015 mean stocking rate did not differ for CS and RS grazing system (Figure 5).

Plant biomass did not differ between CS and RS grazing systems, but there were significant year effects, which were mainly related to the timing and amount of rainfall received. Forage quality naturally changes throughout the season, but it also was impacted by the timing and amount of rainfall. Statistically significant treatment effects were observed for some forage nutritive values, and while these differences were small, when they occurred, they indicated higher nutritive value in RS than in CS treatment The plant nutritional values (Table 4) were determined from clipped biomass (whole plant samples). Nutritive value of forage evaluated from clipped samples is influenced by many factors including climate and ambient temperature, water stress, soil nutrient (fertility) levels, plant species, cultivar, stage of maturity, leaf-to-stem ratio, plant pests and disease, and management [59]. When related to herbage production and nutritive value estimates, there is no clear relationship with animal productivity. This lack of relationship is indicative that other "system" factors (including animal behavior) are impacting animal responses. Animal performance is related to the forage actually selected and consumed by the grazing animal. In this study, cow BCS and calf BW, both taken at weaning, were greater for CS than for RS, indicating greater DM intake [57] and perhaps greater diet selection. Cows on CS had more area from which to select. The nature of management of RS is to force/encourage animals to more uniformly graze the available herbage, and may result in lower diet nutritive value or lower intake.

A collection of papers present additional findings from this experiment to date. Reference [44] reported that the Standardized Precipitation and Evaporation Index (SPEI) has potential for use as a predictive tool for aboveground biomass and thus could provide an early indicator of above- or below-average forage production, which could benefit land and herd management. Findings in this paper that aboveground biomass was not different in CS and RS grazing systems, based on clipping, is similar to results of reference [60], who applied remote sensing techniques to calculate gross primary productivity and the enhanced vegetative index (EVI), which also reported no significant treatment effect for grazing systems. They also reported that variations in EVI among paddocks managed with rotational stocking were small relative to paddocks managed with continuous stocking, suggesting that rotational stocking generated a more uniform grazing pressure on vegetation at annual scale. Overall, climate and inherent pasture conditions are the major drivers of plant growth and productivity.

While individual calf WW was higher in CS grazing systems than in RS grazing systems (Table 5), productivity expressed as calf WW ha^{-1} was not different in the two grazing systems, which is consistent with the findings of [61] that different grazing systems did not affect gain of stocker cattle in South Africa. Economics of forage-based livestock systems are driven by several factors, but productivity and income per unit land area are key for decision-making. Calf rate of gain and WW are determined by genetic influence of both sire and dam, normally defined by mature size, but also by milk production. Both milk production and calf WW were highly related to herbage DM intake by their dams [57], suggesting those management decisions that affect availability and quality of herbage on offer will affect productivity. If RS management can increase forage nutritive value, then animal performance may be improved. However, in this study, the individual calf WW was higher in CS than in RS systems, but total weaned calf weight per unit area did not differ across treatments. Therefore, producers should perhaps look for changes in other response variables such as soil conditions, plant community dynamics, or total forage production as indicators of system effects.

Reference [55] reported that total, particulate, microbial biomass and mineralizable carbon and N fractions were highly stratified with depth and that no significant differences in soil organic carbon and N fractions occurred due to stocking method at any sampling time or depth. Evidence for biological nitrification inhibition in both treatments suggested a mechanism for conservation of available soil N with less opportunity for loss in the prairie soils. In addition, strong association of available N with biologically active carbon indicated slow, but sustained release of N that was strongly coupled to carbon cycling.

Reference [42] reported that effects of stocking methods on macronutrient availability in soil and spatial distribution of macronutrients in soil [43] were variable and often difficult to define. All tested grazing regimes affected soil levels of availability and flux of macronutrients at different times of growing season, and among locations in paddocks. Such responses indicated grazing regimes may not provide uniform distribution of labile macronutrients. The grazing regimes tested appeared to generate localized increases in plant-available macronutrients with change in time of growing season, and among paddock locations, indicating no one system of grazing will prevent hot spots of soil macronutrient availability and flux. This result indicated that organization of the landscape included within the boundaries of individual paddocks played a role in distribution, but actual distances from sampled locations to water sources had less of a material effect on nutrient distributions.

Reference [60] analyzed the phenology and gross primary production of tallgrass pastures in response to CS and RS management systems in drought and non-drought years and found that both the CS and RS grazing systems were resilient to drought, probably due to the conservative stocking rates applied during the study. Spatial patterns of landscape metrics (richness, evenness, and fragmentation) were evaluated for CS and RS grazing system by reference [58], who indicated that shrub encroachment was mainly controlled by the initial status of the pastures rather than the grazing system. Their study also indicated higher proportion of bare soil occurred in RS sub-paddocks that were being grazed or were recently grazed, compared with the CS treatments.

8. Conclusions

In developing the initial study, input from ranchers, conservationists, and others was solicited to identify the most important knowledge gaps, relevant evaluation criteria, and grazing systems of interest. The research was designed in a way to facilitate transfer of knowledge to ranchers. However, due to turnover of key staff, the connection to the producer group was not maintained to the extent envisioned.

Findings to date indicate that plant biomass and animal productivity of the two grazing management systems were similar. There were some indicators that forage nutritive value and soil nutrient content were enhanced in the RS system compared with the CS system, though differences were small and not supported by animal productivity. Analyses of these results prepares us to engage with producers to share the findings. With minimal differences in plant and animal productivity, future research should continue to evaluate changes in soil and plant ecological responses. Additionally, researchers should explore with producers alternative production practices that might increase productivity or profitability and enhance ecological responses. Methods to improve the overall responses to the applied grazing systems might include the use of applied spring burns to manipulate the herbaceous community and eliminate areas of standing dead biomass and provide uniform stands of higher quality biomass, use of patch burning [62] to enhance biodiversity of paddocks, or use of strip grazing to provide more uniform use of paddock and sub-paddock areas. Detailed assessment of vegetative and soil biodiversity may identify responses that plant biomass monitoring and soil nutrient analysis did not detect to date.

Author Contributions: Author contributions are as follows: J.L.S: conceptualization, methodology, original draft preparation, administration, funding acquisition; P.J.S: conceptualization, methodology, formal analysis, writing original draft, and writing and editing; J.P.S.N.: formal analysis, investigation, writing—original draft preparation; B.N. conceptualization, methodology, data curation, writing—review and editing; K.E.T.: forage nutritive value assessment, and writing—review and editing; P.G. supervision, resources, writing—review and editing, project administration; S.C.: methodology, project administration, writing- review and editing; M.B. investigation, supervision, writing—review and editing.

Funding: This research is a contribution from the Long Term Agroecosystem Research (LTAR) network, a project of the USDA-ARS, and was partially supported by funds from the Agriculture and Food Research Initiative Competitive Projects 2012-02355 and 2013-69002-23146 from the USDA National Institute of Food and Agriculture.

Acknowledgments: This project was established to address concerns of ranchers raised at a workshop in 2008 attended by USDA-ARS, USDA-NRCS, the Oklahoma Grazing Lands Conservation Association, the Holistic Management Institute, and Texas AgriLife. The authors gratefully acknowledge the leadership of Dr. Brad Venuto (posthumously) in the establishment of this project. Site management and data collection efforts of Clendon Tucker, Dr. David Von Tungeln, Scott Schmidt, Pat King, Bill Jensen, and many others were critical to this project. Additional researchers have been engaged over time, as evidenced by authorship on this collection of papers.

Conflicts of Interest: The authors declare no conflict of interest.

Disclaimer: Mention of trade names or commercial products is this publication is solely for the purpose of providing specific information and does not imply recommendation or endorsement by the U.S. Department of Agriculture. USDA is an equal opportunity provider and employer.

Appendix A

Appendix A.1 High-Density, Short-Duration (HDSD) Treatments

In addition to the grazing system experiment, an additional study utilized two sets of two small paddocks (0.405 and 0.202 ha) that were established as exclosures within each of the continuous paddocks (see Figure 2) to examine impacts of high-density short-duration (HDSD) stocking on vegetation and soil characteristics. Each year, the herd assigned to each CS paddock were confined within the 0.405 ha paddocks (HDSDx1) for 24 h, returned to the CS paddock overnight, then confined into a 0.202 ha paddocks (HDSDx2) for 24 h (Appendix A). No HDSD occurred in 2012 due to severe drought conditions. The removal of animals for no more than 2 days per year for the embedded stock density experiment was assumed to have a negligible impact on the CS treatment. To avoid unintended plant community shifts caused by high-density grazing in any particular season, the timings of these treatments were sequenced to be different months in different years.

Appendix A.2 Soil Water Content by Depth

The only soil water measurements available for this study were from the USDA-NRCS Soil and Climate Analysis Network (SCAN; https://www.wcc.nrcs.usda.gov/ scan/) site that was installed in the study area in late 1998. Hourly soil volumetric water content (θv), at the 5, 10, 20, 50, and 100 cm (nominally) depths, are measured with a Hydroprobe (Steven's Water Inc., Grants Pass, OR, USA). Due to electronic issues, measurements at the 5 and 10 cm depths were not recorded from November 2014 through June 2015. Measurements from the 20 and 50 cm depths were not recorded from September 2011 through July 2013, and data from the 100 cm depth are missing from November 2014 through June 2015. However, enough data was available to calculate a simple soil profile average for most of the study period.

The averaged measured monthly soil θ_v for the −100 cm profile and 5, 10, 20, 50, and 100 cm depths are shown in Figures A1–A6. At the 5 cm depth (Figure A2), with only two exceptions, θv was typically >15% during the 2008–2010 timeframe. However, due to low precipitation amounts beginning in December of 2010, a steady drying of the soil occurred at this depth. For most of the period from April to November 2011, θv was >15%. The relatively large rainfall events in late 2011 and early 2012 served to bring θv >15%, which lasted through May of 2012. Beginning in June of 2012, however, θv was again <15%, briefly rebounding to 20% in October, and then falling below 15% in November and December 2012. Precipitation in 2013 was above normal, and this is reflected in the 5 cm θv measurements, where θv was >19% in all months and >25% in seven months. Although 2014 was a below-normal rainfall year, the rainfall amounts received enabled θv to stay above 16% in all months for which data were available. The data for the November 2014 through June 2015 was not available, but the precipitation data suggest that θv was likely >15% during this time period. Only four of the 30 months from July 2015 through December 2017 exhibited θv <15%.

Time periods where θv at the 10 cm depth (Figure A3) falls below 15% are largely similar to those observed at the 5 cm depth. Some exceptions include: in 2011, θv at 10 cm is <15% only in three months, whereas at 5 cm, six months were <15%; in 2012, θv was <15% in four months vs. 6 months at 5 cm; no month was below 15% in 2015, 2016, or 2017.

Two 7 to 8 month long time periods were observed where θv <13% was observed at the 50 cm depth (Figure A5): July 2009 through January 2010 and August 2012 through March 2013. One- to two-month-long periods occurred in November of 2011, January of 2015, and December of 2016. Profile averages of θv (Figure A1) indicated July of 2008 and 2009, June of 2010, July through September 2011, January of 2012, and July through September and December of 2012 were ≤13%. The 50 cm (Figure A5) and 100 cm (Figure A6) depths showed similar patterns.

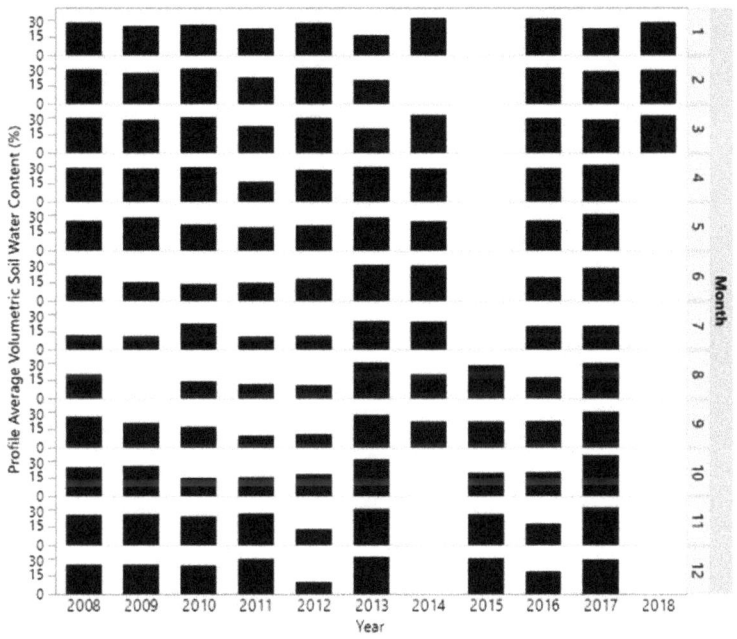

Figure A1. Monthly average volumetric water content (%) for the 0–100 cm profile by year.

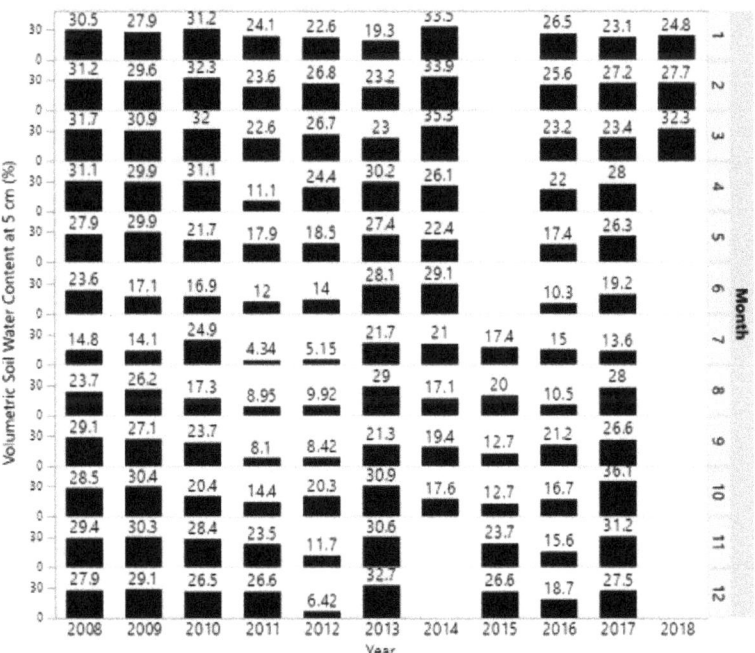

Figure A2. Monthly average volumetric water content (%) at 5 cm depth by year.

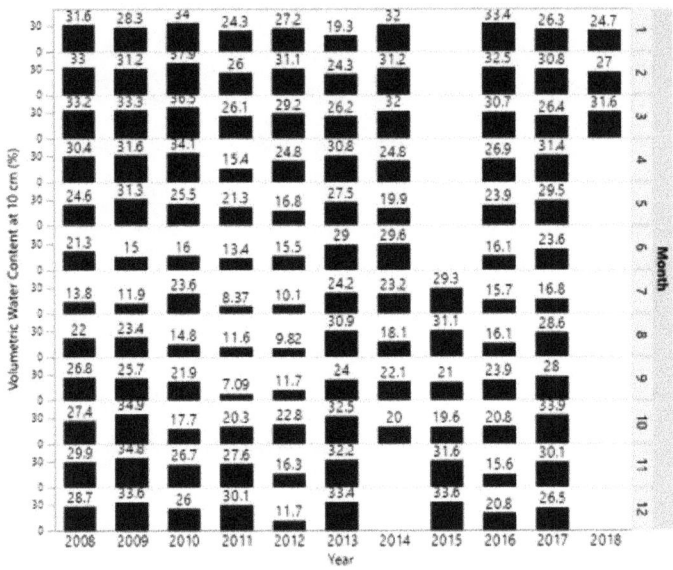

Figure A3. Monthly average volumetric water content (%) at 10 cm depth by year.

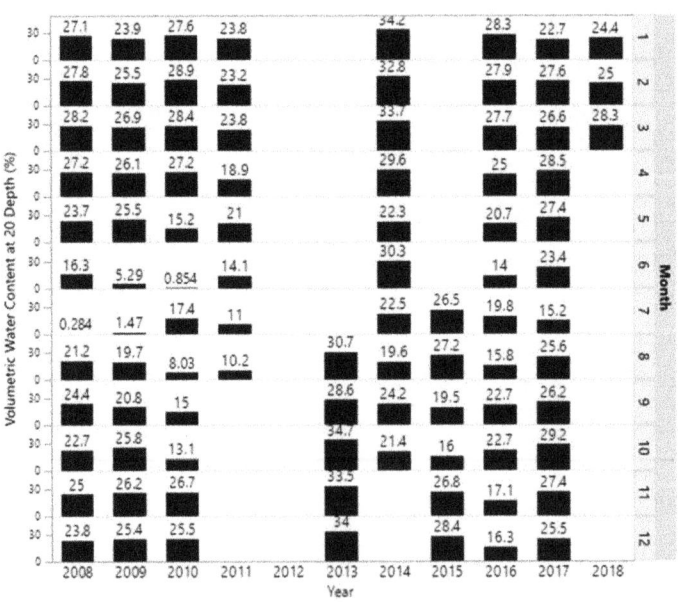

Figure A4. Monthly average volumetric water content (%) at 20 cm depth by year.

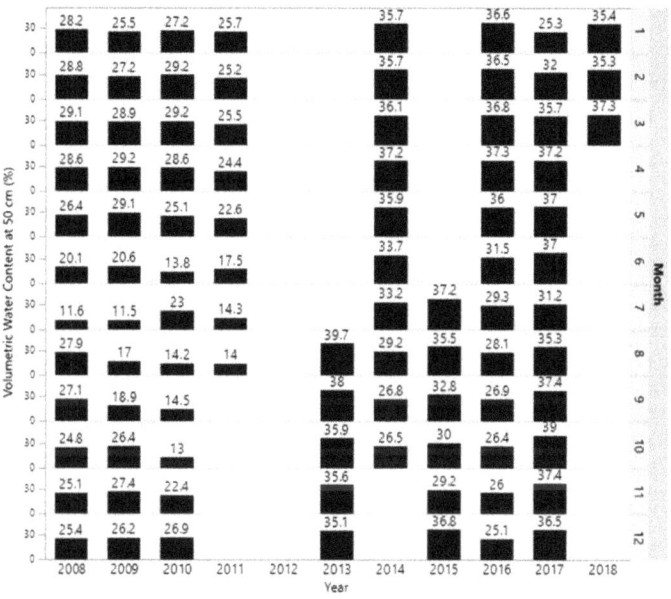

Figure A5. Monthly average volumetric water content (%) at 50 cm depth by year.

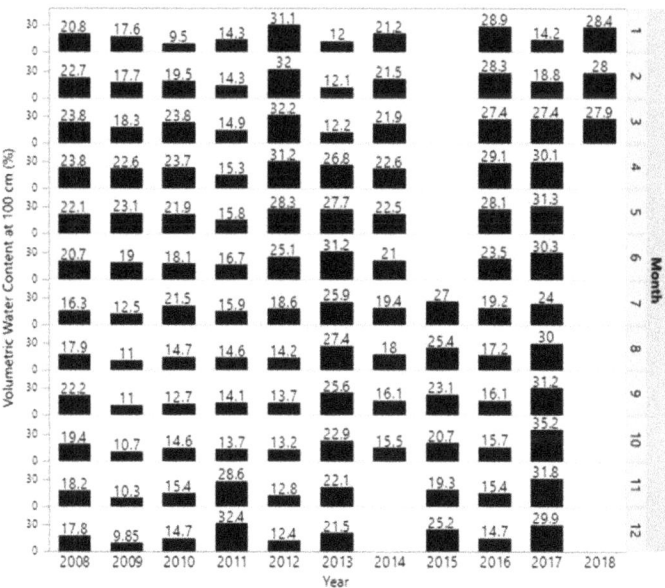

Figure A6. Monthly average volumetric water content (%) at 100 cm depth.

Appendix A.3 Soil Properties

Table A1. Means of percent sand and clay and pH and electrical conductivity (EC) by layer for each treatment.

Treatment [†]	% Sand	% Clay	pH		EC (uS cm^{-1})	
	0–30 cm		0–15 cm	15–30 cm	0–15 cm	15–30 cm
CS	26.7 [A]	23.1 [A]	6.1 [A]	5.92 [A]	329.8 [A]	260.0 [A]
RS	22.6 [A]	25.3 [A]	6.1 [A]	6.02 [A]	292.8 [A]	207.5 [A]

[†] Column means with unlike superscripts differ ($p < 0.05$).

Table A2. Average bulk density for the two soil layers as a function of year × treatment effects.

Year, Treatment [†]	Depth Interval	
	0–15 cm	15–30 cm
	g cm^{-3}	
2009, CS	1.04 [C]	1.21 [B]
2009, RS	1.12 [B]	1.21 [B]
2012, CS	1.28 [A]	1.35 [A]
2012, RS	1.28 [A]	1.33 [A]
2017, CS	1.05 [C]	1.14 [C]
2017, RS	1.07 [BC]	1.16 [C]

[†] Column means with unlike superscript differ ($p < 0.05$).

References

1. Bigelow, D.P.; Borchers, A. *Major Uses of Land in the United States, 2012*; EIB-178; United States Department of Agriculture: Economic Research Service: Washington, DC, USA, 2018.
2. Steiner, J.L.; Franzluebbers, A.J.; Neely, C.; Ellis, T.; Aynekulu, E. Enhancing soil and landscape quality in smallholder grazing systems. In *Soil Management of Smallholder Agriculture. Advances in Soil Science*; Lal, R., Ed.; CRC Press: Boca Raton, FL, USA, 2014; pp. 63–111.
3. Panunzi, E. *Are Grasslands under Threat? Brief Analysis of FAO Statistical Data on Pasture and Fodder Crops*; UN Food and Agriculture Organization: Rome, Italy, 2008.
4. Neely, C.; Bunning, S.; Wilkes, A. *Review of Evidence on Drylands Pastoral Systems and Climate Change*; Land and Water Discussion Paper No. 8; Food and Agriculture Organization of the United Nations: Rome, Italy, 2009.
5. Smith, M.R.; Myers, S.S. Impact of anthropogenic CO_2 emissions on global human nutrition. *Nat. Clim. Chang.* **2018**, *8*, 834–839. [CrossRef]
6. Brussaard, L.; Caron, P.; Campbell, B.; Lipper, L.; Mainka, S.; Rabbinge, R.; Babin, D.; Pulleman, M. Reconciling biodiversity conservation and food security: Scientific challenges for a new agriculture. *Curr. Opin. Environ. Sustain.* **2010**, *2*, 34–42. [CrossRef]
7. Sollenberger, L.E.; Kohmann, M.M.; Dubeux, J.C.B., Jr.; Silveira, M.L. Grassland management affects delivery of regulating and supporting ecosystem services. *Crop Sci.* **2019**, *59*, 441–459. [CrossRef]
8. Boone, R.B.; Conant, R.T.; Sircely, J.; Thornton, P.K.; Herrero, M. Climate change impacts on selected global rangeland ecosystem services. *Glob. Chang. Biol.* **2018**, *24*, 1382–1393. [CrossRef]
9. Hatfield, J.L.; Boote, K.J.; Kimball, B.A.; Ziska, L.H.; Izaurralde, R.C.; Ort, D.; Thomson, A.M.; Wolfe, D. Climate impacts on agriculture: Implications for crop production. *Agron. J.* **2011**, *103*, 351–370. [CrossRef]
10. Izaurralde, R.C.; Thomson, A.M.; Morgan, J.A.; Fay, P.A.; Polley, H.W.; Hatfield, J.L. Climate impacts on agriculture: Implications for forage and rangeland production. *Agron. J.* **2011**, *103*, 371–381. [CrossRef]
11. Steiner, J.L.; Wagle, P.; Gowda, P. Management of water resources for grasslands. In *Improving Grassland and Pasture Management in Agriculture*; Marshall, A., Collins, R., Eds.; Burleigh Dodds Science Publishing: Cambridge, UK, 2017; pp. 265–282.
12. Samson, F.B.; Knopf, F.L.; Ostlie, W.R. Great Plains ecosystems: Past, present, and future. *Wildl. Soc. Bull.* **2004**, *32*, 6–15. [CrossRef]

13. Samson, F.; Knopf, F. Prairie conservation in North America. *BioScience* **1994**, *44*, 418–421. [CrossRef]
14. Conner, R.; Seidl, A.; VanTassell, L.; Wilkins, N. United States Grasslands and Related Resources: An Economic and Biological Trends Assessment. 2002. Available online: http://twri.tamu.edu/media/256592/unitedstatesgrasslands.pdf (accessed on 25 March 2019).
15. Askins, R.; Chavez-Ramirez, F.; Dale, B.; Haas, C.; Herkert, J.; Knopf, F.; Vickery, P.D. *Conservation of Grassland Birds in North America: Understanding Ecological Processes in Different Regions*; The American Ornithologists' Union: Chicago, IL, USA, 2007; pp. 1–46.
16. Comer, P.J.; Hak, J.C.; Kindscher, K.; Muldavin, E.; Singhurst, J. Continent-scale landscape conservation design for temperate grasslands of the Great Plains and Chihuahuan Desert. *Nat. Areas J.* **2018**, *38*, 196–211. [CrossRef]
17. Hill, J.M.; Egan, J.F.; Stauffer, G.E.; Diefenbach, D.R. Habitat availability is a more plausible explanation than insecticide acute toxicity for US grassland bird species declines. *PLoS ONE* **2014**, *9*, e98064. [CrossRef]
18. Derner, J.D.; Smart, A.J.; Toombs, T.P.; Larsen, D.; McCulley, R.L.; Goodwin, J.; Sims, S.; Roche, L.M. Soil health as a transformation change agent for US grazinglands management. *Rangel. Ecol. Manag.* **2018**, *71*, 403–408. [CrossRef]
19. Sarkar, S. Phenology and carbon fixing: A satellite-based study over Continental USA. *Int. J. Remote Sens.* **2018**, *39*, 1–16. [CrossRef]
20. Teague, W.R. Toward restoration of ecosystem function and livelihoods on grazed agroecosystems. *Crop Sci.* **2015**, *55*, 2550–2556. [CrossRef]
21. Park, J.-Y.; Ale, S.; Teague, W.R. Simulated water quality effects of alternate grazing management practices at the ranch and watershed scales. *Ecol. Model.* **2017**, *360*, 1–13. [CrossRef]
22. Wang, J.; Xiao, X.; Qin, Y.; Doughty, R.B.; Dong, J.; Zou, Z. Characterizing the encroachment of juniper forests into sub-humid and semi-arid prairies from 1984 to 2010 using PALSAR and Landsat data. *Remote Sens. Environ.* **2018**, *205*, 166–179. [CrossRef]
23. Caterina, G.L.; Will, R.E.; Turton, D.J.; Wilson, D.S.; Zou, C.B. Water use of Juniperus virginiana trees encroached into mesic prairies in Oklahoma, USA. *Ecohydrology* **2014**, *7*, 1124–1134.
24. Qiao, L.; Zou, C.B.; Stebler, E.; Will, R.E. Woody plant encroachment reduces annual runoff and shifts runoff mechanisms in the tallgrass prairie, USA. *Water Resour. Res.* **2017**, *53*, 4838–4849. [CrossRef]
25. Acharya, B.S.; Hao, Y.; Ochsner, T.E.; Zou, C.B. Woody plant encroachment alters soil hydrological properties and reduces downward flux of water in tallgrass prairie. *Plant Soil* **2017**, *414*, 379–391. [CrossRef]
26. Zhou, Y.; Xiao, X.; Zhang, G.; Wagle, P.; Bajgain, R.; Dong, J.; Jin, C.; Basara, J.B.; Anderson, M.C.; Hain, C.; et al. Quantifying agricultural drought in tallgrass prairie region in the U.S. Southern Great Plains through analysis of a water-related vegetation index from MODIS images. *Agric. For. Meteorol.* **2017**, *246*, 111–122. [CrossRef]
27. Augustine, D.J.; Blumenthal, D.M.; Springer, T.L.; LeCain, D.R.; Gunter, S.A.; Derner, J.D. Elevated CO_2 induces substantial and persistent declines in forage quality irrespective of warming in mixed grass prairie. *Ecol. Appl.* **2018**, *28*, 721–735. [CrossRef]
28. Riitters, K.H.; Wickham, J.D. How far to the nearest road? *Front. Ecol. Environ.* **2003**, *1*, 125–129. [CrossRef]
29. Lal, R. Global food security and nexus thinking. *J. Soil Water Conserv.* **2016**, *71*, 85A–90A. [CrossRef]
30. Holechek, J.L.; Pieper, R.D.; Herbel, C.H. *Range Management Principles and Practices*; Chapter 8 Considerations Concerning Stocking Rate; Pearson, Prentice Hall: Upper Saddle River, NJ, USA, 2004; pp. 216–260.
31. Savory, A.; Butterfield, J. *Holistic Management: A New Framework for Decision Making*, 2nd ed.; Island Press: Washington, DC, USA, 1999; 644p.
32. Teague, W.R.; Apfelbaum, S.; Lall, R.; Kreuter, U.P.; Rowntree, J.; Davies, C.A.; Concer, R.; Rasmussen, M.; Hatield, J.; Wang, T.; et al. The role of ruminants in reducing agriculture's carbon footprint in North America. *J. Soil Water Conserv.* **2016**, *71*, 156–164. [CrossRef]
33. Briske, D.D.; Derner, J.D.; Brown, J.R.; Fuhlendorf, S.D.; Teague, W.R.; Havstad, K.M.; Gillen, R.L.; Ash, A.J.; Willms, W.D. Rotational grazing on rangelands: Reconciliation of perception and experimental evidence. *Rangel. Ecol. Manag.* **2008**, *61*, 3–17. [CrossRef]
34. Briske, D.D.; Sayre, N.F.; Huntsinger, L.; Fernandez-Gimenez, M.; Budd, B.; Derner, J.D. Origin, persistence, and resolution of the rotational grazing debate: Integrating human dimension into rangeland research. *Rangel. Ecol. Manag.* **2011**, *64*, 325–334. [CrossRef]

35. Derner, J.; Briske, D.; Boutton, T. Does grazing mediate soil carbon and nitrogen accumulation beneath C4, perennial grasses along an environmental gradient? *Plant Soil* **1997**, *191*, 147–156. [CrossRef]
36. Fuhlendorf, S.D.; Zhang, H.; Tunnell, T.; Engle, D.M.; Cross, A.F. Effects of grazing on restoration of southern mixed prairie soils. *Restor. Ecol.* **2002**, *10*, 401–407. [CrossRef]
37. Henderson, D.C.; Ellert, B.H.; Naeth, M.A. Grazing and soil carbon along a gradient of Alberta rangelands. *J. Range Manag.* **2004**, *57*, 402–410. [CrossRef]
38. Owensby, C.E.; Ham, J.M.; Auen, L.M. Fluxes of CO_2 from grazed and ungrazed tallgrass prairie. *Rangel. Ecol. Manag.* **2006**, *59*, 111–127. [CrossRef]
39. Follett, R.F.; Reed, D.A. Soil carbon sequestration in grazing lands: Societal benefits and policy implications. *Rangel. Ecol. Manag.* **2010**, *63*, 4–15. [CrossRef]
40. Northup, B.K.; Daniel, J.A. Distribution of soil bulk density and organic matter along an elevation gradient in central Oklahoma. *Trans. ASABE* **2010**, *53*, 1749–1757. [CrossRef]
41. Franzluebbers, A.; Stuedemann, J. Bermudagrass management in the Southern Piedmont USA: VII. Soil-profile organic carbon and total nitrogen. *Soil Sci. Soc. Am. J.* **2005**, *69*, 1455–1462. [CrossRef]
42. Northup, B.K.; Starks, P.J.; Turner, K.E. Stocking methods and soil macronutrient distributions in southern tallgrass paddocks: Are there linkages? *Agronomy* **2019**, *9*, 281. [CrossRef]
43. Northup, B.K.; Starks, P.J.; Turner, K.E. Soil macronutrient responses in diverse landscapes of southern tallgrass to two stocking methods. *Agronomy* **2019**, *9*, 329. [CrossRef]
44. Starks, P.J.; Steiner, J.L.; Neel, J.P.S.; Turner, K.E.; Northup, B.K.; Gowda, P.H.; Brown, M.A. Assessment of the standardized precipitation and evaporation index (SPEI) as a potential management tool for grasslands. *Agronomy* **2019**, *9*, 235. [CrossRef]
45. Spiegal, S.; Bestelmeyer, B.T.; Archer, D.W.; Augustine, D.J.; Boughton, E.H.; Boughton, R.K.; Cavigelli, M.A.; Clark, P.E.; Derner, J.D.; Duncan, E.W.; et al. Evaluating strategies for sustainable intensification of US agriculture through the Long-Term Agroecosystem Research network. *Environ. Res. Lett.* **2018**, *13*, 034031. [CrossRef]
46. Kleinman, P.J.A.; Spiegal, S.; Rigby, J.R.; Goslee, S.C.; Baker, J.M.; Bestelmeyer, B.T.; Boughton, R.K.; Bryant, R.B.; Cavigelli, M.A.; Derner, J.D.; et al. Advancing the Sustainability of US Agriculture through Long-Term Research. *J. Environ. Qual.* **2018**, *47*, 1412–1425. [CrossRef]
47. Garbrecht, J.D.; Rossel, F. Decade-scale precipitation increase in the Great Plains at the end of the 20th century. *J. Hydrol. Eng.* **2002**, *7*, 64–75. [CrossRef]
48. Garbrecht, J.D.; Van Liew, M.; Brown, G.O. Trends in precipitation, streamflow and ET in the Great Plains. *J. Hydrol. Eng.* **2004**, *9*, 360–367. [CrossRef]
49. Garbrecht, J.D.; Zhang, X.C.; Steiner, J.L. Climate change and observed climate trends in the Fort Cobb Experimental Watershed. *J. Environ. Qual.* **2014**, *43*, 1319–1327. [CrossRef]
50. Kunkel, K.E.; Karl, T.R.; Brooks, H.; Kossin, J.; Lawrimore, J.H.; Arndt, D.; Bosart, L.; Changnon, D.; Cutter, S.L.; Doesken, N.; et al. Monitoring and understanding trends in extreme storms: State of Knowledge. *Bull. Am. Meteorol. Soc.* **2013**, *94*, 499–514. [CrossRef]
51. Christian, J.; Christian, K.; Basara, J. Drought and pluvial dipole events within the Great Plains of the United States. *J. Appl. Meteorol. Climatol.* **2015**, *54*, 1886–1898. [CrossRef]
52. Oklahoma Climatological Survey. Climate of Oklahoma. Available online: https://climate.ok.gov/index.php/site/page/climate_of_oklahoma (accessed on 28 March 2019).
53. Omernik, J.M.; Griffith, G.E. Ecoregions of the conterminous United States: Evolution of a hierarchical spatial framework. *Environ. Manag.* **2014**, *54*, 1249–1266. [CrossRef] [PubMed]
54. Day, P.R. Particle fractionation and particle-size analysis. In *Methods of Soil Analysis*; Black, C.A., Ed.; Part I. Agronomy Monograph 9; American Society of Agronomy: Madison, WI, USA; Soil Science Society of America: Madison, WI, USA, 1965; pp. 545–567.
55. Franzluebbers, A.J.; Starks, P.J.; Steiner, P.J. Conservation of soil organic carbon and nitrogen fractions in a tallgrass prairie in Oklahoma. *Agronomy* **2019**, *9*, 204. [CrossRef]
56. FASS. *Guide for Care and Use of Agricultural Animals in Research and Teaching*, 3rd ed.; Federation of Animal Science Societies: Champaign, IL, USA, 2010.
57. Coleman, S.W.; Gunter, S.A.; Sprinkle, J.E.; Neel, J.P. BEEF CATTLE SYMPOSIUM: Difficulties Associated with Predicting Forage Intake by Grazing Beef Cows. *J. Anim. Sci.* **2014**, *92*, 2775–2784. [CrossRef]

58. Ma, S.; Zhou, Y.; Gowda, P.H.; Chen, L.; Starks, P.J.; Steiner, J.L.; Neel, J.P.S. Evaluating the impacts of continuous and rotational grazing on tallgrass prairie landscape using high-spatial-resolution imagery. *Agronomy* **2019**, *9*, 238. [CrossRef]
59. Buxton, D.R. Quality-related characteristics of forages as influenced by plant environment and agronomic factors. *Anim. Feed Sci. Technol.* **1996**, *59*, 37–49. [CrossRef]
60. Zhou, Y.; Gowda, P.H.; Wagle, P.; Ma, S.; Neel, J.P.S.; Kakani, V.G.; Steiner, J.L. Climate effects on tallgrass prairie responses to continuous and rotational grazing. *Agronomy* **2019**, *9*, 219. [CrossRef]
61. Venter, Z.S.; Hawkins, H.; Cramer, M.D. Cattle don't care: Animal behaviour is similar regardless of grazing management in grasslands. *Agric. Ecosyst. Environ.* **2018**, *272*, 175–187. [CrossRef]
62. Cummings, D.C.; Fuhlendorf, S.D.; Engle, D.M. Is altering grazing selectivity of invasive forage species with patch burning more effective than herbicide treatments? *Rangel. Ecol. Manag.* **2007**, *60*, 253–260. [CrossRef]

 © 2019 by the authors. Licensee MDPI, Basel, Switzerland. This article is an open access article distributed under the terms and conditions of the Creative Commons Attribution (CC BY) license (http://creativecommons.org/licenses/by/4.0/).

Article

Soil Quality of Abandoned Agricultural Terraces Managed with Prescribed Fires and Livestock in the Municipality of Capafonts, Catalonia, Spain (2000–2017)

Xavier Úbeda [1,*], Meritxell Alcañiz [1], Gonzalo Borges [1], Luis Outeiro [2] and Marcos Francos [3]

1. Department of Geography, University of Barcelona, Montealegre 6, 08001 Barcelona, Spain; meritxellalpu@gmail.com (M.A.); gonzaloborgespereira@gmail.com (G.B.)
2. Department of Applied Economy, Universidad de Santiago de Compostela, 15705 Santiago de Compostela, Spain; louteiro@gmail.com
3. Departamento de Ciencias Históricas y Geográficas, Universidad de Tarapacá, 18 de Septiembre, 2222, Arica 1010069, Chile; marcosfrancos91@gmail.com
* Correspondence: xubeda@ub.edu; Tel.: +34-93-403-7892

Received: 14 May 2019; Accepted: 21 June 2019; Published: 25 June 2019

Abstract: The abandonment of the economic activities of agriculture, livestock, and forestry since the second half of the 20th century, in conjunction with the exodus of inhabitants from rural areas, has resulted in an increase in the forest mass as well as an expansion of forest areas. This, in turn, has led to a greater risk of forest fires and an increase in the intensity and severity of these fires. Moreover, these forest masses represent a fire hazard to adjacent urban areas, which is a problem illustrated here by the village of Capafonts, whose former agricultural terraces have been invaded by shrubs, and which in the event of fire runs the risk of aiding the propagation of the flames from the forest to the village's homes. One of the tools available to reduce the amount of fuel in zones adjoining inhabited areas is prescribed burns. The local authorities have also promoted measures to convert these terraces into pasture; in this way, the grazing of livestock (in this particular instance, goats) aims to keep fuel levels low and thus reduce the risk of fire. The use of prescribed fires is controversial, as they are believed to be highly aggressive for the soil, and little is known about their long-term effects. The alternation of the two strategies is more acceptable—that is, the use of prescribed burning followed by the grazing of livestock. Yet, similarly little is known about the effects of this management sequence on the soil. As such, this study seeks to examine the impact of the management of the abandoned terraces of Capafonts by means of two prescribed fires (2000 and 2002), which were designed specifically to prevent forest fires from reaching the village. Following these two prescribed burns, a herd of goats began to graze these terraces in 2005. Here, we report the results of soil analyses conducted during this period of years up to and including 2017. A plot comprising 30 sampling points was established on one of the terraces and used to monitor its main soil quality properties. The data were subject to statistical tests to determine whether the recorded changes were significant. The results show modifications to the concentration of soil elements, and since the first prescribed burn, these changes have all been statistically significant. We compare our results with those reported in other studies that evaluate optimum soil concentrations for the adequate growth of grazing to feed goats, and conclude that the soil conditions on the terrace after 17 years are optimum for livestock use.

Keywords: grassland; rural abandonment; goats; soil nutrients; fire risk

1. Introduction

Fire is a widespread phenomenon throughout our planet, and moreover, it is an ecological factor in many ecosystems [1]. In the ecosystem of the Mediterranean basin, the role played by fire is especially important given the region's climate and the inflammable nature of its vegetation. Indeed, some Mediterranean plants have developed reproduction strategies that are dependent on fire. This explains why some Mediterranean fires should be considered part of the natural process [2]. However, today, forest fires present themselves as a problem due to the socioeconomic changes suffered by the region, primarily those associated with the depopulation of rural areas and the abandonment of the silvoagricultural economy over recent decades [3]. Forests have replaced what used to be agricultural fields, increasing plant cover and the spatial continuity of the vegetation, and as a result modifying the fire regime and increasing the risk of forest fires [4]. In short, the accumulation of fuel in Mediterranean forests has become a significant environmental problem [5].

Since the beginning of the 21st century, there has been a need to manage the landscape to avoid large forest fires and minimize their impact at all levels [6]. In this regard, fire management strategies can reduce the fuel load of forests [7]. Two such strategies, prescribed fires (PFs) and prescribed grazing, allow these objectives to be achieved without them having dramatic consequences for the environment. In Catalonia, PFs have been widely employed since 1999; the aims are diverse and dependent on the type of landscape requiring management. Usually, however, the aim has been to reduce the accumulation of fuel in the forest due to the abandonment of forestry, agricultural, and livestock activities over the preceding century [8]. Additionally, the effects of the wind, in combination with certain plant diseases, have led to the accumulation of dead wood in forest areas [9]. In Catalonia, the team responsible for implementing this particular management strategy is the Forest Action Support Group (GRAF in its Catalan acronym), which is part of the Catalan Government's body of firefighters.

A further objective of PFs is the need to reduce fuel loads that have accumulated around inhabited nuclei, especially on former agricultural terraces in the immediate vicinity of villages or towns. Here, the aim is to ensure that the wildfire does not reach the village by eliminating the fuel that can help propagate the fire from the forest to the houses, that is, by creating a firebreak [10].

As stressed above, the abandonment of livestock farming in rural areas has been widespread in many countries. However, these earlier livestock practices served as an efficient control mechanism, as grazing animals not only consumed the vegetation, they also trampled down grasses and plants. In recent years, government agencies have sought to encourage new generations to return to livestock farming—not always, but sometimes, by offering financial aid—for reasons of fire prevention (one such association is known as *Ramats al Bosc* [11]).

As PFs have become more commonplace, their potential effects on a soil's physical, chemical, and biological properties have been widely studied (for a review, see [12]). These effects, in common with those of wildfires, depend on a range of factors and combinations of variables that include fire intensity, soil type, vegetation, land use, and a site's recent management history [13]. Moreover, in the specific case of wildfires, volatilization and the addition of elements, in part due to ash, are important for understanding changes in soil composition [14]. In short, the literature reports a broad range of effects—although not necessarily negative—on the physical, chemical, and biological properties of a soil [12].

Despite the research conducted to date, some questions have yet to be fully addressed. For example, how often can a PF be set in a given ecosystem without causing irreparable damage to its soil parameters? Does the alternation of PF and livestock grazing constitute an effective sequence for managing areas of grazing?

Landscape managers, researchers, and even the public fail to agree on the right frequency of PFs and whether, indeed, this practice is the best management tool that can be applied [15]. Alternatives, most notably mechanical management employing heavy machinery, are expensive, and forest owners are usually unwilling to spend money if they have no guarantee that they can ultimately make a profit.

Here, to understand the scale of the problem, it should be born in mind that in Catalonia, 65% of the territory is forest, and 80% of the property is under private ownership.

The reintroduction of livestock into rural areas has been seen as a natural way to manage the accumulation of forest fuel [16]. In this regard, many actions have been promoted in Catalonia seeking the repopulation of sparsely inhabited areas—including, for example, the opening of a center to train shepherds and the provision of a government subsidy to purchase cattle—while initiatives have been taken to promote rural economies and recover and manage abandoned agricultural terraces and scrubland.

The impact of livestock farming on soils has been quite widely studied in Spain's *dehesa* ecosystems—that is, extensive grassland areas on which large ruminants (primarily, cattle) and pigs are set to feed [17]. More specifically, these studies have examined the impact of trampling on the soil's physical parameters: i.e., compaction, hydrophobicity, bulk density, and erosion problems associated with these surface changes. However, few studies to date have examined the impact of small ruminants—sheep and goats—on a soil's chemical properties. Here, we go some way to rectifying this by identifying the consequences of the reintroduction of livestock on the soil characteristics of the abandoned terraces of the village of Capafonts in a long-term study that spans a period of 17 years.

Capafonts has undergone a process of rural abandonment, its population falling from 400 at the beginning of the 20th century to 100 by the end [18]. This was accompanied by a shift in the predominant economic sector in which its inhabitants were employed, with the primary sector (agriculture–livestock–forestry) losing importance to the tertiary (services, primarily rural tourism) [19]. These social changes have been accompanied by a negative impact on the local environment: the forests have expanded and the tree density has increased, above all, with the abandonment of the terraces on which the almond trees were tended [20].

The village of Capafonts stands on a hill and was formerly surrounded by agricultural terraces. However, today, these have been abandoned to herbs and bushes. The vegetation that now surrounds the village makes it highly vulnerable to fire. Indeed, a wildfire in the forests of the Prades Mountains could easily be propagated to the village as a result of the inflammable vegetation on the terraces (Figure 1). In 2000, to eliminate this risk, firefighters set a PF to burn the vegetation on these terraces and create a firebreak around Capafonts. A second PF was set two years later, in 2002. Since that date, no more interventions have been taken on these terraces.

Figure 1. Photographs of the abandoned terraces and the village of Capafonts and of our study plot with the Prades mountains in the background (2017).

However, the beginning of the 21st century has seen the emergence of a new social and demographic phenomenon: that of, deurbanization or neo-ruralization, in which the urban population has begun to move back to the rural world [21]. Young entrepreneurs believe that the rural world provides opportunities for them to change their lifestyle and put their business ideas into practice. For example, in Capafonts, a young family set up home in the village and began farming goats to make cheese—and significantly, for our study here, the livestock began to graze on the terraces of the former almond groves, which had finally been abandoned in 2004.

The new farmers do not own these terraces, but there are no restrictions on their using them for grazing. Indeed, the original owners appear to have lost any interest in staking claim to them many years ago. In recent years, this farming family has bought various hectares of forest to convert into grazing land for their animals. As a result, the current land use of these terraces is grass pasture.

This study provides novel information about the use of PFs as a long-term management tool. The use of this type of management practice combined with controlled grazing is common in Mediterranean forest environments, such as those found in Catalonia, but few studies of their impact have been carried out to date, and so, few details are available about their effects. The specific objective of this study is to determine the soil conditions on these abandoned terraces after 17 years of fire management—which was designed to create a firebreak around Capafonts—and after more than a decade of goat herding; as such, it represents a significant advance in studies of this type.

2. Materials and Methods

Capafonts, a municipality in the province of Tarragona (NE Spain), provides a classic example of rural abandonment that was initiated in the final decades of the 20th century, a process not unlike that which affected many towns and villages in the developed world. The traditional forestry practices involving the extraction of wood and charcoal from the oak woodlands were discontinued, and the farmers gradually abandoned their rain-fed crops and livestock grazing (primarily, goats in this area).

The municipality of Capafonts lies in the Prades Mountains in the province of Tarragona (NE Iberian Peninsula). It has a population of 102 inhabitants (2017) and occupies an area of 13.27 km^2. Our study plot is at an altitude of 740 m a.s.l. and was burnt for the first time on 25 February 2000.

Capafonts has an average annual rainfall of 712 mm, with autumn and spring maxima, and an average annual temperature of 12 °C. The substrate is calcareous and the soil is Xerorthents [22]. Until the first half of the 20th century, the municipality's terraces were planted with almond trees. Following their abandonment, the vegetation that emerged was a grass species (*Brachypodium phoenicoides* L.), which reached heights of around one meter. The plant cover was homogeneous and quite dense, and represented a considerable fire risk, as it could serve to propagate flames from the forest to the village.

The physical characteristics of the soil in terms of its particle sizes are as follows: 34.14% sands, 31.84% silt, and 34.02% clays, with 53.6% of the soil comprising coarse elements (>2 mm).

The sampling design was based on a 4 m × 18 m grid, with 30 sampling points set at 2-m intervals. This experimental design without replicates, including that of the control area, was previously used and described by Francos et al. [9] and Alcañiz et al. [15]. The large number of samples taken at each sampling ensures the reliability of our statistical outcomes. Sampling was conducted before and just after the prescribed fire. An additional sampling was completed one year after the prescribed fire of 2000. A second prescribed fire was set in 2002 two years after the first PF. We sampled the soil to determine the impact of repeating the fire management exercise. In 2004, four years after, another sampling was conducted to determine the changes that had taken place in the soil and, from that year onwards, a herd of goats was set to graze on this plot. A final sampling was conducted in 2017, 17 years after the first prescribed fire.

The sampling design used in this study is the same as that described for the Montgrí plot (see [15]). Soil samples were taken from the top layer (0–5 cm) at each of the 30 sampling points using a small pick. Then, ash was removed, and the sample was air dried in the laboratory and sieved to obtain a <2 mm fraction. Soil pH [1:2.5] was analyzed following extraction with deionized water and measured with a pH meter [23]. Soil organic carbon (OC) was measured using the loss-on-ignition (LOI) method described in [24]. Total N was determined by elemental analysis (NaA2100 Protein Nitrogen Analyser). Available P was analyzed using the Olsen Gray method [25]. Exchangeable cations, Ca^{2+}, Mg^{2+}, and K⁺ were analyzed by ammonium acetate extraction [26] and determined by atomic absorption spectrophotometry. CaCO$_3$ was analyzed with a Bernard calcimeter.

Before conducting the statistical analyses, data normality and homogeneity were checked using a Shapiro–Wilk and Levene test. All the data followed a Gaussian distribution and presented homogeneity of variance. Thus, we applied a one-way ANOVA test, and significant differences were identified at $p < 0.05$. Tukey's post-hoc test identified significant differences between sampling moments. A redundancy analysis (RDA) was carried out to identify the variations between the different sampling moments. The soil properties included were pH, OC, N, Ca^{2+}, Mg^{2+}, K⁺, and CaCO$_3$. These analyses were carried out using SPSS 20.0 and Canoco for Windows 4.5.

3. Results

We show the soil chemical results from the six sampling periods in Table 1, and also report statistically significant increases or decreases in these properties by means of the *p*-values from a one-way ANOVA.

Almost all the soil parameters analyzed here increased after the first PF; however, a year later, these values had fallen. The second PF caused slight increases in soil calcium, magnesium, potassium, and phosphorus, but organic carbon and nitrogen levels did not rise. At the end of the study in 2017, there was considerable variation in the soil parameters compared to their initial values, and our objective here is to determine (see Section 4.6 for details) whether the soil quality is optimal for sustaining grazing.

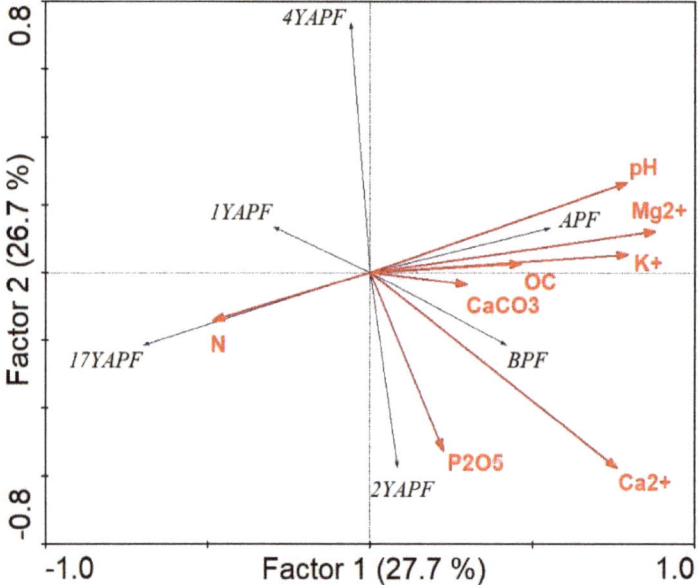

Figure 2. Redundancy analysis (RDA) showing the relation between factors 1 and 2. pH, organic carbon (OC), nitrogen (N), Ca^{2+}, Mg^{2+}, K⁺, P$_2$O$_5$ and CaCO$_3$. Before the prescribed fire (BPF), after the prescribed fire (APF), one year after the prescribed fire (1YAPF), two years after the prescribed fire (2YAPF), four years after the prescribed fire (4YAPF), and 17 years after the prescribed fire (17YAPF)

Table 1. Minimum, maximum, mean (in bold), Prescribed Fire (PF), standard deviation (SD), variance, and standard error (SE) values before the prescribed fire (BPF), after the prescribed fire (APF), one year after the prescribed fire (1YAPF), two years after the prescribed fire (2YAPF), four years after the prescribed fire (4YAPF), and 17 years after the prescribed fire (17YAPF). Different letters represent significant differences at a $p < 0.05$. * $p < 0.05$, ** $p < 0.01$, *** $p < 0.001$. $N = 30$.

Soil Properties	Statistics	Before PF	After PF	1 Year	2 Years and Immediately after Second PF	4 Years Goats Introduced	17 Years
pH	Min	7.37	7.73	7.63	7.60	7.29	7.07
	Max	8.11	8.23	7.91	7.79	8.24	7.63
	Mean	**7.83**	**8.00**	**7.77**	**7.71**	**7.76**	**7.37**
	SD	0.14	0.13	0.07	0.05	0.18	0.13
	Variance	0.02	0.02	0.01	0.00	0.03	0.02
	***	b	a	bc	c	bc	d
OC (%)	Min	3.55	4.32	6.54	7.77	4.54	1.10
	Max	21.98	24.74	14.18	12.17	12.98	12.51
	Mean	**10.32**	**11.53**	**11.42**	**10.09**	**8.56**	**5.39**
	SD	4.56	4.04	1.70	1.10	2.00	2.85
	Variance	20.76	16.34	2.90	1.21	3.76	8.13
	***	ab	a	a	ab	b	c
N (%)	Min	0.26	0.32	0.29	0.23	0.20	0.32
	Max	0.53	0.64	0.64	0.50	0.50	0.94
	Mean	**0.36**	**0.46**	**0.48**	**0.34**	**0.32**	**0.63**
	SD	0.08	0.24	0.11	0.07	0.08	0.14
	Variance	0.01	0.06	0.01	0.00	0.01	0.02
	***	c	b	b	c	c	a
Ca^{2+} (ppm)	Min	16,710	13,830	5120	15,138	4578	9014
	Max	29,900	31,280	13,420	21,731	9066	12,632
	Mean	**22,130**	**20,784**	**8836**	**18,179**	**6469**	**10,177**
	SD	4329	4121	2037	1668	869	790
	Variance	18,748,186	16,990,130	4,149,975	2,783,349	756,751	625,038
	***	a	a	c	b	d	c
Mg^{2+} (ppm)	Min	614	960	320	533	474	325
	Max	1583	1444	830	838	886	640
	Mean	**1006**	**1244**	**507**	**627**	**713**	**446**
	SD	254	139	130	67	78	73
	Variance	63,476	19,543	17,043	4522	6173	5446
	***	b	a	d	c	c	d
K^+ (ppm)	Min	697	900	459	606	563	460
	Max	1732	1816	928	1003	956	865
	Mean	**1132**	**1326**	**665**	**738**	**771**	**591**
	SD	319	246	134	95	96	98
	Variance	101,881	60,900	18,183	9155	9273	9617
	***	b	a	cd	c	c	d
P_2O_5 (ppm)	Min	49.12	5.98	78.95	105.20	41.20	45.00
	Max	138.01	245.72	253.16	320.44	142.54	143.55
	Mean	**84.70**	**132.24**	**137.25**	**216.53**	**65.24**	**69.61**
	SD	24.33	42.62	36.59	48.89	20.28	19.39
	Variance	591.74	1816.65	1338.91	2390.59	411.39	375.89
	***	c	b	b	a	c	c
$CaCO_3$ (%)	Min	21.48	19.57	11.01	20.60	21.00	21.60
	Max	43.61	32.97	28.76	30.91	29.05	29.13
	Mean	**28.72**	**24.34**	**20.80**	**24.28**	**25.87**	**24.54**
	SD	4.30	3.23	3.86	2.04	1.99	1.67
	Variance	18.50	10.45	14.95	4.16	4.03	2.81
	***	a	b	c	b	b	b

The RDA allows us to determine differences in soil properties over time. Factor 1 in the RDA explains 27.7% of the variance, and Factor 2 explains 26.7% of the variance, that is, a combined total of 54.4% of the variance. The variables with the highest explanatory capacity are Ca^{2+}, Mg^{2+}, and pH, while those with the lowest capacity are $CaCO_3$ and OC. The RDA clearly separates the prescribed fire events that played an important role—that is before the prescribed fire (BPF), after the prescribed fire (APF), and two years after the prescribed fire (2YAPF)—from those that happened some time after the prescribed fire. Ca^{2+} could be associated with BPF, Mg^{2+}, pH, K^+, OC, and $CaCO_3$ were closely associated with APF, and P_2O_5 was associated with 2YAPF (Figure 2).

4. Discussion

4.1. pH

Changes in pH after a fire can be attributed to the oxidation of the organic matter and the incorporation of cations produced by combustion [27]. According to the literature, after a PF, different outcomes are found: some studies report no changes in the pH in their soils [28,29], some observe no significant changes [30], while others report an increase [31,32].

The reasons for these changes, significant or otherwise, as well as for the lack of alterations in soil pH, after prescribed burns appear to be related to any treatments conducted prior to fuel combustion [30], and, according to Alcañiz et al. [15], the intensity at which the fire burned at the surface and in the first few centimeters of the soil layer. Various authors, including [15], stressed that the recurrence of prescribed fires at short time intervals can also have a differential impact on soil pH. This conclusion is ratified by Muqaddas et al. [33], who reported the effects of periodic PFs with frequency regimes of two and four years over a 35-year period.

Here, we found a significant increase in pH after the first burn, but no significant change was recorded after the second. This behavior can be attributed to there being more fuel to burn in the first PF as different species of bush had colonized the area, whereas during the second burn, what was burnt was largely dry grasses. Moreover, in line with the findings of Muqaddas et al. [33], repeated PFs at the same site result in different impacts on a soil's chemical qualities.

Few studies have examined the pH changes in soils subject to grazing. However, Teague et al. [34] reported significant pH changes and variations depending on whether grazing is intense ('heavy continuous') or more moderate ('light continuous'). Here, we found a statistically significant decrease in pH values 13 years after goats were introduced (i.e., 17 years after the first PF). This decrease is not unexpected if we consider (see below for more details) that the cation levels responsible for higher pH values also undergo a decrease.

4.2. Organic Carbon and Total Nitrogen

Organic carbon (OC) is one of the most widely studied parameters in this context, given that it is critical for understanding soil quality, whether it be for forestry or agriculture [35]. Most studies report an increase in OC since with combustion, part of the organic matter that burns is partially pyrolyzed [14]. In PFs of both high and low intensity, this causes OC levels to increase [36,37]. However, in line with other parameters, the recurrence of PFs at short time intervals can reduce the magnitude of these changes [33]. Here, we found that the changes after the first PF were more noticeable than those after the second, when OC levels actually fell.

In the case of grazing, Teague et al. [34] reported that plots subject to heavy continuous grazing had the lowest levels of organic carbon. Qasim et al. [38] reviewed 12 studies from around the globe comparing non-grazed plots with those used for grazing, and reported a fall in OC levels in all cases, with just one of the studies not reporting a significance greater than 95%. In our study, we also recorded a statistically significant decrease in OC levels 4YAPF and the introduction of livestock grazing, which is logical if we consider that there is less vegetation after the reintroduction of goats.

Most studies reported an increase in nitrogen after a prescribed burn [15,39]; however, some did not find any significant change [40]. Úbeda et al. [10] associated this increase to the incorporation of micro ash particles produced by combustion into the soil.

In Capafonts, while there was no significant increase in nitrogen after the first PF, we did record a significant rise after the second burn. Similar responses have been recorded by both Muqaddas et al. [33] and Blankenship and Arthur [41]. In these two studies, plots were also burned every two years, suggesting that such frequent burning does not provide sufficient time for the stabilization of certain soil parameters.

In the case of grazing, some studies reported a decrease in soil nitrogen following the passage of cattle [38] compared to soils on which cattle had not been allowed to graze. However, in Capafonts, nitrogen increased considerably, which may be due to the type of vegetation grown on this plot. *Brachypodium phoenicoides* is a legume and, apart from having high nitrogen content [42], it has been shown to be a good nitrogen fixer in the soil [43].

4.3. Cations

Various studies indicate that PFs act to trigger a change in levels of calcium, magnesium, and potassium, as well as changes in the distribution of these cations [44]. Some report an increase after a PF [31,39], as occurred at the Capafonts plot after the first burn, although not after the second. The small quantity of ash produced by the first fire could account for this increase, while the washing or leaching of ash following the second fire might explain why cation values fell. Several studies have also examined the duration of these increases, with some reporting long-term effects [32] and others finding only very ephemeral changes [45]. Alcañiz et al. [15] found that after burning, and once the vegetation had begun to recover (after a year), there was a very significant reduction in potassium levels due to leaching and the high consumption of this cation by plants. Here, we recorded a similar trend. Overall, Alcañiz et al. [15] concluded that PFs are beneficial because they increase nutrient availability.

In line with our findings here, Teague et al. [34] found a statistically significant decrease in the content of calcium and potassium in grass grazing. These authors attribute part of the decline in these cations to erosive processes in places of heavy grazing. However, in Capafonts, the decrease must be attributable to the consumption of these cations by plants, given that our plot is a terrace, and as such provides no slope for erosive processes to occur.

4.4. Phosphorous

In the case of available phosphorus, we found an increase after both PFs. McKee [46] claimed that PFs accelerate the phosphorus cycle, and that in places where limited phosphorus is available, PFs can trigger levels of this nutrient. However, it is not advisable to set frequent PFs, because according to Brye [28] in the United States, annual PFs over a 12-year period significantly reduced the soil phosphorus content.

Teague et al. [34] found no changes in phosphorus levels in association with heavy or light grazing. Likewise, our data showed no statistically significant change in phosphorus content, rising only slightly from 65.24 mg/kg to 69.61 mg/kg between 2005–2017.

However, Aarons et al. [47] and Sharpley and Moyer [48] reported an increase in assimilable phosphorus in the soil, which they attribute to cattle droppings. However, here, given that the number of goats grazing the plot was small, phosphorous levels in the soils of Capafonts did not increase following the reintroduction of livestock on the terraces.

4.5. Calcium Carbonates

Various studies have attributed the changes in soil $CaCO_3$ content after a fire to the incorporation of ash. Pereira et al. [14], in studies conducted both in Portugal and Catalonia, showed how the ash resulting from a very severe combustion (white ash) contains a high percentage of $CaCO_3$. However, here, we recorded an overall fall in $CaCO_3$ content, which could indicate that the PFs were not severe enough to create intense burning.

In the case of grazing, $CaCO_3$ content is typically reported as increasing. Stavi et al. [49] found that there was an increase of $CaCO_3$ content with livestock grazing, although they detected the importance of plot orientation in relation to the predominant vegetation type. This suggests that the vegetation available to the livestock is determined by the soil's calcium carbonate content.

4.6. Multivariable Analysis and Soil Quality in 2017

The multivariable analysis shows that the groups of variables, and the changes in soil properties, are largely determined by the two prescribed burns, and as such, this type of management practice produces greater changes than those resulting from controlled grazing. This is particularly true of the first prescribed burning, around which a large number of variables are grouped.

The soil properties recorded at the Capafonts site in 2017 were compared with the reference values published in various studies conducted on soils dedicated to the cultivation of grazing to provide grazing for different types of livestock. INIA [50], Espinoza et al. [51], and Sela [52] established a range of reference values for P_2O_4 from >30 to 100 mg/kg. The corresponding value for Capafonts stands at 69.61 mg/kg, which means it lies within these limits. In the case of the cations (i.e., Ca, Mg, and K) the values recorded at Capafonts (see Table 1) are above the minimum values specified as references by Sela [52], namely 2000 mg/kg in the case of calcium, 180 mg/kg in that of magnesium, and between 200–800 mg/kg for potassium. Espinoza et al. [51] also established an optimal percentage value for OC in soils for grazing. The minimum value of 1.16% is well below that of 5.39% recorded at Capafonts.

Therefore, we can conclude that exposure to prescribed burnings and 13 years of goat herding have not degraded the soils' chemical properties with the exception of a slight magnesium deficiency, according to the recommended limits identified by Landon [53]. However, the other references consulted do not consider this level unsuitable for grazing. Likewise, AQM [54] recommended a 1% nitrogen level for grazing soils, while our sample has a level of 0.63%, albeit that nitrogen levels at our site have increased in recent years after falling as low as 0.32% (see Table 1). Overall, the soil quality data indicated that the soil of the Capafonts plot is of optimum quality and has not suffered degradation over time.

5. Conclusions

Prescribed burns are a good tool for managing the accumulation of forest fuel, but repeating PFs every two years can result in soil nutrient loss.

The soil parameters analyzed herein present a normal variation if we consider their exposure to prescribed fires and goat browsing. While the changes in the soil's chemical properties are significant, this does not mean that PFs and livestock farming should be shunned. Indeed, on the basis of the data reported herein, the state of the soil today, 17 years after the first PF, can be considered optimal.

A recommendation that could be made to GRAF firefighters when setting prescribed burns is to ensure that the fires do not acquire any great intensity, as it has been shown that high-intensity PFs reduce soil phosphorus content. Therefore, the next step is to sit the GRAF firefighters, the local council, and livestock farmers down at the same table to consider the best timing for the next PF in Capafonts.

Author Contributions: X.Ú. is the head of the research group and wrote the paper; M.A. did the fieldwork and wrote the paper; G.B. did the laboratory analysis; L.O. is the co-supervisor of M.A. PhD Thesis and did the revision of the paper; M.F. did the fieldwork and data treatment.

Funding: Spanish Ministry of Economy and Competitiveness and the European Union via European Funding for Regional Development (FEDER) CGL2013-47862-C2-1-R; CGL2016-75178-C2-2-R [AEI/FEDER, UE] and Agència de Gestió d'Ajuts Universitaris i de Recerca de la Generalitat de Catalunya 2017SGR1344.

Acknowledgments: This study was made possible thanks to Projects CGL2013-47862-C2-1-R and POSTFIRE_CARE Project (CGL2016-75178-C2-2-R [AEI/FEDER, UE]) sponsored by the Spanish Ministry of Economy and Competitiveness and the European Union via European Funding for Regional Development (FEDER). We also thank the FPU Program (FPU13/00139) promoted by the Ministry of Economy, Culture and Sports. Financial support from the Postdoctoral Program I2C, organized by the *Xunta de Galicia*, is gratefully acknowledged for helping in the preparation of this manuscript. We also enjoyed the benefits of grant 2017SGR1344 awarded by the *Agència de Gestió d'Ajuts Universitaris i de Recerca de la Generalitat de Catalunya*, which served to support the activities of the research groups (SGR2017-2019). We thank the members of the GRAF team for providing support in the field and helping in completing the project. Finally, we would like to thank the Scientific and Technological Centers at the University of Barcelona (CCiTUB) for undertaking analyses of soil chemical parameters.

Conflicts of Interest: The authors declare no conflict of interest.

References

1. Doerr, S.H.; Santín, C. Global trends in wildfire and its impacts: Perceptions versus realities in a changing world. *Philos. Trans. R. Soc. B: Biol. Sci.* **2016**, *371*, 20150345. [CrossRef] [PubMed]
2. Pausas, J.G.; Llovet, J.; Rodrigo, A.; Vallejo, R. Are wildfires a disaster in the Mediterranean basin?—A review. *Int. J. Wildland Fire* **2009**, *17*, 713–723. [CrossRef]
3. Vélez, R. *La Defensa Contra Incendios Forestales: Fundamentos y Experiencias*; McGraw Hill Interamericana De España S.A.U: Madrid, Spain, 2000.
4. Moreira, F.; Rego, F.C.; Ferreira, P.G. Temporal (1958–1995) pattern of change in a cultural landscape of northwestern Portugal: Implications for fire occurrence. *Landsc. Ecol.* **2001**, *16*, 557–567. [CrossRef]
5. Pausas, J.G.; Fernández-Muñoz, S. Fire regime changes in the Western Mediterranean Basin: From fuel-limited to drought-driven fire regime. *Clim. Chang.* **2012**, *110*, 215–226. [CrossRef]
6. Fernandes, P.M.; Davies, G.M.; Ascoli, D.; Fernández, C.; Moreira, F.; Rigolot, E.; Molina, D. Prescribed burning in southern Europe: Developing fire management in a dynamic landscape. *Front. Ecol. Environ.* **2013**, *11*, e4–e14. [CrossRef]
7. Verkerk, P.J.; de Arano, I.M.; Palahí, M. The bio-economy as an opportunity to tackle wildfires in Mediterranean forest ecosystems. *For. Policy Econ.* **2018**, *86*, 1–3. [CrossRef]
8. Galán, M.; Lleonart, S. Plans de gestió de grans incendis forestals. In *Incendis Forestals, Dimensió Sociambiental, Gestió Del Risc i Ecologia Del Foc*; Plana, E., Ed.; XCT2001-00061; Xarxa ALINFO: Solsona, Spain, 2004; pp. 50–55.
9. Francos, M.; Pereira, P.; Alcañiz, M.; Mataix-Solera, J.; Úbeda, X. Impact of an intense rainfall evento on soil properties following a wildfire in a Mediterranean environment (North-East Spain). *Sci. Total Environ.* **2016**, *572*, 1353–1362. [CrossRef] [PubMed]
10. Úbeda, X.; Lorca, M.; Outeiro, L.; Bernia, S.; Castellnou, M. Effects of prescribed fire on soil quality in Mediterranean grassland (Prades Mountains, northeast Spain). *Int. J. Wildland Fire* **2005**, *14*, 379–384. [CrossRef]
11. Available online: http://ramatsalbosc.org/ (accessed on 15 May 2019).
12. Alcañiz, M.; Outeiro, L.; Francos, M.; Úbeda, X. Effects of prescribed fires on soil properties: A review. *Sci. Total Environ.* **2018**, *613*, 944–957. [CrossRef]
13. Outerio, L.; Asperó, F.; Úbeda, X. Geostatistical methods to study spatial variability of soil cations after a prescribed fire and rainfall. *Catena* **2008**, *74*, 310–320. [CrossRef]
14. Pereira, P.; Úbeda, X.; Martin, D. Fire severity effects on ash chemical composition and water-extractable elements. *Geoderma* **2012**, *141*, 105–114. [CrossRef]
15. Alcañiz, A.; Outeiro, L.; Francos, M.; Farguell, F.; Úbeda, X. Long-term dynamics of soil chemical properties after a prescribed fire in a Mediterranean forest (Montgri Massif, Catalonia, Spain). *Sci. Total Environ.* **2016**, *572*, 1329–1335. [CrossRef]
16. Lovreglio, R.; Meddour-Sahar, O.; Leone, V. Goat grazing as a wildfire prevention tool: A basic review. *iFor. Biogeosci. For.* **2014**, *7*, 260–268. [CrossRef]
17. Pulido, M.; Schnabel, S.; Lavado Contador, J.F.; Lozano-Parra, J.; González, F. The impact of heavy grazing on soil quality and pasture production in rangelands of SW Spain. *Land Degrad. Dev.* **2018**, *29*, 219–230. [CrossRef]
18. Institut d'Estadística de Catalunya. Available online: www.idescat.cat (accessed on 29 June 2018).
19. Capafonts Local Council. *Butlletí d'informació Municipal (2011–2015)*; Capafonts Local Council: Capafonts, Spain, 2015; 32p.
20. FAO. *State of the World's Forests*; FAO: Rome, Italy, 2011; p. 179.
21. Badia, A.; Valldeperas, N. El valor histórico y estético del paisaje: Claves para entender la vulnerabilidad de la interfaz urbano-forestal frente a los incendios. *Scr. Nova* **2015**, *19*, 1–26.
22. USDA. *Claves para la Taxonomía de Suelos*; United States Department of Agriculture: Washington, DC, USA, 1999; p. 410.
23. MAPA (Ministerio de Agricultura Pesca y Alimentación). *Métodos Oficiales de Análisis Vol III*; Secretaría Técnica General: Madrid, Spain, 1996.
24. Heiri, O.; Lotter, A.F.; Lemcke, G. Loss on ignition as a method for estimating organic and carbonate content in sediments: Reproducibility and comparability of results. *J. Paleolomnol.* **2001**, *5*, 101–110. [CrossRef]

25. Olsen, S.R.; Cole, C.V.; Frank, S.W.; Dean, L.A. *Estimation of Available Phosphorus in Soils by Extraction with Sodium Bicarbonate*; USDA Circular No 939; US Government Printing Office: Washington, DC, USA, 1954.
26. Knudsen, D.; Petersen, G.A. Lithium Sodium and potassium. In *Methods of Soil Analysis*; Soil Science Society of America: Madison, WI, USA, 1986; Volume 2, pp. 225–246.
27. Certini, G. Effects of fire on properties of forest soils: A review. *Oecologia* **2005**, *143*, 1–10. [CrossRef] [PubMed]
28. Brye, K.R. Soil physicochemical changes following 12 years of annual burning in a humid-subtropical tallgrass prairie: A hypothesis. *Acta Oecol.* **2006**, *30*, 407–413. [CrossRef]
29. Valkó, O.; Deak, B.; Magura, T.; Torok, P.; Kelemen, A.; Tóth, K.; Horvarth, R.; Nagy, D.D.; Debnar, Z.; Zsigrai, G.; et al. Supporting biodiversity by prescribed burning in grasslands-A multi-taxa approach. *Sci. Total Environ.* **2016**, *572*, 1377–1384. [CrossRef]
30. Switzer, J.M.; Hope, G.D.; Grayston, S.J.; Prescott, C.E. Changes in soil chemical and biological properties after thinning and prescribed fire for ecosystem restoration in a Rocky Mountain Douglas-fir forest. *For. Ecol. Manag.* **2012**, *275*, 1–13. [CrossRef]
31. Arocena, J.M.; Opio, C. Prescribed fire-induced changes in properties of sub-boreal forest soils. *Geoderma* **2003**, *113*, 1–16. [CrossRef]
32. Lavoie, M.; Starr, G.; Mack, M.C.; Martin, T.; Gholz, H.L. Effects of a prescribed fire on understory vegetation, carbon pools, and soil nutrients in a longleaf pine-slash pine forest in Florida. *Nat. Area J.* **2010**, *30*, 82–94. [CrossRef]
33. Muqaddas, B.; Zhou, X.; Lewis, T.; Wild, C.; Chen, C. Long-term frequent prescribed fire decreases surface soil carbon and nitrogen pools in wet sclerophyll forest of Southeast Queensland, Australia. *Sci. Total Environ.* **2015**, *536*, 39–47. [CrossRef] [PubMed]
34. Teague, W.R.; Dowhower, S.L.; Baker, S.A.; Haile, N.; DeLaune, P.B.; Conover, D.M. Grazing management impacts on vegetation, soil biota and soil chemical, physical and hydrological properties in tall grass prairie. *Agric. Ecosyst. Environ.* **2011**, *141*, 310–322. [CrossRef]
35. Gónzalez-Pérez, J.A.; González-Vila, F.J.; Almendros, G.; Knicker, H. The effects of fire on soil organic matter—A review. *Environ. Int.* **2004**, *30*, 855–870. [CrossRef] [PubMed]
36. Soto, B.; Díaz-Fierros, F. Interactions between plant ash leachates and soil. *Int. J. Wildland Fire* **1993**, *3*, 207–216. [CrossRef]
37. Scharenbroch, B.C.; Nix, B.; Jacobs, K.A.; Bowles, M.L. Two decades of low-severity prescribed fire increases soil nutrient availability in Midwestern, USA oak (*Quercus*) forest. *Geoderma* **2012**, *183*, 89–91. [CrossRef]
38. Qasim, S.; Gul, S.; Shah, M.H.; Hussain, F.; Ahmad, S.; Islam, M.; Rehman, G.; Yaqoob, M.; Shah, S.Q. Influence of grazing exclosure on vegetation biomass and soil quality. *Int. Soil Water Conserv. Res.* **2017**, *5*, 62–68. [CrossRef]
39. Shakesby, R.A.; Bento, C.P.M.; Ferreira, C.S.S.; Ferreira, A.J.D.; Stoof, C.R.; Urbanek, E.; Walsh, R.P.D. Impacts of prescribed fire on soil loss and soil quality: An assessment based on an experimentally-burned catchment in central Portugal. *Catena* **2015**, *128*, 278–293. [CrossRef]
40. Roaldson, L.M.; Johnson, D.W.; Miller, W.W.; Murphy, J.D.; Walker, R.F.; Stein, C.M.; Glass, D.W. Prescribed fire and timber harvesting effects on soil carbon and nitrogen in a pine forest. *Soil Sci. Soc. Am. J.* **2014**, *78*, S48–S57. [CrossRef]
41. Blankenship, B.A.; Arthur, M.A. Soil nutrient and microbial response to prescribed fire in an oak-pine ecosystem in eastern Kentucky. In Proceedings of the 12th Central Hardwood Forest Conference, Lexington, KY, USA, 1–2 March 1999; Stringer, J., Loftis, D., Eds.; Gen. Tech. Rep. SRS 24. USDA: Asheville, NC, USA, 1999; pp. 39–50.
42. Canals, R.M.; Pedro, J.; Rupérez, E.; San Emeterio, L. Nutrient pulses after prescribed Winter fires and preferential patterns of N uptake may contribute to the expansion of B rachypodium pinnatum (L.) P Beauv. in highland grasslands. *Appl. Veg. Sci.* **2014**, *17*, 419–428. [CrossRef]
43. Bowen, G.D.; Danso, S. Investigación sobre el nitrógeno en los cultivos perennes. *OIEA Boletín* **1987**, *2*, 5–8.
44. McNabb, D.H.; Cromack, K., Jr. Effects of prescribed fire on nutrients and soil productivity. In *Natural and Prescribed Fire in Pacific Northwest Forests*; Walstad, J.D., Radosevich, S.R., Sandberg, D.V., Eds.; Oregon State University Press: Corvallis, OR, USA, 1990; pp. 125–141.
45. Afif, E.; Oliveira, P. Efectos del fuego prescrito sobre el matorral en las propiedades del suelo. Investigaciones Agrarias. *Sistema de Recursos Forestales* **2006**, *15*, 262–270.

46. McKee, W.H. *Changes in Soil Fertility Following Prescribed Burning on Costal Pine Sites*; United States Department of Agriculture Forest Service Research Paper SE-234; Southern Forest Experiment Station: Asheville, NC, USA, 1982.
47. Aarons, S.R.; Hosseini, H.M.; Dorling, L.; Gourley, C.J.P. Dung decomposition in temperate dairy pastures. II contribution to plant-available soil phosphorus. *Aust. J. Soil Res.* **2004**, *42*, 115–123. [CrossRef]
48. Sharpley, A.; Moyer, B. Phosporous forms in manure and compost and their release during simulated rainfall. *J. Environ. Qual.* **2000**, *29*, 1462–1469. [CrossRef]
49. Stavi, I.; Ungar, E.D.; Lavee, H.; Sarah, P. Grazing-induced spatial variability of soil bulk density and content of moisture, organic carbon and calcium carbonate in a semi-arid grassland. *Catena* **2008**, *75*, 288–296. [CrossRef]
50. INIA (Instituto Nacional de Investigación y Tecnología Agraria y Alimentaria). *Interpretación de Análisis de Suelos*; Serie Actas 4; INIA: Madrid, Spain, 2016; pp. 10–14.
51. Espinoza, L.; Slaton, N.; Mozaffari, M. *Cómo Interpretar Los Resultados de Los Análisis de Suelos*; Division of Agriculture Research & Extension, University of Arkansas System: Little Rock, AR, USA, 2007; Available online: https://www.uaex.edu/publications/PDF/FSA-2118SP.pdf (accessed on 15 May 2019).
52. Sela, G. *Guía de Interpretación de Análisis de Suelos*; Smart Fertilizer Management: London, UK, 2018; pp. 1–4.
53. Landon, S. *Introducción al análisis de suelos*; CIAT: Cali, Colombia, 1983.
54. AQM. Available online: http://aqmlaboratorios.com (accessed on 5 June 2018).

© 2019 by the authors. Licensee MDPI, Basel, Switzerland. This article is an open access article distributed under the terms and conditions of the Creative Commons Attribution (CC BY) license (http://creativecommons.org/licenses/by/4.0/).

Article

Distance from Night Penning Areas as an Effective Proxy to Estimate Site Use Intensity by Grazing Sheep in the Alps

Simone Ravetto Enri [1,*], Alessandra Gorlier [2], Ginevra Nota [1], Marco Pittarello [1], Giampiero Lombardi [1] and Michele Lonati [1]

1. Department of Agricultural, Forest and Food Sciences, University of Torino, largo Braccini 2, 10095 Grugliasco (TO), Italy; ginevra.nota@unito.it (G.N.); marco.pittarello@unito.it (M.P.); giampiero.lombardi@unito.it (G.L.); michele.lonati@unito.it (M.L.)
2. School of Natural Resources and the Environment, University of Arizona, Tucson, AZ 85721, USA; agorlier@email.arizona.edu
* Correspondence: simone.ravettoenri@unito.it

Received: 23 May 2019; Accepted: 17 June 2019; Published: 21 June 2019

Abstract: Livestock site use intensity can vary widely across a grazing area due to several factors such as topography and distance from sheds and water sources. However, an accurate approximation of animal site use should be assessed for each part of the grazing area to apply effective management strategies. In the Alps, shepherds manage sheep through lenient supervision during the day and confining the animals in temporary night penning areas (TNPA) at night. In our case study, we assessed sheep site use over the grazing area with global positioning system (GPS) collars and calculated the sums of inverse distances from all TNPA (unweighted and weighted on the number of penning nights) and from all water sources, as well as the slope, on 118 sample points. We assessed the relative importance of these variables in affecting site use intensity by animals using different sets of models. Both the unweighted and weighted distances from TNPA were found to be the most important factors. The best fitting model accounted for the weighted distance from TNPA and the distance from water, but the latter showed a lower relative importance. Our study suggests that using the distance from TNPA, preferably weighted on the number of penning nights, is an effective proxy to estimate the spatial variability of sheep stocking rate during grazing in the Alps.

Keywords: drinking sources; GIS; grazing behavior; pastures; spatial distribution; stocking rate

1. Introduction

Livestock grazing is useful for the implementation of management strategies that address the restoration, improvement, or maintenance of grassland vegetation [1]. As a basic criterion, the effectiveness of such strategies depends on livestock stocking rate, which can strongly affect nutrient availability, plant species diversity, and vegetation dynamics [2,3]. Stocking rate can be quantified for the entire grazing area as the overall average number of animals per hectare and time unit. However, the overall stocking rate dismisses animal site use intensity, which can have wide variability over the grazing area due to differences in forage quality and quantity, topography (e.g., slope), animal behavior, and the presence of attractive points such as drinking troughs, sheds, and milking areas [4–9]. Animal site use can be measured directly at each site of the grazing area through global positioning system (GPS) collars, visual observations, etc. Otherwise, it can be estimated through indirect measures (proxies) such as the distance from congregation areas like sheds or water sources. These proxies usually assume that animal site use decreases with increasing distances from congregation areas [10]. More specifically, several authors [6,11,12] proved that inverse distances from congregation areas were linearly related to animal site use. However, the reliability of such proxies was rarely validated with direct measurements [5,13] and, to date, no comparative studies have been conducted to assess the different proxies.

In the Alps, a reliable proxy to estimate site use intensity by animals would be particularly useful since pastures are characterized by a high spatial heterogeneity due to changes in topography and vegetation. In these environments, sheep flocks are commonly managed by shepherds in a daily routine, which entails lenient supervision during the day and confinement in temporary areas during the night (temporary night penning areas—TNPA) [14]. TNPA confine flocks to areas of about 1–3 m²/sheep by means of electrified fences, and they are moved over the grazing area every one-four days, generally located in sites with homogeneous topographic conditions and limited presence of rocks. TNPA help to prevent wolf attacks and, occasionally, contrast shrub encroachment and improve grassland vegetation due to livestock trampling and dung deposition [15,16]. TNPA, as well as water sources, gentle terrains, and milking areas, can therefore be considered the main congregation areas that affect grazing sheep site use in the Alps.

Our study aims to implement a method that uses a GPS/GIS assessment to determine the relative importance of distance from TNPA, distance from water sources, and slope in affecting sheep site use intensity during grazing.

2. Materials and Methods

The study was conducted in the northwestern Italian Alps (45°08′ N, 7°06′ E) in the Site of Community Interest 'Oasi xerotermiche della Valle di Susa—Orrido di Chianocco e Foresto' (SCI IT1110030), an area characterized by a xerothermic and sub-Mediterranean climate with an average annual temperature of 11 °C and an average annual precipitation of 670 mm [17]. Slopes ranged from 4° to 65° (average 28°) and the elevation ranged from 510 to 1260 m a.s.l. The grazing area was characterized by homogeneous seminatural dry grasslands dominated by *Stipa pennata* L., *Bromus erectus* Hudson, and *Festuca ovina* s.l.

From 15 April to 16 May 2015, a flock of 250 Bergamasca (meat breed) sheep grazed over a 45 ha area. Fourteen TNPA (average area: 737 ± 74.0 m²) were progressively set out over the area and each was used to fence the sheep in for two to three consecutive nights (2.3 ± 0.47; mean ± standard deviation). During the period, four water sources homogeneously distributed over the grazing area were also made available to the sheep.

Ten randomly selected sheep were equipped with GPS collars (Model Corzo, Microsensory SLL, Fernàn Nùñez, Spain; 5 m accuracy) and tracked at 15 min intervals for the entire duration of the experiment. The tracked sheep were dry ewes, two to four years old, weighing approximately 70 kg, and regularly fed on Alpine pastures during the summer. During the study, the flock experienced this specific grazing area for the first time. We assumed the 10 selected sheep as representative of the entire flock, since sheep are a livestock species characterized by a highly cohesive grazing behavior.

We randomly generated 160 sample points over the grazing area and assessed the number of GPS fixes within a 30 m buffer zone around each of them as a direct measurement of the site use intensity by grazing sheep [13,16]. The 30 m distance was considered to encompass a zone with homogeneous vegetation and topographic conditions. The sample points were spaced 60 m apart to avoid overlaps between buffers. When a buffer zone exceeded the grazing area, the number of GPS fixes included was weighted by the within-the-grazing-area portion and rounded to the nearest integer value. Forty-two sample points were excluded from further analysis, as they exceeded the boundaries of the grazing area for more than 25% of their buffer zone, so 118 sample points were retained.

According to the following formulas, for each sample point we calculated:

(i) the sum of inverse distances from all TNPA (hereafter 'unweighted distance from TNPA')

$$\text{Unweighted } d_{TNPA} = \sum_{i=1}^{i=14}\left(\frac{1}{d_i}\right) \quad (1)$$

where d_i is the distance from each TNPA;
(ii) the sum of inverse distances from all TNPA weighted on the number of consecutive penning nights for each of them (hereafter 'weighted distance from TNPA')

$$\text{Weighted } d_{TNPA} = \sum_{i=1}^{i=14} \left(\frac{n_i}{d_i}\right) \quad (2)$$

where d_i is the distance from each TNPA and n_i is the corresponding number of nights;
(ii) the sum of inverse distances from all water sources (hereafter 'distance from water')

$$d_{water} = \sum_{j=1}^{j=4} \left(\frac{1}{d_j}\right) \quad (3)$$

where d_j is the distance from each water source;
(iv) the slope, which is assessed as the average value of the buffer zone using a 10 m resolution digital terrain model [18].

Geographical analyses were conducted using Quantum GIS version 2.18.26 [19].

To assess the relative importance of (i) distance from TNPA (weighted and unweighted), (ii) distance from water, and (iii) slope in predicting the actual site use intensity by the sheep during grazing, we ran 11 generalized linear models (GLMs). We set the sheep site use intensity (i.e., the count of GPS fixes within each buffer zone) as response variable and set the distances from TNPA and from water as well as the slope in all possible combinations as explanatory variables. We specified a negative binomial error distribution for the GPS count and a logarithmic link function [20]. All explanatory variables were standardized (Z-scores) before performing GLMs to allow for the analysis of effect size by scrutinizing model parameters (β coefficients). Autocorrelation was tested using Pearson's correlation before running the GLMs. Residual deviance, percent of explained deviance (D^2), Akaike information criterion with small-sample correction (AICc), and Bayesian information criterion (BIC) were used to compare the goodness of the model fit. D^2 was calculated according to the following formula:

$$D^2 = \frac{\text{null deviance} - \text{residual deviance}}{\text{null deviance}}$$

where null deviance is the deviance of an intercept-only GLM and residual deviance is the deviance that remains unexplained after the model fit. Statistical analyses were performed using SPSS 25 (SPSS Inc., Chicago, IL, USA).

3. Results

The daily acquisition rate of the GPS devices refers to the total potential of daily fix acquisitions, and was 44.2 ± 2.54% (mean ± standard error). Explanatory variables showed a not significant ($p \geq 0.05$) or weak ($R \leq 0.25$) autocorrelation and all of them were retained in the models. Average values for the response and explanatory variables in buffer zones are provided in Table 1. According to the performed GLMs, the sheep site use was significantly related to the selected predictors (Table 2). More particularly, it was always positively affected by both unweighted and weighted distances from TNPA and the distance from water sources, but negatively by the slope. However, the slope effect was not significant when the distances from TNPA were weighted on the number of penning nights (M7 and M8). Among the explanatory variables, the distance from TNPA had the most influence (highest β coefficients) in all the models, followed by the distance from water and then by the slope. β coefficients increased for the distance from water and the slope in M9, M10, and M11, but their effect size was lower than those of the distances from TNPA.

Table 1. Summary statistics for the dependent and explanatory variables used in the models. Values apply to the 30 m buffer zone around the 118 random points. Numbers in brackets refer to formulas detailed in the Methods section. TNPA refers to temporary night penning areas.

Variable	Minimum	Mean	Maximum
Site use intensity (global positioning system (GPS) count)	0.00	35.14	434.00
Distance from TNPA—unweighted (m^{-1}) (1)	0.01	0.04	0.14
Distance from TNPA—weighted (m^{-1}) (2)	0.03	0.08	0.29
Distance from water (m^{-1}) (3)	0.02	0.10	1.67
Slope (°)	10.92	28.41	43.87

Table 2. Summary of generalized linear models (GLMs) of sheep site use intensity by different sets of explanatory variables. Numbers in brackets refer to formulas in the Methods section. β coefficients and significance levels are provided for each variable as results of the related GLM. Goodness of model fit (best values are highlighted in bold): D^2, explained deviance; AICc, Akaike's information criterion with small-sample correction; BIC, Bayesian information criterion. ***, $p < 0.001$; **, $p < 0.01$; *, $p < 0.05$; ns, $p \geq 0.05$.

Generalized Linear Model	Distance from TNPA	Distance from Water	Slope	Residual Deviance	D^2 %	AICc	BIC
M1: Distance from TNPA (unweighted) (1)	1.43 ***	-	-	111.5	63.1	892.7	898.1
M2: Distance from TNPA (unweighted) (1) + distance from water (3)	1.41 ***	0.18 **	-	102.6	66.0	885.9	894.0
M3: Distance from TNPA (unweighted) (1) + slope	1.40 ***	-	−0.20 *	109.2	63.9	892.5	900.6
M4: Distance from TNPA (unweighted) (1) + distance from water (3) + slope	1.36 ***	0.21 **	−0.21 *	100.4	66.8	885.8	896.5
M5: Distance from TNPA (weighted) (2)	1.47 ***	-	-	107.4	64.5	888.6	894.0
M6: Distance from TNPA (weighted) (2) + distance from water (3)	1.43 ***	0.21 **	-	99.2	67.2	**882.5**	**890.6**
M7: Distance from TNPA (weighted) (2) + slope	1.44 ***	-	−0.12 ns	106.5	64.8	889.7	897.8
M8: Distance from TNPA (weighted) (2) + distance from water (3) + slope	1.39 ***	0.23 **	−0.15 ns	98.2	67.5	883.6	894.4
M9: Distance from water (3)	-	0.76 ***	-	282.6	6.5	1063.8	1069.2
M10: Slope	-	-	−0.62 ***	267.8	11.4	1048.9	1054.4
M11: Distance from water (3) + slope	-	0.96 ***	−0.63 ***	243.3	19.5	1026.6	1034.7

Lower AICc and BIC scores were obtained in models including distances from TNPA and specifically, in models based on weighted distances (M5, M6, M7, and M8). The lowest values in terms of residual deviance and D^2 were performed by M8, which considered weighted distance from TNPA, distance from water, and slope, although this latter variable was not significant. The same model deprived of slope (M6) showed the best fit according to AICc and BIC values.

4. Discussion

All GPS devices worked as expected with a satisfactory acquisition rate. However, the harsh morphology of the study area (i.e., very rocky, rough, and a steep mountainous environment) had a negative effect on the accuracy of GPS fix acquisition. This determined that the signal bounced off of a considerable proportion of GPS fixes out of the study area borders, so these were excluded from the analyses.

The present research highlighted the remarkable relationships that exist among site use intensity by grazing sheep and specific environmental/management predictors, namely, distance from night penning areas, distance from water, and slope. As expected, site use intensity was inversely related to the slope and directly related to the distance from TNPA and water sources in all GLMs we performed [5,21]. Specifically, the models that included the distance from TNPA (both unweighted and weighted, i.e., from M1 to M8) explained remarkable percentages of deviance (>63%, higher than shown in Putfarken et al. and in Dorji et al. [13,22]), which proved the pivotal role of TNPA in affecting sheep distribution during grazing. Instead, the distance from water sources and the slope showed a weaker influence, as demonstrated by their lower β coefficients. This was also observed in models based only on these variables (M9, M10, and M11), which had the lowest explained deviances, confirming the findings of other authors [13,22]. Nevertheless, unlike our study, previous studies did not compare different regressive models that consider environmental and management variables.

The models that included distances from TNPA (M1 to M8) also achieved the best fitting results in terms of AICc and BIC, which varied within a range of 10 points. Therefore, according to Burnham and Anderson [23], all of them can be considered as having comparable reliability. The limited differences among the models including the unweighted and weighted distances from TNPA may be due to the low variability in the number of penning nights among TNPA used in Equation (2), which resulted in distributions with a high similarity. Nevertheless, M6 (weighted distance from TNPA + distance from water) can be considered as the best fitting model as it presented the lowest AICc and BIC scores. Moreover, in this model, the relative importance of distance from water was lower (β coefficient was sevenfold smaller) than that of the weighted distance from TNPA, suggesting that the implementation of predictive models that include the distance from water sources could be of limited effectiveness. This finding was in contrast to the results of the study by Putfarken et al. [13], which highlighted a higher relative importance of the distance from the drinking trough as compared to the distance from the sheep shed. Nevertheless, in their trial study, they tested the effects of only one drinking trough and one sheep shed in different management conditions (i.e., higher stocking rate, longer grazing season, lowland mesotrophic grasslands, and with a cow and sheep mixed grazing system). Moreover, the distance of a given site from all available water sources could be difficult to assess in some situations, e.g., when linear water sources like mountain streams are available for livestock over the grazing area. Therefore, according to our results, the distance from TNPA, preferably weighted on the number of penning nights (Equation (2)), can be reasonably considered as the main driver and a suitable and easily measured proxy to estimate the spatial variability of sheep stocking rate during grazing.

Future research should avoid some shortcomings still evident in our study, such as (i) the short duration of the experiment, (ii) the low grassland forage variability related to the occurrence of one vegetation community, (iii) the limited number of tracked animals (avertible by selecting rotating collared sheep), and (iv) the lack of information about animal behavior activity (i.e., resting, grazing, and traveling categories). Nonetheless, the approach we propose, which is based on a comparison among different models including environmental and management predictors, could also be applied for other livestock species and categories, shepherding managements, vegetation types, and environments.

Author Contributions: Conceptualization, M.L.; Methodology, S.R.E., M.L.; Data Gathering and Preparation, S.R.E., A.G., G.N., M.P.; Writing—Original Draft Preparation, S.R.E., A.G., G.N., M.P., G.L., M.L.; Supervision, G.L., M.L.

Funding: Research was funded by the EC-LIFE program, project LIFE12 NAT/IT/000818 Xero-grazing (Principal Investigator Giampiero Lombardi).

Acknowledgments: The authors gratefully thank the 'Ente di gestione delle aree protette delle Alpi Cozie' (Coordinating Beneficiary) and the Franco Pia farm for their constant support and the provision of the flock.

Conflicts of Interest: The authors declare no conflict of interest.

References

1. Metera, E.; Sakowski, T.; Słoniewski, K.; Romanowicz, B. Grazing as a tool to maintain biodiversity of grassland—A review. *Anim. Sci. Pap. Rep.* **2010**, *28*, 315–334.
2. Pittarello, M.; Probo, M.; Lonati, M.; Lombardi, G. Restoration of sub-alpine shrub-encroached grasslands through pastoral practices: Effects on vegetation structure and botanical composition. *Appl. Veg. Sci.* **2016**, *19*, 381–390. [CrossRef]
3. Perotti, E.; Probo, M.; Pittarello, M.; Lonati, M.; Lombardi, G. A 5-year rotational grazing changes the botanical composition of sub-alpine and alpine grasslands. *Appl. Veg. Sci.* **2018**, *21*, 647–657. [CrossRef]
4. Bailey, D.W.; Gross, J.E.; Laca, E.A.; Rittenhouse, L.R.; Coughenour, M.B.; Swift, D.M.; Sims, P.L. Mechanisms That Result in Large Herbivore Grazing Distribution Patterns. *J. Range Manag.* **1996**, *49*, 386. [CrossRef]
5. Svoray, T.; Shafran-Nathan, R.; Ungar, E.D.; Arnon, A.; Perevolotsky, A. Integrating GPS technologies in dynamic spatio-temporal models to monitor grazing habits in dry rangelands. In *Recent Advances in Remote Sensing and Geoinformation Processing for Land Degradation Assessment*; Taylor and Francis: Leiden, The Netherlands, 2009; pp. 301–312.
6. Manthey, M.; Peper, J. Estimation of grazing intensity along grazing gradients – the bias of nonlinearity. *J. Arid Environ.* **2010**, *74*, 1351–1354. [CrossRef]
7. Russell, M.L.; Bailey, D.W.; Thomas, M.G.; Witmore, B.K. Grazing Distribution and Diet Quality of Angus, Brangus, and Brahman Cows in the Chihuahuan Desert. *Rangel. Ecol. Manag.* **2012**, *65*, 371–381. [CrossRef]
8. Probo, M.; Lonati, M.; Pittarello, M.; Bailey, D.W.; Garbarino, M.; Gorlier, A.; Lombardi, G. Implementation of a rotational grazing system with large paddocks changes the distribution of grazing cattle in the south-western Italian Alps. *Rangel. J.* **2014**, *36*, 445–458. [CrossRef]
9. Pittarello, M.; Probo, M.; Lonati, M.; Bailey, D.W.; Lombardi, G. Effects of traditional salt placement and strategically placed mineral mix supplements on cattle distribution in the Western Italian Alps. *Grass Forage Sci.* **2016**, *71*, 529–539. [CrossRef]
10. Tarhouni, M.; Ben Salem, F.; Ouled Belgacem, A.; Neffati, M. Acceptability of plant species along grazing gradients around watering points in Tunisian arid zone. *Flora Morphol. Distrib. Funct. Ecol. Plants* **2010**, *205*, 454–461. [CrossRef]
11. Fernandez-Gimenez, M.; Allen-Diaz, B. Vegetation change along gradients from water sources in three grazed Mongolian ecosystems. *Plant Ecol.* **2001**, *157*, 101–118. [CrossRef]
12. Wesuls, D.; Pellowski, M.; Suchrow, S.; Oldeland, J.; Jansen, F.; Dengler, J. The grazing fingerprint: Modelling species responses and trait patterns along grazing gradients in semi-arid Namibian rangelands. *Ecol. Indic.* **2013**, *27*, 61–70. [CrossRef]
13. Putfarken, D.; Dengler, J.; Lehmann, S.; Härdtle, W. Site use of grazing cattle and sheep in a large-scale pasture landscape: A GPS/GIS assessment. *Appl. Anim. Behav. Sci.* **2008**, *111*, 54–67. [CrossRef]
14. Lombardi, G. Optimum management and quality pastures for sheep and goat in mountain areas. *Options Méditerranéennes. Série A Séminaires Méditerranéens* **2005**, *67*, 19–29.
15. Espuno, N.; Lequette, B.; Poulle, M.-L.; Migot, P.; Lebreton, J.-D. Heterogeneous Response to Preventive Sheep Husbandry during Wolf Recolonization of the French Alps. *Wildl. Soc. Bull.* **2004**, *32*, 1195–1208. [CrossRef]
16. Pittarello, M.; Gorlier, A.; Lonati, M.; Perotti, E.; Lombardi, G. *Temporary Night Penning as Effective Tool to Improve Plant Diversity in Nutrient-Poor Dry Grasslands. Proceedings of the 19th Symposium of the European Grassland Federation, Alghero, Italy, 7–10 May 2017*; Wageningen Academic Publishers: Wageningen, The Netherlands, 2017; Volume 22, pp. 381–383.
17. Biancotti, A.; Bellardone, G.; Bovo, S.; Cagnazzi, B.; Giacomelli, L.; Marchisio, C. *Distribuzione Regionale delle Piogge e Temperature*; Collana Studi Climatologici del Piemonte; Regione Piemonte: Torino, Italy, 1998; Volume 1.

18. Regione Piemonte Digital Terrain Model with 10 Meters Resolution. Available online: http://www.geoportale.piemonte.it/geonetworkrp/srv/ita/metadata.show?id=2486&currTab=rndt (accessed on 12 February 2019).
19. QGIS Development Team. *QGIS Geographic Information System*; Open Source Geospatial Foundation: Beaverton, OR, USA, 2016.
20. McCullagh, P.; Nelder, J.A. *Generalized Linear Models*; Chapman and Hall: London, UK, 1983; ISBN 978-0-412-23850-5.
21. Amiri, F. A model for classification of range suitability for sheep grazing in semi-arid regions of Iran. *Livest. Res. Rural Dev.* **2009**, *21*, 241–266.
22. Dorji, T.; Totland, Ø.; Moe, S.R. Are Droppings, Distance from Pastoralist Camps, and Pika Burrows Good Proxies for Local Grazing Pressure? *Rangel. Ecol. Manag.* **2013**, *66*, 26–33. [CrossRef]
23. Burnham, K.P.; Anderson, D.R. *Model Selection and Multimodel Inference: A Practical Information-Theoretic Approach*, 2nd ed.; Springer-Verlag: New York, NY, USA, 2002; ISBN 978-0-387-95364-9.

© 2019 by the authors. Licensee MDPI, Basel, Switzerland. This article is an open access article distributed under the terms and conditions of the Creative Commons Attribution (CC BY) license (http://creativecommons.org/licenses/by/4.0/).

Article

Soil Macronutrient Responses in Diverse Landscapes of Southern Tallgrass to Two Stocking Methods

Brian K. Northup *, Patrick J. Starks and Kenneth E. Turner

United States Department of Agriculture-Agricultural Research Service, 7207 West Cheyenne St., El Reno, OK 73036, USA; Patrick.starks@ars.usda.gov (P.J.S.); ken.turner@ars.usda.gov (K.E.T.)
* Correspondence: brian.northup@ars.usda.gov; Tel.: +1-(405)-262-5291

Received: 22 March 2019; Accepted: 15 June 2019; Published: 20 June 2019

Abstract: Macronutrient (N, P, S, K, Ca, and Mg) availability and distribution in soils of grassland ecosystems are affected by diverse factors, including landscape position, climate, and forms of management. This study examined flux in plant-available macronutrients in production-scale (60 to 80 ha) paddocks of southern tallgrass prairie of central Oklahoma, United States, managed (2009–15) under two contrasting stocking methods (continuous yearlong; rotational stocking among 10 sub-paddocks). Macronutrient availability within the 0–7.5 cm and 7.5–15 cm soil depths were determined with sets of anion-cation exchange membrane probes at 16 locations within paddocks, oriented along transects from water sources to far corners. No clear overall effect related to stocking method was recorded for all macronutrient distributions. The only significant stocking method × location interaction occurred for K ($p = 0.01$). All other macronutrients displayed significant ($p < 0.08$) location effects that were common across stocking methods. Effects relatable to stocking method occurred in interactions with soil depth or time of year ($p < 0.10$), but responses of macronutrient flux to stocking method in these interactions varied. Higher flux occurred in available S, Ca, and Mg in proximity (<24 m) to water sources, which may be related to grazing, but local features of the landscape may also have been involved. More attention to landscape features included within paddocks, and standardized organization of water and other features within paddocks, would improve the potential to define grazing effects on macronutrient distribution.

Keywords: exchange membranes; grazing management; soil macronutrients

1. Introduction

Macronutrients are present in a range of forms within soils of grasslands, and are related to the nature of underlying parent materials [1]. Macronutrient availability and distributions are also affected by type and productivity of plant communities, position within landscape, and climate, and varies with time [1–3]. Livestock grazing has been considered an important component of altering availability of macronutrients within grassland ecosystems. Cattle consume plant biomass, and redistribute macronutrients within consumed biomass through removal by grazing, movement of nutrients off-paddock in body weight, and recycling through excreta that is returned to the landscape [3–6]. Re-distribution of macronutrients in excreta tends to be non-uniform, and can result in high concentrations within localized areas of paddocks [7,8]. Increased amounts of labile N, P, K, and S were reported in areas adjacent to watering facilities, corners, and other paddock structures [6,9,10].

Methods of stocking cattle (grazing systems) are thought to be capable of altering distribution of grazing and excreta within grazed landscapes, and hence improve the distribution of labile macronutrients [3,10,11]. Dividing larger land units into smaller sub-paddocks and applying rotational stocking to shift timing and frequency of grazing among paddocks is thought to be capable of achieving more uniform levels of grazing, and use of paddock areas [10,12,13]. However, both availability and distribution of macronutrients in production-scale paddocks are both primarily related to local soil properties, at pedon (m^2) through catena (100's m^2) scales of organization [2,14,15]. Macronutrient availability and distributions are further affected by type of plant community, position within landscape, and climate [1–3,16–18]. Such features tend to result in variable distribution patterns without overlaying any effects related to grazing [2,16–18].

United States southern tallgrass prairies are diverse native ecosystems (up to 120 herbaceous and woody species within 1.0 ha areas) that can be productive components of livestock production in the region, with proper management and adequate precipitation [19]. They are definable as low-input–low-output ecosystems and function without fertilizer or irrigation applications [2,12]. However, the tallgrasses still require adequate amounts of macronutrients to be productive during growing seasons, and grazing by cattle can re-distribute important resources within the landscape of paddocks [2,8]. A companion paper to this study [20] reported a lack of uniform responses in the distribution of 8 macronutrients among a limited set of contrasting locations in paddocks of southern tallgrass prairie managed under four different stocking methods. The most notable effect [20] was that levels of plant-available macronutrients either changed among locations, or times of growing season, within paddocks under the different applied stocking methods. Local landscape conditions within paddocks were noted as possible drivers of responses [20], and results indicated the need for a more comprehensive examination of macronutrient distributions within larger production-scale paddocks. This result, combined with the dichotomy related to management effects versus inherent natural variability of landscapes, points to the question of whether grazing applied in different stocking methods influences distribution of excreta, and hence plant-available macronutrients, within production-scale paddocks of rangeland.

This study was undertaken to define the degree of flux in plant-available macronutrients (NO_3^-, NH_4^+, P, S, K, Ca, and Mg) in soils within larger areas of southern tallgrass prairie encompassed in production-scale paddocks under two contrasting stocking methods. The working null hypotheses were: (1) no difference would occur in distributions and levels of 8 plant-available macronutrients at a range of locations within landscapes of production-scale paddocks, and (2) no difference would occur among times of growing seasons or soil depths, in response to stocking methods.

2. Materials and Methods

2.1. Study Site

This study was conducted within sets of production-scale paddocks of tallgrass prairie at the United States Department of Agriculture–Agricultural Research Service, Grazinglands Research Laboratory (35°33'29" N, 98°1'50" W) in central Oklahoma, United States. The entire site was located in a rolling upland landscape, with a range of landscape features present. Included were local easterly and westerly-facing slopes of 3 to 6% on riser positions, and toe and tread slope positions with 0 to 2% slope, bordering the risers [21]. The long-term average (±1 standard deviation (s.d.)) precipitation (long-term average (LTA); 1977 to 2012) during calendar years was 941 (±174) mm. Annual amounts ranged from 1468 mm (2007) to 646 mm (2011), with a bimodal distribution pattern. Maxima occurred during April through June (334 ± 58 mm), and September through October (175 ± 54 mm). Long-term minimum and maximum monthly temperatures occurred during January (2.8 (±2.7) °C; −4.0 to 8.6 °C) and July (28.1 (±1.5) °C; 25.6 to 32.4 °C).

The ecosystem of the study area (346 ha) was defined as southern tallgrass prairie, and was identified as a Loamy Prairie ecological site [21]. These perennial grasslands are remnant plant communities of the original southern tallgrass ecosystems that existed in central Oklahoma, United States, prior to European settlement. They were never cultivated, nor replanted to native species following cultivation. The dominant species were the perennial warm-season tallgrasses big bluestem (*Andropogon gerardii*), Indiangrass (*Sorghastrum nutans*), and little bluestem (*Schizachyrium scoparium*). These three species provide an average (±1 s.d.) of 70% (±10%) of the total annual biomass produced at the study area [21]; annual productivity ranges from 1 to 5 Mg ha^{-1}, depending on amount and timing of precipitation [20–23]. The area has historically (1970's to 2009) supported cowherds that were managed to produce calves for research on growth and production of yearling stocker cattle [22,23]. Stocking methods applied throughout the historical period were not consistent. They were changed regularly to meet management requirements for calves used in research projects. Included were periods of management under either, or both, continuous and rotational stocking during growing seasons (April to September) at lower densities (2.0 to 3.0 ha cow/calf pair^{-1} year^{-1}).

A range of soil series belonging to different families and subgroups of the Mollisol order have been recorded in the area, based on landscape position within the site [24,25]. All listed soils evolved from parent material that was Permian-aged Dog Creek shale, a reddish-brown shale containing thin inter-beds of sandstones and siltstones [24]. Three sub-types of Norge series silt loams (Fine-silty, mixed, active thermic Udic Paleustolls) situated on riser (mid-slope) positions of the landscape were the most-common [21]. Kirkland or Renfrow silt loams (Fine, mixed superactive, thermic Udertic Paleustolls) situated on tread positions (summit locations) and Port silt loams (Fine-silty, mixed superactive, thermic Cumulic Haplustolls) at toe positions bounded the Norge series. Six additional related soils exist as inclusions, or complexes, within boundaries of each of these primary soils [20]. The surface soils (upper 30 cm) of these series have variable, but near-neutral pH (6.7 (±0.6)), cation exchange capacity of 13.5 (±3.8) cmol kg^{-1} soil, low water-holding capacities (3 (±1) mm cm^{-1} soil), and variable rates of permeability (33 (±17) mm h^{-1}) [20].

2.2. Experimental Design

The study area included sets of paddocks assigned to different stocking methods [25]. Two replicate, 61 (±2) ha paddocks were managed by continuous yearlong stocking as controls. The remaining area included two ~80 ha sites that were each sub-divided into sets of 10 sub-paddocks of rotationally stocked rangeland. These sub-paddocks were managed by application of shorter grazing periods, with timing and frequency changed annually to mimic an adaptive system of rotational stocking [12,13]. Paddocks were managed under their assigned stocking methods from 2009 through 2015, to define impacts of stocking methods on plant communities and soil properties. Grazing pressure of the different treatments were achieved with herds of cow-calf pairs (~600 kg cows and ~249 kg calves at weaning, hereafter defined as animal units (AU)), that were assigned to each replicate paddock. Annual (2009–2015) herd sizes in the continuous and groups of rotational-stocked sub-paddocks were 18 (±3.4) and 26 (±2.5) head, respectively. Animal units assigned to rotationally stocked paddocks during 2009 to 2015 grazed sub-paddocks in 7- to 10-day grazing bouts, 2 to 4 times annually, with timing of bouts varied annually. Daily allotment of forage per AU was 17 kg day^{-1}, or ~6.2 Mg AU^{-1} year^{-1}. Total grazing pressure applied to the continuous and rotational-stocked paddocks was, respectively, 108 (±20) and 119 (±11) animal unit days (AUD) ha^{-1} year^{-1}.

2.3. Data Collection

Data were collected twice during the 2015 growing season—mid-March and early-August. These two periods represented (1) the time when growth by native grasses initiated during growing seasons (March), and (2) the time when peak living biomass occurs (August) for southern tallgrass prairies during late summer [21]. Availability of macronutrients within the 0 to 7.5 cm and 7.5 to 15 cm depth increments of soil were determined on each date, at a series of 16 locations oriented along transects from water sources (W) to far corners (FC), and three additional points, within paddocks (Figure 1). Sampled locations included a set ($n = 7$) within paddocks that were at the same physical distances, between 1.5 to 37 m between sampled locations and water sources (W) within paddocks, and at W. Past research at other US sites noted increased amounts of macronutrients in soils in "zones" close to water or shade, but not specific distances [8,10]. The intense spacing of these 8 locations was to determine if exact areas in proximity to water exist where high amounts of macronutrients might occur.

Data were also collected from eight additional locations (Figure 1) that were identified as potential high- and low-traffic areas [7,8,10,11]. Included were nearest and furthest corners (NC and FC, respectively), nearest and furthest fence lines (NF and FF, respectively), paddock midpoint (PMP), and 25%, 37%, 50%, and 75% of distances between W and PMP (0.25 PMP, 0.37 PMP, 0.50 PMP, and 0.75 PMP, respectively). Distances from W to locations were defined by measuring wheel, and elevation of locations defined by hand-held Global Positioning System (GPS) units. The sampled locations for NF and FF were at closest points along fences that were immediately north or south of water sources. While similar in geographic terms, the actual location of these eight positions and FC within paddocks and sub-paddocks varied in distance from W (Table 1). This variance in distance for the same position within paddocks was due to differences in size, shape, and dimensions of paddocks, and locations of water within paddocks, assigned to the applied stocking methods [25].

Table 1. Mean (±1 standard deviation (s.d.)) distances (meters) from 9 locations where macronutrient availability was measured to water sources within paddocks and sub-paddocks assigned to stocking methods.

Stocking Method	Location †								
	Near Corner	Near Fence	0.25 PMP	0.37 PMP	0.50 PMP	0.75 PMP	PMP	Far Fence	Far Corner
Continuous	14 (11)	77 (13)	112 (5)	167 (8)	222 (8)	338 (9)	450 (10)	976 (69)	912 (43)
Rotational	55 (14)	35 (11)	58 (17)	82 (11)	110 (13)	166 (13)	220 (10)	423 (14)	455 (43)

† PMP = paddock mid-point.

Macronutrient availability (hereafter also noted as flux) within soils was determined with Plant Root Simulator^{TM} probes (Western Ag Innovations Inc., Saskatoon, SK, Canada). The probes are comprised of paired sets of anion and cation exchange membranes encased in plastic housings. The probes provide estimates of nutrient uptake by plant roots, and represent a rate-based process (μg uptake cm^{-2} probe surface area 14 days^{-1} soil depth^{-1}) of the soil–plant interface [26,27]. At each location within paddocks, two sets of probes with anion and cation membranes were located at each depth. The probes were buried in situ during the March sampling for 14-d incubation periods, as soil moisture approximated field capacity (21.5% (±3%) volumetric water) in response to precipitation events immediately before, and early in, incubation periods.

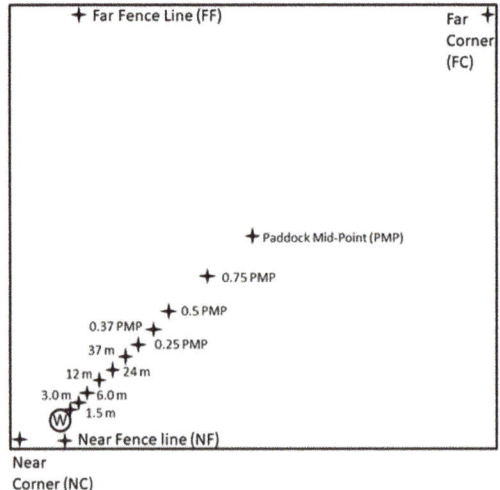

Figure 1. Illustration of locations sampled within production-scale paddocks.

Soil moisture during the August sampling approximated permanent wilt point (12.5% (±1%) volumetric water) due to a summer drought. Therefore, replicate ($n = 4$) cores (5.38 cm diameter) of soil at each location and soil depth were collected by plunging hammer and core tubes. These cores were removed from core tubes in a laboratory, wetted with deionized water to field capacity, and dissected into two longitudinal sections. Replicate ($n = 2$) anion and cation membrane probes were sandwiched within dissected cores for 14-d incubation periods at a temperature of 23 °C.

The probes used during both incubation periods were removed from soil post-incubation, lightly washed with deionized water to remove soil, packaged in groups, refrigerated, and sent to the probe manufacturer for analyses. The manufacturer used colorimetric analyses by automated flow injection to determine NO_3-N and NH_4-N, and all other macronutrients (ions) were measured using inductively-coupled plasma spectrometry.

Two sets of soil samples were collected by plunging hammer and 5.38 cm diameter core tubes on each sampling date at each location within paddocks, to define physical properties of soils for the two depth increments. Collected samples were placed in soil tins for transport and storage. Moist bulk densities of one set of samples were defined [28], followed by analyses for particle fractions by hydrometer methods in a sodium hexa-metasulfate solution [29]. The second set of soil samples was passed through a 2.0 mm sieve and used to define estimates of soil organic matter based on low temperature loss on ignition [30].

2.4. Statistical Analyses

Probe-measured availability of total mineral N ($NH_4^+ + NO_3^-$), NO_3^--N, NH_4-N, P, S, K, Ca, and Mg in soils were explored (Table 2) to determine whether data transformations were required for statistical tests [31]. The natural logarithm (Ln) transform was applied, as required, to improve cumulative distribution functions of the populations of observations to more closely fit a normal distribution. Transformed levels of flux of total minerals of N, NO_3, NH_4, P, S, and K, and the raw values for Ca and Mg were analyzed in SAS 9.3 (SAS Institute, Cary, NC, USA).

Table 2. Descriptive statistics and distribution functions of populations ($n = 256$) of macronutrient fluxes in the upper 15 cm of soils in paddocks managed by different stocking methods.

	Macronutrients							
	Mineral N	NO_3^--N	NH_4^+-N	P	S	K	Mg	Ca
Distribution								
Skewness	3.7	3.6	7.9	3.6	3.2	1.7	0.4	<0.1
Kurtosis	17.8	16.7	82.6	18.5	10.1	3.8	0.2	−0.6
K-S normality [†]	0.25	0.26	0.31	0.26	0.32	0.15	0.04	0.05
K-S p [†]	<0.01	<0.01	<0.01	<0.01	<0.01	<0.01	>0.15	0.10
c.v. [†]	144	156	204	157	190	74	44	43
	----------	----------	----------	(μg cm^{-2} probe)	----------	----------	----------	
Statistics								
Mean	8	7	0.8	1.2	15	34	39	189
s.d. [†]	12	12	1.6	1.7	28	25	17	8
Median	4	3	0.4	0.4	5	28	38	192
Minimum	<1	<1	0.1	0.1	<1	<1	4	21
Maximum	87	83	20.0	15.0	154	155	93	399
Ln Transform	Y	Y	Y	Y	Y	Y	N	N

[†] K-S = Kolmogrov-Smirnov normality test; $p \geq 0.10$ approximate normality; c.v. = coefficient of variation (%); s.d. = standard deviation.

Data were analyzed by longitudinal (repeated) measures analyses [32] within mixed models (PROC MIXED). Grazing regime, soil depth, and time of growing season were main effects in analyses, while sampled locations within paddocks were the longitudinal element. Preliminary analyses of variance applied to macronutrients attempted to utilize different physical attributes of soils and distances between paddock locations and water sources as covariates to improve the function of statistical models. However, all attempts to use these attributes as covariates were non-significant ($p > 0.13$) and failed to improve statistical tests. An examination of correlation coefficients of the relationship between macronutrient fluxes and physical attributes showed weak relationships in all cases (Table 3). The most effective variance-covariance matrix to account for covariance and autocorrelation among the varied locations within paddocks would be the unstructured (UN) procedure [33]. However, there were not enough degrees of freedom (d.f.) to produce stable models with this, or other more complex (i.e., power, exponential) matrix structures [32,33]. Therefore, the compound symmetry (CS) structure was used to account for covariance and autocorrelation among locations within paddocks. The limited number of d.f. also forced analyses to be restricted to main effects and 2-way interactions between stocking methods, and remaining main effects and longitudinal factor [32]. Reported means for significant main and interaction effects were back-transformed to original scales [31]. Level of significance of statistical tests was set at $p = 0.10$.

Table 3. Correlation coefficients (r) of relationships between soil properties and physical features of paddock locations with plant-available flux in 8 macronutrients.

Soil Properties	Macronutrients							
	NO_3	NH_4	Total N	P	S	K	Mg	Ca
Clay	0.04	−0.08	0.04	−0.08	0.50 #	−0.36	0.17	0.24
Silt	−0.37	−0.01	−0.38	−0.05	−0.39	0.12	−0.19	−0.16
Sand	0.40 #	0.06	0.40 #	0.30	0.18	0.05	0.13	0.05
Bulk Density	−0.09	−0.04	−0.10	−0.38	0.01	−0.29	0.07	−0.16
Organic matter	−0.01	−0.01	−0.01	0.11	0.20	0.14	0.01	0.11
Elevation	−0.15	−0.25	−0.13	−0.02	−0.14	0.17	−0.26	−0.12
DTW †	−0.17	−0.01	−0.16	−0.02	−0.25	−0.09	−0.09	−0.11

† Distance (m) to water source from sampled locations; # indicates significance of $p \leq 0.10$.

3. Results and Discussion

3.1. Soil Properties

The means and standard deviations (s.d.) of the measured physical attributes are presented here to provide estimates of soil properties at the sampled locations within paddocks (Table 4). Soils on the study site showed a degree of variability in physical attributes among paddocks assigned to the two stocking methods. Percentages of different particle fractions of soils in paddocks assigned to continuous stocking were less consistent than in sub-paddocks assigned to rotational stocking, which was likely due to the greater number of soils encountered within these larger units [25]. Paddocks receiving continuous stocking had greater percentages of sand and lower amounts of silt than in rotational-stocked sub-paddocks, while percentages of clay in soils were similar. Moist bulk densities of soils within paddocks receiving rotational stocking were lower than in paddocks receiving continuous stocking, most likely due to the higher percentage of sand in soils of continuously stocked paddocks [34]. Amounts of soil organic matter showed distinct differences among stocking methods. Paddocks managed by continuous stocking had lower concentrations than were recorded in paddocks managed under rotational stocking. However, organic matter in soils was more variable in paddocks receiving rotational stocking (c.v. = 29% vs. 24%).

A review of these properties (Table 4) indicates the two stocking methods may have affected some of the physical properties of soil, as has been noted in responses to both longer- and shorter-term applications of stocking methods [13,35,36]. However, the amount of natural variability (e.g., standard deviation) present in the measured attributes, which is also a component of the landscapes enclosed within paddock boundaries [1], indicated the means were not consistent within different paddocks assigned to the same stocking method. The location of paddocks within the landscape of the study area likely affected these properties. The continuous stocked paddocks had predominantly western exposures, while rotationally stocked sub-paddocks had largely easterly exposures. Such differences in exposure can have large effects on catena-based soil development within landscapes [1,2,34]. This difference in predominant exposure likely contributed to the higher amounts of sand, corresponding greater bulk densities, and lower amounts of silt, noted in the continuously stocked paddocks.

Table 4. Mean and standard deviations of particle fractions, bulk density, and organic matter of 7.5 cm increments of surface soils in paddocks managed by two stocking methods.

Stocking Method	Soil Attributes				
	Particle Fractions			Bulk Density	Organic Matter
	Clay	Silt	Sand		
	---------- (%) ----------			(g cm^{-3})	(g kg^{-1})
Continuous	23 (6)	35 (12)	42 (9)	1.17 (0.15)	38 (9)
Rotational	22 (1)	46 (4)	33 (3)	1.09 (0.16)	43 (12)

Stocking methods have some capacity to result in changes in soil organic matter and bulk density in the US Southern Great Plains (SGP) over time. Research in south-central Oklahoma, United States [35], noted increasing bulk densities of soils of perennial grassland in response to 10 years of different animal densities applied by rotational stocking, compared to no grazing. However, the higher bulk density of soils in the continuously stocked paddocks of the current study may be related to the greater amounts of sand that were recorded [34]. Concentrations of soil organic matter were lower in the continuously stocked paddocks, though amounts were variable within paddocks managed under both stocking methods. However, the distribution of soil properties and the native grasslands they support varies within landscapes without grazing [2]. For example, variable patterns were recorded in the spatial distribution of bulk density and organic matter of soils in 1.6 ha paddocks of southern tallgrass prairie within 2 km of this study site, after 25 years of three sustained stocking methods [15]. Distribution patterns noted within those paddocks were definable both within and across multiple paddocks, often within 20 m of spatial scale, which defined the catenae-scale organization of the landscape. Such patterns indicate soils in the area of the current study, which were defined as members of different families and subgroups of Mollisols, likely had similar non-uniform distributions, and variable, fine-scale patterns of spatial distribution.

3.2. Mineral N

Main effects related to stocking method ($0.15 < p < 0.76$) and stocking interactions with paddock locations ($0.21 < p < 0.34$) were not significant for NO_3^-, NH_4^+, or total N. This lack of effect by stocking method in general, and their effects on N fluxes at paddock locations, was unexpected given the number of years of applied treatments and broad range of sampled locations ($n = 16$). These locations included a series with a low likelihood of animal visitation, such as the middle area of paddocks under continuous stocking under the low animal densities that were applied. In comparison, studies in introduced perennial grasslands reported significantly greater amounts of NO_3^- and total mineral N with increase in length of time cattle grazed rotationally stocked sub-paddocks, but found no differences between effects of longer rotational or continuous stocking [10,37]. Studies on the effects of stocking methods on soil properties of native prairie in northcentral Texas, United States, also reported no differences in NO_3^--N concentrations between continuous and rotationally stocked native prairie [13].

Stocking methods interacted with soil depth for NO_3^- ($F_{1, 194} = 3.4$; $p = 0.07$) and total N ($F_{1, 194} = 3.4$; $p = 0.07$). The stocking method × with time of year interaction on NH_4^+ flux was also significant ($F_{1, 194} = 16.6$; $p < 0.01$). Main effects related to paddock location were also significant ($0.01 < p < 0.08$) for NO_3^-, NH_4^+ and total N. The highest amount of total mineral and NO_3^-N within the stocking method × soil depth interactions (Table 5) were noted for the upper 7.5 cm of the profile under rotational stocking, with the second-highest amounts recorded for the upper 7.5 cm under continuous stocking. The lowest amounts of NO_3^- flux were recorded in the upper soil depth under rotational stocking. Within the stocking method × time of year interaction on NH_4^+, the highest and second-highest amounts of flux were recorded, respectively, in response to rotational and continuous stocking during March. The lowest amounts of flux were noted for responses to rotational stocking during August. The greater total and NO_3-N fluxes recorded in the upper 7.5 cm of soil was similar to results noted in studies on other warm-season grasslands [10]. The higher amounts noted during March were likely related to cooler soil temperatures that occur during winter and early-spring in the region, which result in lower amounts of mineralization of NH_4^+ to NO_3^- [3]. Research in paddocks of other warm-season grasses in Florida, United States, noted no differences in concentrations of NH_4^+ among different forms of rotational and continuous stocking [10].

Table 5. Stocking method × soil depth interactions in flux of total mineral N and NO_3, and stocking method × time of year in NH_4 flux within soils. [†]

	Macronutrients					
	Total N		NO3-N		NH4-N	
	Soil Depth		Soil Depth		Time of Year	
Stocking Method	0–7.5 cm	7.5–15 cm	0–7.5 cm	7.5–15 cm	March	August
	------------------------------------ ($\mu g\ cm^2$ probe 14 d^{-1})------------------------------------					
Continuous	3.8 ab	3.1 bc	3.0 ab	2.3 bc	0.5 b	0.3 c
Rotational	4.6 a	2.3 c	3.7 a	1.6 c	0.7 a	0.2 d
Diff	1.3		1.2		0.1	

[†] Diff is statistical differences for means tests; numbers within groups of columns with the same letter were not different at $p = 0.10$.

The greatest amounts of flux in NO_3^- among paddock locations occurred at locations 37 m from water sources, with the second-greatest amounts noted from 3 to 6 m from water (Figure 2A). Alternatively, the lowest amounts of flux were recorded at locations near 0.25 PMP, 0.5 PMP, PMP, and FC. Locations with the second-lowest amounts of flux in NO_3^- were noted at water sources and 1.5 m from water. Amounts of flux at remaining locations belonged to groups with intermediate amounts. Overall, there was a general decline in amounts of NO_3^- flux from 3 m from water sources to the far corner of paddocks (FC), with the notable exception of 37 m.

The distribution of amounts of NH_4^+ flux recorded at paddock locations differed from the distribution noted for NO_3^- (Figure 2B). The greatest amounts of flux in NH_4^+ among paddock locations occurred along fence lines furthest from water sources (FF). Amounts of flux (0.3 to 0.5 μg NH_4^+ cm^{-2}) at all remaining sampled locations were ≤31% of this level, and belonged to the same means group. The distribution of NH_4^+ within paddocks was unusual, given the 3- to 7-fold greater amounts at FF, across stocking methods, compared to the other locations. Studies in paddocks of tame warm-season grasses noted greater amounts of NH4+ in zones of paddocks that were closest and furthest from water, relative to intermediate locations [37]. The driver for high amounts of NH_4^+ flux at FF in the current study was not clear, as these locations were not consistent in terms of landscape positions, or distances from water. The FF locations were situated at both tread and toe positions of the landscapes of the two continuously stocked and four rotationally stocked units that were sampled. However, the amounts of NH_4^+ flux recorded in differences among locations were small, covering a range of 1.1 $\mu g\ cm^{-2}$ of surface areas of probes.

The distribution of amounts of flux in total mineral N recorded at paddock locations were somewhat similar to the distribution pattern noted for NO_3^- (Figure 2C). This similarity was related to the high amounts of NO_3^- relative to the total amount; 65% to 88% of total mineral N at 14 of 16 locations was NO_3^-. The greatest amounts of flux in total N among paddock locations occurred at locations 37 m from water sources, with the second-greatest amounts noted from 3.0 from water, and at sampled locations along the fence furthest from water sources. Alternatively, the lowest amounts of flux were recorded at locations near 0.25 PMP and FC. Locations with the second-lowest amounts of flux in total N were noted within 1.5 m of water sources, 0.37 PMP, 0.5 PMP, and PMP. Amounts of flux at remaining locations belonged to different groups with intermediate amounts. As with NO_3^-, there was a general decline in amounts of flux in total N from 3.0 m from water sources to the far corner of paddocks, with the notable exception of 37 m.

Other studies have noted hotspots of mineral N in soils at specific locations within paddocks, especially in closer proximity to water, corners, and along fences, due to uneven distribution of animal use of paddock areas [38]. For example, research [8] reported that cattle grazing shortgrass rangeland in northeast Colorado, United States, spent ~27% of the time on paddocks at locations near water sources and corners, which represented 2.5% of the total area of paddocks, resulting in shifts in N distributions. Studies undertaken to define redistribution of N in tamegrass paddocks in Florida, United States, noted greater amounts of mineral N in zones closer to water and shade, relative to the remainder of paddock areas [10,37]. These distribution patterns were similar across different forms of rotational and continuous stocking applied to bermudagrass (*Cynodon dactylon*) and bahiagrass (*Paspalum notatum*) paddocks [10,37]. In comparison, the distribution of flux in plant available NO_3^- and total mineral N in the current study showed variability among sampled locations, but higher amounts occurred at locations other than in immediate proximity to water. Further, fluxes in plant available N at other (supposed) high-traffic locations (i.e., NC and FC) were among the lower groups of mean responses that were recorded. Other studies on native rangeland noted positive correlations between heterogeneity in paddock use by cattle and size of paddocks [11]. Given the variable distributions of mineral N in the current study, the large sizes of paddocks and sub-paddocks may have limited uniform use of paddock (and sub-paddock) areas by cattle.

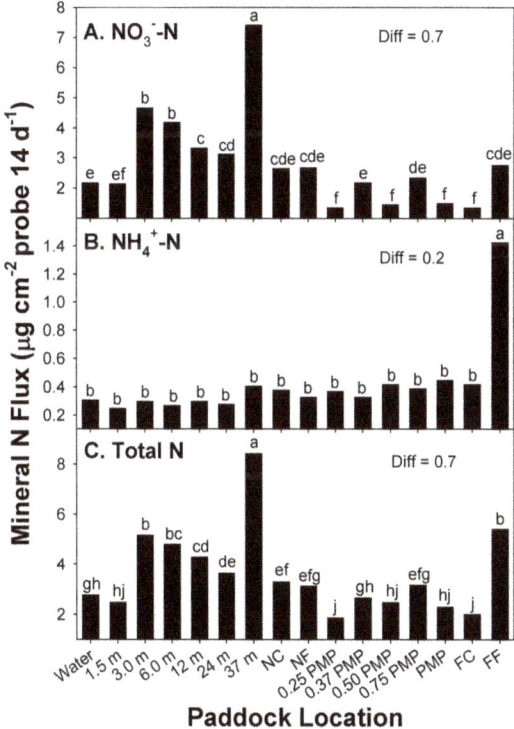

Figure 2. Location effects on flux in plant-available soil (**A**) NO_3^-, (**B**) NH_4^+, and (**C**) total mineral N; columns within panels with the same letter were not different at $p = 0.10$.

3.3. Phosphorus and Magnesium

There were significant soil depth ($F_{1, 194} = 86.8$; $p < 0.01$) and paddock location ($F_{15, 194} = 5.3$; $p < 0.01$) main effects on flux of available P. The interaction between stocking method and time of year was also significant ($F_{1, 194} = 24.8$; $p < 0.01$) for flux in available P. Among effects of soil depth, greater amounts of P flux were noted within the upper 7.5 cm depth than the 7.5–15 cm increment (0.75 versus 0.28 µg P cm^{-2}; Diff = 0.45 µg). Greater amounts of flux in plant-available P within the uppermost segment of the soil profile was not unexpected, as such responses have been reported elsewhere for a range of different grasslands [3]. Larger concentrations in P were recorded in the upper 15 cm of soil than in deeper increments of bermudagrass paddocks in Florida, United States, managed under continuous and rotational stocking methods, with similar responses across stocking methods [10].

Within the stocking method × time of year interaction (Table 6), the largest fluxes in plant-available P occurred in response to rotational stocking during March, while the second-largest occurred under continuous stocking during August. The lowest flux in available P within the interaction was recorded for responses to rotational stocking during August. Differences among mean amounts of plant-available P during the growing season are driven by amounts present in soil solution at the start of growing seasons, plus amounts that become soluble during growing seasons [3]. Responses to continuous stocking indicated a degree of consistency between these factors and uptake by plants (Table 6), while declines in P flux under rotational stocking in summer may indicate depletion of available pools. However, the concentration of plant-available P in soils is generally low [3], so the effects of stocking methods during different times of the growing season are unclear. It is difficult to assess what such low fluxes in plant-available P represent, as it is only a portion of the entire P pool. Other research on P distributions within paddocks of grassland under rotational and continuous stocking noted differences in amounts related to lengths of applied grazing [10,37].

Table 6. Stocking method × time of year interaction effects on flux in plant-available P and Mg in soil [†].

	Macronutrients			
	P		Mg	
Time of Year	Continuous	Rotational	Continuous	Rotational
	---------------- (µg cm^{-2} probe 14 d^{-1}) ----------------			
March	0.50 ab	0.63 a	25.2 bc	37.7 a
August	0.58 a	0.24 b	35.5 ab	18.6 c
Diff	0.26		6.6	

[†] Diff was statistical difference of means tests, and numbers within groups of columns with the same letter were not different at $p = 0.10$.

Among effects related to paddock location, the greatest amounts of flux in P were recorded at 6 m from water, while the second-greatest amounts occurred at 12 m from water (Figure 3A). In contrast, locations with the lowest amounts of P flux were noted at 0.25 PMP through 0.5 PMP. Locations with the second-lowest amounts of P flux occurred at NF and FF. Flux in P at all other locations belonged to different means groups, with intermediate amounts recorded. The potential of the two applied stocking methods to affect distribution of plant-available P within different areas of paddocks (sub-paddocks) was not entirely clear. While there were definite areas with high and low P flux, these zones did not translate to locations normally considered high-traffic areas, such as water sources, corners, or fences [37,38]. Cattle could have loafed roughly 6 to 12 m from water in all sampled paddocks and sub-paddocks, but that would have also resulted in higher flux in other neighboring locations.

The primary sources of plant-available P in soils during growing seasons is derived from recycling of P in soil organic matter via microbes, and through inputs and decomposition of livestock feces, which are slow processes [3,10]. Small amounts are also derived by weathering of parent materials of soils. In this instance, a likely contributing factor for the lower amounts of flux in available P at the more centrally-located positions in paddocks could be a lack of use of these areas by cattle [10], or a lack of deposition of excreta in these areas when frequented by cattle. Cattle grazing larger areas of rangeland can travel up to 1.6 km day^{-1} while searching for forage, though areas frequented in such travel may not be grazed or used for loafing [38,39]. Other research on P distributions within paddocks of grasslands under rotational and continuous stocking noted higher amounts of inorganic P closer to water or shade compared to paddock centers and more distant locations (78–130 [10], and 8–17 [40] ppm), or different times of growing seasons (11–27 ppm [9]), but no effects of stocking method on distributions, as was recorded in the current study.

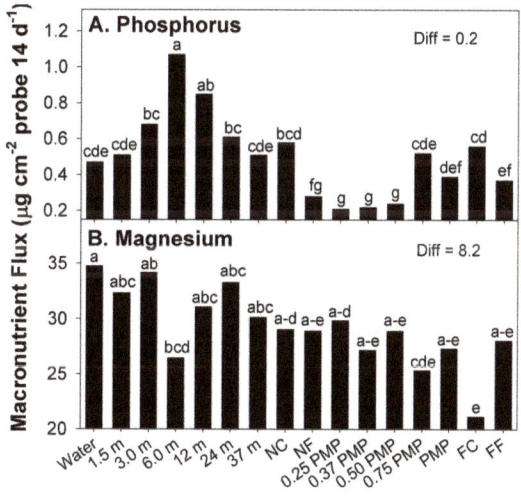

Figure 3. Location effects on flux in plant-available (**A**) phosphorus and (**B**) magnesium in soils; columns within panels with the same letter were not different at $p = 0.10$.

Among effects on plant available Mg, main effects related to paddock location ($F_{15, 195} = 1.8$; $p = 0.03$) and stocking method × time of year interactions ($F_{1, 194} = 12.2$; $p < 0.01$) were significant. All other main and interaction effects were not ($0.11 < p < 0.60$). Within stocking method × time of year interactions (Table 6), the largest fluxes in plant-available Mg occurred in response to rotational stocking during March, while the second-largest occurred under continuous stocking during August. The lowest flux in Mg occurred in response to rotational stocking during August.

Differences in mean responses in this interaction varied compared to results of other studies. Studies on both native prairie and tame pasture have reported higher amounts of Mg in soils under rotational than continuous stocking [3,13], as was noted in the current study, in addition to changes in distribution patterns related to stocking methods [37]. Research has also recorded declines in Mg concentrations in soils through leaching under grazing [3,39], as in the decline in flux recorded under rotational stocking between March and August. In contrast, the increase in Mg over the growing season under continuous stocking may be related to inputs via feces, which is the primary input source, due to yearlong residency on paddocks [3,37]. Soil type can also have some effects on flux in available Mg, with lower amounts recorded in sandier soils. However, the primary sources of plant-available Mg are found in animal excreta, and primarily (>75% of total) in feces [3], so the longer grazing time afforded by continuous stocking provided a greater opportunity for Mg enrichment. The current study noted higher amounts under rotational stocking in March, but continuous stocked paddocks, dominated by sandier soils, generated similar high fluxes in August.

The greatest amounts of flux in available Mg among paddock locations, across stocking methods, occurred at water sources, with the second-greatest amounts noted at 3 m from water (Figure 3B).

Alternatively, the lowest amounts of flux were recorded at FC locations. Locations with the second-lowest amounts of flux in Mg were noted at 0.75 PMP. Amounts of Mg flux at remaining locations belonged to groups with intermediate amounts, with some mean fluxes belonging to 5 means groups. Therefore, there was a degree of similarity in amounts of Mg flux at different locations along transects. A total of 13 means belonged to the group containing the largest response, while 7 responses belonged to the means group with the lowest amounts of flux. Overall, there was a general decline in Mg flux from water sources to FC, but the degree of variability present in flux among paddock locations makes identifying the presence of hot spots difficult. Research on paddocks of introduced perennial grasses managed by different stocking methods reported greater amounts of Mg in larger zones of paddocks closer to water or shade [10,37]. A similar effect was noted in the current study in the lower means of Mg flux with increasing distance from water, though the test for differences noted a degree of similarity between locations near water through FF, which were distances of 423 and 976 m from water.

3.4. Calcium and Sulfur

Main effects of time of growing season ($F_{1,\,194} = 155.7$; $p < 0.01$) and paddock location ($F_{15,\,194} = 2.3$; $p < 0.01$) on flux in available Ca were significant; all other main and interaction effects were not ($0.11 < p < 0.90$). The lack of effects related to stocking methods was unexpected. Earlier research on native prairie in north-central Texas, United States, noted greater amounts of Ca in surface soils of rotational stocked sub-paddocks than in larger, continuously stocked paddocks [13]. Among times of growing season, higher flux in available Ca was noted at time of sampling during August, compared to March (181 vs. 116 µg Ca cm^{-2} probe; Diff = 5.4 µg). Among effects related to paddock location, the greatest amounts of flux in Ca were recorded at water sources 3.0 and 6.0 m from water (Figure 4A). Locations with the second greatest amounts of Ca flux were noted at 1.5 m from water, and 0.75 PMP. Locations with the lowest and second-lowest amounts of Ca flux were noted at far corners from water sources, and 0.25 PMP, respectively. Flux in Ca at all other locations belonged to different means groups with intermediate amounts recorded.

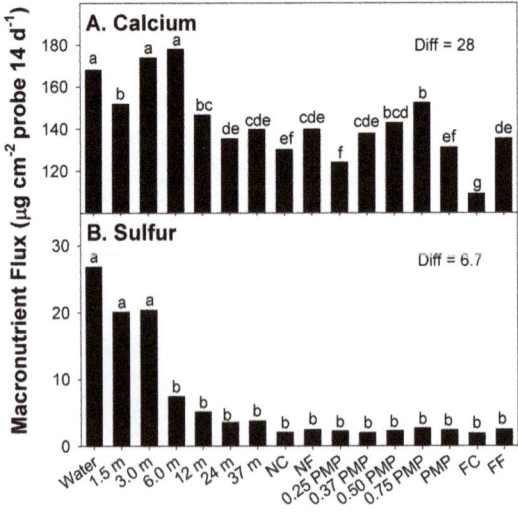

Figure 4. Location effects on flux of plant-available (**A**) calcium and (**B**) sulfur in soils; columns within panels with the same letter were not different at $p = 0.10$.

Calcium in excreta from animals is primarily found in feces, with <3% of animal inputs to soil found in urine. Fecal sources of Ca are also more significant drivers of Ca flux in grassland soils than amounts derived from decomposition of plant residues, or breakdown of parent materials of soils [3]. However, reports of Ca inputs to grassland soils via cattle excreta varies widely. Further, Ca movement from dung into available pools in soils is not well defined, particularly in native rangelands of the United States. A study on soil responses of native prairie ecosystems in northcentral Texas, United States, reported higher concentrations of Ca in soils of paddocks managed by rotational than continuous stocking, with both concentrations lower than was recorded under no grazing [13]. While the current study did show greater amounts of Ca flux in close proximity to water across stocking methods, other locations that would be considered high-traffic zones for cattle (i.e., corners and fences) tended to show lower flux, despite the application of stocking methods for 6 years.

Differences in amounts of flux of plant-available Ca during different seasons have been previously reported. Dickinson and Craig [40] reported Ca in soils increased with amount of precipitation received. However, the current study showed lower amounts of Ca flux during spring, under wetter conditions in conjunction with precipitation events, compared to flux recorded in summer. Precipitation received during individual calendar years in the US SGP, including the study site, is highly variable in amount and timing of precipitation [41]. Such variability in received precipitation would result in some level of variability of plant-available Ca and other macronutrients. Larger precipitation events than were encountered during this study may be required. As such, Ca movement from dung into soil without such events likely occurs at rates that are similar to amounts derived from organic matter in dung [42,43].

The main effects of time of growing season ($F_{1, 194}$ = 5.4; p = 0.02) and paddock location ($F_{15, 194}$ = 2.0; p = 0.02) on flux in available S were significant; all other main and interaction effects were not (0.34 < p < 0.98). Among times of growing season, higher flux in available S was recorded at time of sampling during August than March (4.8 vs. 3.7 µg S cm^{-2} probe; Diff = 0.7 µg). Among effects of paddock location, the greatest amounts of flux in available S were recorded within 3.0 m of water sources (Figure 4B). Thereafter, amounts of S flux at 6.0 m from water to FC belonged to the same means group and displayed low and consistent amounts of flux. Flux of available S at these locations ranged from 6 to 28% of amounts recorded closer to water sources.

The majority (>90%) of plant-available S within the uppermost sections of soil profiles in grazed temperate grasslands are present in labile organic forms derived from animal excreta and plant residues [44]. Inputs of plant-available S from grazing animals is related to relationships between retention and throughput of consumed forage in animals. Amounts of S in urine and feces varies with type of grassland, location within landscape where deposited, and form of management. Research in Australia and New Zealand reported 50 to 70% of S excreted by cattle was labile forms of sulfate in urine [44,45]. Alternatively, S concentration in dung tends to be small (~0.3% of dung dry matter) and generally in organic forms that mineralize at slow rates, similar to soil organic matter [3,44].

Cattle retain roughly 25% of S within biomass of consumed forage, with the remainder excreted, primarily in urine [3]. The current study showed redistribution of S within grazed paddocks was limited, and largely occurred in proximity (≤3.0 m) to water, compared to other locations. Further, other high traffic areas (corners and fences) showed low amounts of flux compared to water sources. In comparison, research in Florida, United States, [10] reported no significant effects of stocking method or paddock locations related to S distribution within grazed bermudagrass paddocks.

3.5. Potassium

There were significant differences in main effects related to time of year ($F_{1, 194}$ = 37.4; p < 0.01) and soil depth ($F_{1, 194}$ = 66.9; p < 0.01) on plant-available K, as was the stocking method × paddock location interaction ($F_{15, 194}$ = 2.3; p = 0.01). All other main and interaction effects were not significant (0.14 < p < 0.96). Greater amounts of flux were noted during March than were recorded in August (24.4 vs. 14.7 µg K cm^{-2} probe; Diff = 7.4 µg). Among soil depths, greater amounts of flux were noted within the upper 7.5 cm depth of profile than the 7.5 to 15 cm increment (26.1 vs. 13.7 µg K cm^{-2} probe; Diff = 7.4 µg). Main effects of time of year and soil depth were similar to results reported in the broader literature, with greater amounts of K flux noted during periods with greater amounts soil water and in sections of soil profiles near the surface [3,15]. Sampling during March of the current study occurred after a series of precipitation events that allowed moisture in the upper sections of soil profile to approximate field capacity. Such conditions are important in movement of K from plant residues and excreta into soil solution [3]; drought conditions prior to sampling in August would limit such movement, and reduce amounts of plant-available K in soils.

Within the stocking method × paddock location interaction in flux of available K, the greatest and second greatest amounts occurred, respectively, under rotational stocking at 12 and 24 m from water sources (Figure 5). Amounts at 37 m from water in continuous stocked paddocks had some similarity to amounts at 24 m locations in rotational stocked paddocks. In contrast, the lowest flux in available K were recorded at 0.37 PMP under both stocking methods, and FC in rotational stocked paddocks. All remaining means within the interaction belonged to ranges of means groups, with many values belonging to 7 means groups within the interaction. Amounts of flux in plant available K under both stocking methods displayed undulations among low and high amounts with increasing distance from water, though at different locations.

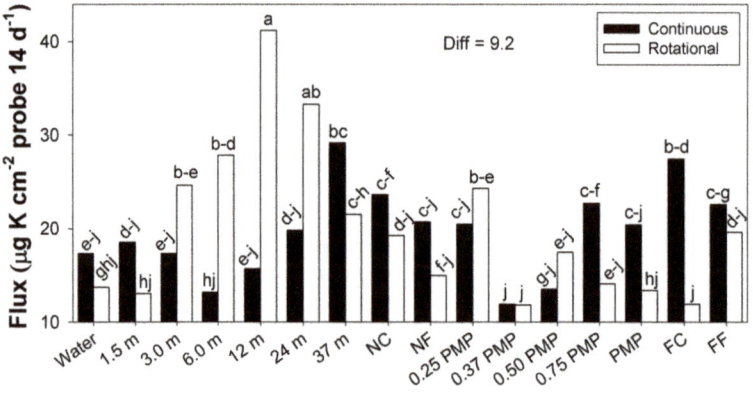

Figure 5. Stocking method × paddock location interaction in flux of plant-available K in soils; columns with the same letter were not different at $p = 0.10$.

The interaction between stocking method and paddock location on K flux (Figure 5) was the only occurrence of stocking method affecting the distribution of a macronutrient within paddock space. However, this impact is not clear and displayed a degree of variability, as was recorded for most of the macronutrients. Potassium is highly labile and can be easily leached from excreta and grassland soils, particularly with occurrences of high rainfall [3], which are uncommon in the US SGP [41]. Cattle excreta (particularly urine) is a primary source of plant-available K, and amounts in urine patches tend to exceed uptake requirements of plants [3,35]. Therefore, amounts of K flux could be related to some combination of conservation of K in soils by drought and local distribution of urine patches, though many of these locations did not translate to areas normally noted for high animal use [37,38]. Effects related to stocking method and paddock location on K flux showed both similarities and differences from results of other studies. Experiments testing distributions of macronutrients within paddocks of bahiagrass reported no difference among rotational and continuous stocking for soil K, but higher concentrations of K in zones near water and shade [37]. In comparison, experiments on stocking methods applied to bermudagrass paddocks also reported no difference in K concentrations in soils among stocking methods [10]. Similarly, K concentrations in soils of native prairie in northcentral Texas, United States, reported no differences among continuous and rotational stocking [13]. In contrast, amounts of flux in available K in the current study showed differences among stocking methods that occurred as undulations across paddocks and sub-paddocks.

4. Conclusions

Responses noted during this study did not provide clear definitions of the impact of stocking methods on amounts and distributions of flux in most plant-available macronutrients within paddocks. There was some evidence of redistribution in the higher fluxes of S, Ca, and Mg close to water sources, but little evidence related to effects of individual stocking methods, despite the differences between rotational (multiple grazing periods per year) and continuous (year-round) stocking. There were effects recorded for amounts of macronutrient flux among times of year and soil depths, with both stocking methods generating hot spots of flux, as was noted in other research. However, flux of 7 of the macronutrients at different locations within paddocks and sub-paddocks assigned to the stocking methods were similar, despite the large differences in paddock (and sub-paddock) sizes (and transect lengths) under the two stocking systems. One factor that likely contributed to the lack of effects of stocking methods on distributions of fluxes within paddocks and sub-paddocks was the non-uniform distribution and orientation of different features of landscapes within paddocks in relation to water sources [46], and differences in the primary exposure of landscapes within these paddocks and sub-paddocks.

Such variability in landscape features is not unusual in paddocks at production-scales, compared to smaller experimental paddocks, and can affect distribution of grazing and pasture use by cattle [47]. One feature that defines the occurrence of non-uniform landscapes within production-scale paddocks of the US SGP is related to the organization of land ownership by the Public Land Survey System. This system results in regular grids organized by township and range, on a sectional (259 ha) and sub-sectional basis. Therefore, paddocks and sub-paddocks applied to rangelands are largely organized without regard to positions within the larger landscape, resulting in multiple landscape features with multiple soils, and local plant communities, within paddock boundaries. This variability, in turn, limits the capacity of research to determine if stocking methods influence macronutrient distributions through landscape use, and placement of excreta, in production-scale paddocks and sub-paddocks. Therefore, if large production-scale paddocks are to be used in research, attention to how landscape features and locations of water sources are organized within paddocks is required [48]. Such an approach would improve the capacity of studies to define effects of stocking methods on macronutrient distributions, and landscape use by grazing cattle [49].

Author Contributions: Conceptualization, P.J.S. and K.E.T.; methodology and experimental design, B.K.N., P.J.S. and K.E.T.; statistical analysis, B.K.N.; resource management and project administration, P.J.S., K.E.T. and B.K.N.; data curation, P.J.S. and B.K.N.; writing—original draft preparation, B.K.N.; review and editing, B.K.N., K.E.T. and P.J.S.; funding acquisition, P.J.S.

Funding: This research was partially supported by funds from the Agriculture and Food Research Initiative Competitive Projects 2012-02355 and 2013-69002-23146, from United States Department of Agriculture (USDA), National Institute of Food and Agriculture.

Acknowledgments: This research is a contribution from the Long-Term Agroecosystem Research (LTAR) network. The authors wish to recognize Agricultural Research Service technicians Kory Bollinger and Jeff Weik for their assistance in managing the experiment. Mention of trademarks, proprietary products, or vendors does not constitute guarantee or warranty of products by USDA and does not imply its approval to the exclusion of other products that may be suitable. All programs and services of the USDA are offered on a nondiscriminatory basis, without regard to race, color, national origin, religion, sex, age, marital status, or handicap.

Conflicts of Interest: The authors declare no conflicts of interest.

References

1. Jenny, H. *Factors of Soil Formation: A System for Quantitative Pedology*; Dover Publications: Mineola, NY, USA, 1994.
2. Archer, S.; Smeins, F.E. Ecosystem-level processes. In *Grazing Management: An Ecological Perspective*; Heitschmidt, R.K., Stuth, J.W., Eds.; Timber Press: Portland, OR, USA, 1991; pp. 109–139.
3. Whitehead, D.C. *Nutrient Elements in Grassland: Soil-Plant-Animal Relationships*; CABI Publishing: New York, NY, USA, 2000.
4. Norman, M.J.T.; Green, J.O. The local influence of cattle dung and urine upon yield and botanical composition of permanent pasture. *J. Br. Grassl. Soc.* **1958**, *13*, 39–45. [CrossRef]
5. Sheldrick, W.; Syers, J.K.; Lindgard, J. Contribution of livestock excreta to nutrient balances. *Nutr. Cycl. Agroecosyst.* **2003**, *66*, 119–131. [CrossRef]
6. Orwin, K.H.; Bertram, J.E.; Clough, T.J.; Condron, L.M.; Sherlock, R.R.; O'Callaghan, M. Short-term consequences of spatial heterogeneity in soil nitrogen concentrations caused by urine patches of different sizes. *Appl. Soil Ecol.* **2009**, *42*, 271–278. [CrossRef]
7. Schnyder, H.; Locher, F.; Auerswald, K. Nutrient redistribution by grazing cattle drives patterns of topsoil N and P stocks in a low-input pasture ecosystem. *Nutr. Cycl. Agroecosyst.* **2010**, *88*, 183–195. [CrossRef]
8. Augustine, D.J.; Milchunas, D.G.; Derner, J.D. Spatial redistribution of nitrogen by cattle in semiarid rangeland. *Rangel. Ecol. Manag.* **2013**, *66*, 56–62. [CrossRef]
9. Saunders, W.H.M. Effects of cow urine and its major constituents on pasture properties. *N. Z. J. Agric. Res.* **1982**, *25*, 61–68. [CrossRef]
10. Mathews, B.W.; Sollenberger, L.E.; Nair, V.D.; Staples, C.R. Impact of grazing management on soil nitrogen, phosphorus, potassium, and sulfur distribution. *J. Environ. Qual.* **1994**, *23*, 1006–1013. [CrossRef]
11. Barnes, M.K.; Norton, B.E.; Maeno, M.; Malachek, J.C. Paddock size and stocking density affect spatial heterogeneity of grazing. *Rangel. Ecol. Manag.* **2008**, *61*, 380–388. [CrossRef]
12. Briske, D.D.; Derner, J.D.; Brown, J.R.; Fuhlendorf, S.D.; Teague, W.R.; Havstad, K.M.; Gillen, R.L.; Ash, A.J.; Willms, W.D. Rotational grazing on rangelands: Reconciliation of perception and experimental evidence. *Rangel. Ecol. Manag.* **2008**, *61*, 3–17. [CrossRef]
13. Teague, W.R.; Dowhower, S.L.; Baker, S.A.; Haile, N.; DeLaune, P.B.; Conover, D.M. Grazing management impacts on vegetation, soil biota and soil chemical, physical and hydrological properties in tall grass prairie. *Agric. Ecosyst. Environ.* **2011**, *141*, 310–322. [CrossRef]
14. Williams, R.D.; Ahuja, L.R.; Naney, J.W.; Ross, J.D.; Barnes, B.B. Spatial trends and variability of soil properties and crop yield in a small watershed. *Trans. ASAE* **1987**, *30*, 1653–1660. [CrossRef]
15. Northup, B.K.; Daniel, J.A. Distribution of soil bulk density and organic matter along an elevation gradient in central Oklahoma. *Trans. ASABE* **2010**, *53*, 1749–1757. [CrossRef]
16. Oades, J.M. The role of biology in the formation, stabilization, and degradation of soil structure. *Geoderma* **1993**, *56*, 377–400. [CrossRef]
17. Burke, I.; Lauenroth, W.; Vinton, M.; Hook, P.; Kelly, R.; Epstein, H.; Agular, M.; Robles, M.; Aguilera, M.; Murphy, M.; et al. Plant-soil interactions in temperate grasslands. *Biogeochemistry* **1998**, *42*, 121–143. [CrossRef]
18. Hook, P.; Burke, I.C. Biogeochemistry in a shortgrass landscape: Control by topography, soil texture and microclimate. *Ecology* **2000**, *81*, 2686–2703. [CrossRef]
19. Northup, B.K.; Phillips, W.A.; Daniel, J.A.; Mayeux, H.S. Managing southern tallgrass prairie: Case studies on grazing and climatic effects. In *Proceedings: 2nd National Conference on Grazing Lands, Nashville, TN, USA*; Theurer, M., Peterson, J., Golla, M., Eds.; Omnipress Inc.: Madison, WI, USA, 2003; pp. 834–890.
20. Northup, B.K.; Starks, P.J.; Turner, K.E. Effects of stocking methods on soil macronutrient availability at contrasting locations in tallgrass paddocks. *Agronomy* **2019**, *9*, 281. [CrossRef]
21. USDA-NRCS. *Soil Survey of Canadian County, Oklahoma. Supplement Manuscript*; US Department of Agriculture–Natural Resource Conservation Service, and Oklahoma Agricultural Experiment Station: Stillwater, OK, USA, 1999.
22. Phillips, W.A.; Coleman, S.A. Productivity and economic return of three warm-season grass stocker systems of the southern Great Plains. *J. Prod. Agric.* **1995**, *8*, 334–339. [CrossRef]

23. Phillips, W.A.; Northup, B.K.; Mayeux, H.S.; Daniel, J.A. Performance and economic returns of stocker cattle on tallgrass prairie under different grazing management strategies. *Prof. Anim. Sci.* **2003**, *19*, 416–423. [CrossRef]
24. Goodman, J.M. Physical environments of Oklahoma. In *Geography of Oklahoma*; Morris, J.W., Ed.; Oklahoma Historical Society: Oklahoma City, OK, USA, 1977; pp. 9–25.
25. Zhou, Y.; Gowda, P.H.; Wagle, P.; Ma, S.; Neel, J.P.S.; Kakani, V.G.; Steiner, J.L. Climate effects on tallgrass prairie responses to continuous and rotational grazing. *Agronomy* **2019**, *9*, 219. [CrossRef]
26. Szillery, J.E.; Fernandez, I.J.; Norton, S.A.; Rustad, L.E.; White, A.S. Using ion-exchange resins to study soil response to experimental watershed acidification. *Environ. Monit. Assess.* **2006**, *116*, 383–398. [CrossRef]
27. Dick, W.A.; Culman, S.W. Biological and biochemical tests for assessing soil fertility. In *Soil Fertility Management in Agroecosystems*; Chatterjee, A., Clay, D., Eds.; ASA, CSSA, and SSSA: Madison, WI, USA, 2016; pp. 134–147.
28. Grossman, R.B.; Reinsch, T.G. Bulk density and linear extensibility. In *Methods of Soil Analysis: Part 4. Physical Methods*; Dane, J.H., Topp, G.C., Eds.; SSSA: Madison, WI, USA, 2002; pp. 201–228.
29. Gee, G.W.; Bauder, J.W. Particle-size analysis. In *Methods of Soil Analysis, Part I. Physical and Mineralogical Methods*; Klute, A., Campbell, G.S., Jackson, R.D., Mortland, M.M., Nielsen, D.R., Eds.; ASA and SSSA: Madison, WI, USA, 1986; pp. 383–411.
30. Nelson, D.W.; Sommers, L.E. Total carbon, organic carbon, and organic matter. In *Methods of Soil Analysis: Part 3. Chemical Methods*; Sparks, D.L., Page, A.L., Helmke, P.A., Loeppert, R.H., Soltonpour, P.N., Tabatabai, M.A., Johnston, C.T., Sumner, M.E., Eds.; SSSA and ASA: Madison, WI, USA, 1996; pp. 961–1010.
31. Steel, R.G.D.; Torrie, J.H. *Principles and Procedures of Statistics: A Biometrical Approach*, 2nd ed.; McGraw-Hill: New York, NY, USA, 1980.
32. Littel, R.C.; Milliken, G.A.; Stroup, W.W.; Wolfinger, R.D. *SAS Systems for Mixed Models*; SAS Institute Inc.: Cary, NC, USA, 1996.
33. Patetta, M. *Longitudinal Data Analysis with Discrete and Continuous Responses: Course Notes for Instructor-Based Training*; SAS Institute Inc.: Cary, NC, USA, 2005.
34. Buol, S.W.; Hole, F.D.; McCracken, R.J. *Soil Genesis and Classification*, 2nd ed.; Iowa State University Press: Ames, IA, USA, 1980.
35. Daniel, J.A.; Potter, K.N.; Altom, W.; Aljoe, H.; Stevens, R. Long-term grazing density impacts on soil compaction. *Trans ASAE* **2002**, *45*, 1911–1915. [CrossRef]
36. Wheeler, M.A.; Trilica, M.J.; Fraser, G.W.; Reeder, J.D. Seasonal grazing effects soil physical properties of a montane riparian community. *J. Range Manag.* **2002**, *55*, 49–56. [CrossRef]
37. Dubeux, J.C.B., Jr.; Sollenberger, L.E.; Vendramini, J.M.B.; Interrante, S.M.; Lira, M.A., Jr. Stocking methods, animal behavior, and soil nutrient redistribution: How are they linked? *Crop Sci.* **2014**, *54*, 2341–2350. [CrossRef]
38. Bailey, D.W.; Gross, J.E.; Laca, E.A.; Rittenhouse, L.R.; Coughenor, M.B.; Swift, D.M.; Sims, P.L. Mechanisms that result in large herbivore grazing patterns. *J. Range Manag.* **1996**, *49*, 386–400. [CrossRef]
39. Arnold, G.W.; Dudzinski, M.L. *Ethology of Free-Ranging Domestic Animals*; Elsevier Scientific Publ.: New York, NY, USA, 1979.
40. Dickinson, C.H.; Craig, J. Effects of water on the decomposition and release of nutrients from cow pats. *New Phytol.* **1990**, *115*, 139–147. [CrossRef]
41. Schneider, J.M.; Garbrecht, J.D. A measure of usefulness of seasonal precipitation forecasts for agricultural applications. *Trans. ASAE* **2003**, *46*, 257–267. [CrossRef]
42. Steele, K.W.; Judd, M.J.; Shannon, P.W. Leaching of nitrate and other nutrients from grazed pasture. *N. Z. J. Agric. Res.* **1984**, *27*, 5–12. [CrossRef]
43. Underhay, V.H.S.; Dickinson, C.H. Water, mineral and energy fluctuations in decomposing cow pats. *J. Br. Grassl. Soc.* **1978**, *33*, 189–196. [CrossRef]
44. Nguyen, M.L.; Goh, K.M. Sulphur cycling and its implications on sulphur fertilizer requirements of grazed grassland ecosystems. *Agric. Ecosyst. Environ.* **1994**, *49*, 173–206. [CrossRef]
45. Haynes, R.J.; Williams, P.H. Nutrient cycling and soil fertility in the grazed pasture ecosystem. *Adv. Agron.* **1993**, *49*, 119–199.
46. Stuth, J.W. Foraging behavior. In *Grazing Management: An Ecological Perspective*; Heitshcmidt, R.K., Stuth, J.W., Eds.; Timber Press: Portland, OR, USA, 1991; pp. 65–83.

47. Wallis-DeVries, M.F.; Schippers, P. Foraging in a landscape mosaic, selection for energy and minerals in free-ranging cattle. *Oecologia* **1994**, *100*, 107–117. [CrossRef] [PubMed]
48. Cook, C.W.; Stubbendieck, J. *Range Research: Basic Problems and Techniques*; Society for Range Management: Denver, CO, USA, 1986.
49. Long, J.W.; Medina, A.M. Consequences of ignoring geologic variation in evaluating grazing impacts. *Rangel. Ecol. Manag.* **2006**, *59*, 373–382. [CrossRef]

© 2019 by the authors. Licensee MDPI, Basel, Switzerland. This article is an open access article distributed under the terms and conditions of the Creative Commons Attribution (CC BY) license (http://creativecommons.org/licenses/by/4.0/).

Article

Stocking Methods and Soil Macronutrient Distributions in Southern Tallgrass Paddocks: Are There Linkages?

Brian K. Northup *, Patrick J. Starks and Kenneth E. Turner

United States Department of Agriculture-Agricultural Research Service, El Reno, OK 73036, USA; patrick.starks@ars.usda.gov (P.J.S.); ken.turner@ars.usda.gov (K.E.T.)
* Correspondence: brian.northup@ars.usda.gov; Tel.: +1-405-262-5291

Received: 22 March 2019; Accepted: 16 May 2019; Published: 31 May 2019

Abstract: Broad ranges of factors (parent materials, climate, plant community, landscape position, management) can influence macronutrient availability in rangeland soils. Two important factors in production-scale paddocks are the influences of location in space and land management. This study examined plant-available macronutrients (total mineral and nitrate-N, P, S, K, Ca, and Mg) in soils, with paired sets of probes (anion and cation exchange membranes) that simulate uptake by plant roots. Data were collected from sets of paddocks of southern tallgrass prairie in central Oklahoma, managed by four stocking methods during the 2015 growing season (mid-March, growth initiation by native grasses, and early-August, time of peak living plant biomass). Macronutrient availability in the 0–7.5 cm and 7.5–15 cm depths were determined at locations in close proximity to water (water tanks and 25% of the distance between tanks and paddock mid-points (PMP)), and distances near the mid-points of paddocks (70% of the distance between water and mid-points (0.7 PMP), and PMP). All of the tested stocking methods affected levels of availability of macronutrients at different times of the growing season, and among different locations within paddocks. Such responses indicated stocking methods may not result in uniform distributions of flux in plant-available macronutrients. The overall exposure of landscapes and arrangement of features within paddocks also appeared to influence macronutrient distributions.

Keywords: exchange membranes; range management; soil macronutrients

---

1. Introduction

A diverse range of factors influence the availability of macronutrients in soils of native grasslands, often with a large degree of interaction among factors. The availability of macronutrients, and the forms of macronutrients, within soils of native grasslands generally reflects the nature of parent materials from which the soils evolved, and is further affected by plant communities, climate, position within landscapes, and management [1–3]. These factors result in variable spatial and temporal distribution patterns for plant-available macronutrients in soils of ecosystems dominated by perennial grasses [1,3]. Hence, the importance of macronutrients to landscape organization and biological function of grasslands [4–6].

Cattle can affect the distribution of plant-available macronutrients in grassland ecosystems by grazing plant biomass and; removing macronutrients from paddocks in weight gain (via consumption, rumination and conversion of macronutrients to animal mass), and by redistributing unincorporated macronutrients in excreta within paddocks in heterogeneous patterns [3,7–9]. The behavior of cattle and their preferential use of different zones of the landscape within paddocks are factors that can drive redistribution of macronutrients within paddocks [1,10,11]. The redistribution of macronutrients through excreta can be non-uniform, resulting in high concentrations of more labile forms in localized areas of paddocks [12,13]. There have been reported increases in amounts of labile forms of N, P, K, and S in areas near watering facilities, corners, and other structures within paddocks of grazed rangeland and perennial grasslands [9,14,15].

Some systems of stocking are thought to be capable of altering how cattle redistribute more labile forms of macronutrients, primarily N, P, K, and S within grazed landscapes [3,15,16]. This impact is of particular interest in cases where perennial grasslands are managed by different methods of rotational stocking, which may prevent high and disproportionate loadings in certain areas of paddocks [15]. However, the natural distribution of macronutrients, and physical properties, of soils in production-scale paddocks tend to be variable without overlaying the effects of applied grazing systems [1].

Given the naturally variable landscapes within production-scale paddocks, an important question is whether stocking methods can influence distributions of plant-available macronutrients within managed rangeland landscapes. The objective of this study was to apply a simple test for differences in levels of plant-available macronutrients in soils at two contrasting locations within paddocks (near water sources, near paddock centers) of southern tallgrass prairie managed under different methods of stocking. The working hypotheses were, that there would be no difference in the level of availability of eight macronutrients at these locations within paddocks, or differences in amounts at different times of growing seasons, in response to four stocking methods.

2. Materials and Methods

This study was conducted at the USDA-ARS Grazinglands Research Laboratory (35°33′29″ N, 98°1′50″ W; 435 m elevation) in central Oklahoma, USA. The entire site was located in a rolling upland landscape, with a range of features present. Included were local easterly and westerly-facing slopes (3 to 6%) on riser positions, and toe and tread slope positions (0 to 2% slopes), bordering the risers [17]. The long-term average (LTA; 1977 to 2012) precipitation [±1 standard deviation [SD]) during calendar years was 941(±174) mm. The majority of precipitation received annually had a bimodal distribution with maxima in April through June (334 ± 58 mm), and September through October (175 ± 54 mm). Long-term monthly minimum and maximum temperatures (±1 s.d.) were recorded in January [2.8(± 2.7) °C; −4.0 to 8.6 °C] and July [28.1(± 1.5) °C; 25.6 to 32.4 °C].

The study was located within production-scale paddocks on 346 ha of native grassland defined as southern tallgrass prairie. The grasslands within the paddocks are remnants of the original southern tallgrass ecosystem that existed in central Oklahoma prior to settlement, and were never cultivated or replanted to native grasses following cultivation. The plant community was identified as a Loamy Prairie ecological site ([17], ecological site number 080AY056OK). Dominant grasses were the perennials big bluestem (*Andropogon gerardii*), little bluestem (*Schizachyrium scoparium*), and Indiangrass (*Sorghastrum nutans*). These three indigenous species generate 60% to 80% of the total above-ground biomass produced annually within the area that was included in the study [18]. The area was historically (1977 to 2009) used to support cow herds that were managed to generate calves of different breed-types and crosses for research projects on production systems applied to yearling stocker cattle [19,20]. Stocking methods that were applied during the historical period were changed regularly to meet management requirements for yearling cattle used in research projects. Included were periods of management under either, or both, continuous and rotational stocking applied to portions of the area during growing seasons (April to September) at low densities (2.0 to 3.0 ha cow/calf pair^{-1} growing season^{-1}).

A range of soils exists within the area [17,21], based on position within the landscape. Norge series silt loams (fine-silty, mixed, active thermic Udic Paleustolls), situated on riser (mid-slope) positions of the landscape were the most common soil [17]. The Norge series has three sub-types based on the location along risers, and degree of slope. Renfrow or Kirkland silt loams (fine, mixed-superactive, thermic Udertic Paleustolls) on tread positions (summit locations) and Port silt loams (fine-silty, mixed-superactive, thermic Cumulic Haplustolls) at toe positions bounded the Norge series within the study site. Up to six related soils are present as inclusions within the boundaries of each of these soils [17]. The surface soils of these series have near-neutral but variable pH levels (6.7 ± 0.6), a water-holding capacity of 3 ± 1 mm cm^{-1} soil, and permeability rates of 33 ± 17 mm h^{-1} [17]. These soils evolved from parent material, defined as Permian-aged Dog Creek shale, comprised of reddish-brown shale with thin inter-beds of siltstone and sandstone [21].

Two replicate 61(± 2) ha paddocks, managed by continuous yearlong stocking, were included in the study as controls. Additional sets of sub-paddocks that were components of two replicate 80(± 2) ha paddocks (10 sub-paddocks each) of rotationally-stocked rangeland were also included. The design and organization of paddocks and sub-paddocks used in this study were variable in size, shape, and dimensions. Two sets of additional small paddocks (0.4 and 0.2 ha) were also established within the areas of the continuous paddocks, with one set each at toe and tread positions. These small units were used to mimic the application of high-density short-duration rotational stocking (HDRG), known as mob stocking [22], and their impacts on plant communities and soil properties.

Grazing pressure was achieved by herds of cow–calf pairs (~500 kg cows and up to 249 kg calves at weaning) as animal units (AU) that were assigned to replicate experimental units receiving continuous and rotational stocking. Daily forage allotments per AU were 14 kg d^{-1}, or ~5.1 Mg AU^{-1} yr^{-1}. Herd sizes in the continuous paddocks and sets of rotational-stocked sub-paddocks were 18(± 3.4) and 26(± 2.5) AU yr^{-1} respectively, during 2009–2015. The total grazing pressure applied to the continuous and rotational paddocks averaged roughly 108 and 119 animal unit days (AUD) ha^{-1} yr^{-1}. The cow herds assigned to continuous-stocked paddocks were used to apply the high-density short-duration stocking treatments, which are also known as mob stocking. These paddocks were grazed once per year for 24-h, with the timing of grazing varied annually. Cattle from the two continuous-stocked paddocks were randomly applied to one of the HDRG treatments on each replicate for 24 h, returned to the continuous paddocks for 48 h, and then applied to the remaining HDRG treatment for 24 h. The mob-stocked paddocks served as examples of two levels of high-density rapid-rotation stocking [18(± 3.4 head) for 24 h on 0.8 ha (HDRG-1×) and 0.4 ha (HDRG-2×) units, which applied 23 and 45 AUD ha^{-1} yr^{-1} of grazing pressure] that have garnered producer interest in Oklahoma, USA, in recent years. Cow–calf pairs assigned to the rotational-stocked paddocks grazed sub-paddocks in 7 to 10-day grazing bouts, up to four times per year, with the number and timing of bouts varied annually. Therefore, the applied rotational and HDRG systems were representative of different adaptive forms of rotational stocking [22]. All paddocks in this study were managed under their assigned stocking methods from 2009 through 2015.

Data were collected at two times during the 2015 growing season: mid-March at the initiation of growth by native grasses, and early August, when peak living biomass occurs for southern tallgrass prairie during late summer [18]. Sub-paddocks within the rotational stocked units that were sampled in March 2015 were last grazed in mid-August 2014, while the mob-stocked paddocks were previously grazed in early September 2014. The ~7-month delay in sampling of the rotational and mob-stocked paddocks in March 2015 occurred because there was only one set of mob-stocked paddocks per replicate, and their planned timing of grazing for 2015 was late summer. Sampling of the rotational sub-paddocks and mob-stocked units in August 2015 occurred ~7 days after grazing by cattle. This coordination of sampling within the rotational and HDRG-stocked paddocks allowed for more effective comparisons among different forms of rotational stocking.

At each sampling date, the availability of macronutrients within the 0 to 7.5 cm and 7.5 to 15 cm depth increments of soil were determined at four locations within the paddocks at four locations within two replicate paddocks (or sub-paddocks) per stocking method. Locations were oriented along straight lines from water sources to centers of units. All water sources within paddocks or sub-paddocks of this study were located near paddock corners. Locations were: (1) 1.0 to 2.0 m distance from water tanks, (2) 25% of the distance between tanks and the mid-point of paddocks (0.25 PMP), (3) 70% of the distance between water and the mid-point of paddocks (0.7 PMP), and (4) at the mid-points of paddocks (PMP). The distances of 0.25 PMP, 0.7 PMP, and PMP locations from water within paddocks varied with size and shape of the individual paddocks and sub-paddocks that were sampled (Table 1). Therefore, all locations represented random locations within each paddock or sub-paddock due to pasture size, shape, and dimensions. These locations represented areas within rangeland paddocks with two divergent capacities to attract cattle: (1) areas where animals are guaranteed to interact with the soil and plant community (locations closer to water), and (2) areas at the furthest point from water and other attractant features (fence lines, corners) within paddocks [23].

Table 1. Paddock sizes and distances (±1 SD) between water and locations within paddocks.

Stocking Method	Paddock Size	Paddock Locations			
		Water	0.25 PMP	0.7 PMP	PMP
	(ha)	(m)			
Continuous	61.0 (±2)	2(±1)	109(±7)	301(±9)	435(±8)
Rotational	6.7 (±2)	2(±1)	43(±17)	119(±33)	174(±29)
HDRG-1×	0.4	2(±1)	10(±2)	26(±2)	37(±2)
HDRG-2×	0.2	2(±1)	9(±1)	23(±2)	35(±2)

Availability of macronutrients within soils was determined with Plant Root Simulator ^{TM} probes (Western Ag Innovations Inc., Saskatoon, SK, Canada). The probes consist of paired sets of anion and cation exchange membranes encased in plastic housings. At each sampled location within paddocks, two sets of probes with anion and cation membranes were co-located within 30 cm areas around paddock locations, at each soil depth. The probes were buried in situ during the March sampling for a 14-day incubation period. Soil moisture approximated field capacity in March [22(±3)% volumetric water] due to precipitation events that occurred prior to incubation periods.

Soil moisture at the time of the August sampling approximated permanent wilt-point [12(±1)% volumetric water] due to a prolonged dry period during summer. Therefore, replicate ($n = 4$) 5.38 cm diameter soil cores for each location and soil depth were collected and removed intact to a laboratory. Collected cores were wetted to field capacity with deionized water, dissected into two sections along the long axis of cores, and probes with anion and cation membranes were each sandwiched within two separate cores for incubation. The probes were removed following incubation periods for both sampling dates, lightly washed with deionized water to remove soil, packaged in groups by paddocks, refrigerated, and sent to the probe manufacturer for analyses. The manufacturer used colorimetric analyses via automated flow injection to determine NO_3-N and NH_4-N, and the other macronutrients (ions) were measured using inductively-coupled plasma spectrometry.

Two additional sets of soil cores were collected at times of sampling to define physical properties of soils, within a 50 cm diameter area as the probes, at the different sampled locations. Moist bulk densities of one set of samples ($n = 8$ soil depth^{-1} stocking method^{-1}) were defined [24], followed by analyses of particle fractions by hydrometer methods in a sodium hexa-metasulfate solution [25]. A second set of soil samples ($n = 8$ soil depth^{-1} stocking method^{-1}) were used to define estimates of soil organic matter based on low temperature [380 °C for 16 h on oven-dried (105 °C for 24 h) samples] loss on ignition [26].

Exploratory analyses were applied to amounts of macronutrients absorbed by the probes (Table 2) to determine whether data transformations were required for statistical tests [27]. We applied natural logarithm (Ln) transforms, as required, to improve the cumulative distribution functions of the populations of observations so they more closely fit a normal distribution. Transformed levels of flux of total mineral N, NO_3, NH_4, P, S, and K, and the raw values for Ca and Mg were analyzed in SAS 9.3 (SAS Institute, Cary, NC) by longitudinal (repeated) measures analyses [28] within mixed-models (PROC MIXED). Grazing regime, soil depth, and time of growing season were main effects in analyses, while sampled location within paddock was the longitudinal effect. Particle fractions, bulk density, organic matter, and distances between water sources and sampled locations were tested as covariates, to improve function and tests of statistical models. However, all physical attributes of soil were reported as non-significant for effects on analyses (0.23 < P < 0.88). Distances between water and sampled locations were retained as a covariate, as they differed among individual paddocks assigned to stocking methods, and tests indicated there was some influence on the function of statistical models [29]. Therefore, means and standard deviations (SD) of particle fractions, bulk density, and organic matter are reported to provide some estimates of soil properties at the sampled locations within paddocks.

Compound symmetry (CS) variance–covariance matrices [28] were used to account for covariance and autocorrelation among locations within paddocks as there were not enough degrees of freedom (d.f.) to utilize more complex matrix structures [29]. All analyses were restricted to 2-way interactions among main factors and the longitudinal factor due to a lack of d.f. required to test higher-order interactions [27]. The PDIFF procedure of SAS [29] was used to test for differences among means of significant main and interaction effects. Reported means were back-transformed to original scales of flux [27]. The level of significance of statistical tests was set at $P = 0.10$.

Table 2. Descriptive statistics and distribution functions of populations of plant-available macronutrients in soils of paddocks managed under different stocking methods.

	Macronutrients						
	Mineral N	NO_3^--N	P	S	K	Mg	Ca
	——($\mu g\ 10\ cm^{-2}$ probe 7.5 cm soil depth^{-1} 14 d^{-1})——						
Statistics							
Mean	57	53	4.2	93.9	203	1444	292
Std. deviation	78	74	6.4	191.6	142	602	111
Median	32	30	2.1	27.4	190	1465	307
Minimum	2	1	0.2	0.7	10	113	26
Maximum	481	426	44.8	1124.1	781	2770	563
Distribution							
Skewness	3.4	3.1	4.0	3.7	1.2	−0.1	−0.2
Kurtosis	13.9	11.2	19.5	14.4	2.3	−0.6	−0.5
K-S normality	0.3	0.3	0.3	0.3	0.1	0.1	0.1
K-S P	<0.01	<0.01	<0.01	<0.01	>0.2	>0.2	<0.2
Ln Transform	Y	Y	Y	Y	Y	N	N

3. Results and Discussion

3.1. Physical Properties

Soils on the study site showed a degree of similarity for some physical attributes among paddocks assigned to the different stocking methods, while other attributes were more variable (Table 3). Particle fractions of soils in paddocks were not constant among paddocks assigned to the different stocking methods, based on SD. Soils in paddocks assigned to continuous stocking had proportions of sand and silt with coefficients of variation (c.v.) that ranged from 16% to 41%, while soils in the rotational and HDRG-stocked paddocks varied from 4% to 14%. Soils in the HDRG-2× and rotational-stocked paddocks had greater, and more uniform percentages of silt. Alternatively, soils in paddocks managed by continuous and HDRG-1× stocking had higher mean percentages of sand than was noted for rotational or HDRG-2× stocked paddocks. Percentages of clay in soils of all paddocks were relatively similar.

Moist bulk density of soils in paddocks managed under all stocking methods showed greater variance in the upper 7.5 cm (c.v. = 15% to 21%) than the 7.5 to 15 cm (c.v. = 7% to 12%) increment (Table 3). Bulk density was lower for the upper 7.5 cm increment in rotational-stocked paddocks compared to densities in paddocks assigned to the other stocking methods. Moist bulk density in all sampled paddocks increased with soil depth. Bulk density of the 7.5 to 15 cm increment was 36%, 75%, 39%, and 49% higher than the upper 7.5 cm increment in the continuous, rotational, HDRG-1×, and HDRG-2×-stocked paddocks, respectively.

Amounts of soil organic matter showed distinct differences within paddocks assigned to the different stocking methods. Paddocks managed by HDRG stocking had lower mean concentrations in the upper 7.5 cm of the profile than in paddocks managed by continuous or rotational stocking (Table 3). Soil organic matter in paddocks managed by HDRG and rotational stocking were more uniform among soil depths, while continuous grazing had lower concentrations in the 7.5 to 15 cm increment. Overall, soil organic matter was more variable in paddocks managed by continuous (c.v. = 25%) and rotational (c.v. = 29%) stocking than under HDRG-1× (c.v. = 23%) and HDRG-2× (c.v. = 8%) stocking, which was primarily due to the larger size of paddocks and broader ranges of soils encompassed within the paddocks [17].

Table 3. Mean (±1SD) particle fractions, bulk density, and organic matter of two soil depths (0–7.5 and 7.5–15 cm) in paddocks managed by different stocking methods; 8 samples depth^{-1} method^{-1}.

Stocking Method [†]	Soil Properties									
	Particle Fractions						Bulk Density		Organic matter	
	Clay		Silt		Sand					
	0–7.5	7.5–15	0–7.5	7.5–15	0–7.5	7.5–15	0–7.5	7.5–15	0–7.5	7.5–15
	(%)						(Mg m^{-3})		(g kg^{-1})	
Continuous	20(5)	26(6)	34(14)	36(10)	46(9)	38(6)	0.99(0.15)	1.35(0.15)	42(10)	34(9)
Rotational	20(2)	23(2)	45(5)	46(2)	35(5)	31(2)	0.79(0.17)	1.39(0.15)	43(13)	43(12)
HDRG-1×	17(2)	21(2)	41(4)	37(4)	42(4)	42(4)	0.93(0.12)	1.29(0.15)	33(8)	35(7)
HDRG-2×	18(2)	22(2)	55(6)	53(3)	27(6)	25(2)	0.92(0.19)	1.37(0.10)	35(2)	35(4)

[†] HDRG-1× and HDRG-2× were high-density rapid-rotation stocking (62 and 123 AU ha^{-1} for 24 h respectively, each year).

The review of these properties indicated the physical attributes of soils within the paddocks included in the study were variable among paddocks assigned to stocking methods. Stocking methods may have affected some soil properties after sustained application from 2009 to 2015, particularly under rotational stocking. However, the amount of variability (c.v.) in the measured attributes indicated that such means were not consistent among the different larger-paddocks and sub-paddocks assigned to continuous and rotational stocking. The level of variability present in most of the physical attributes also declined with the size of the paddock. Therefore, the size, organization, and location of paddocks within the larger landscape of the study area likely had some effects on soil properties [1]. For example, the rotational-stocked paddocks had largely easterly exposures, while the continuous-stocked paddocks had predominantly westerly exposures [17]. Both the continuous- and rotational-stocked paddocks also included multiple slope positions, from toe slope through to tread positions. Such organizational features of the landscape within paddocks can have large effects on localized, catena-based development of soils, and their physical (and chemical) properties [1,2,30]. In contrast, the paddocks assigned to HDRG stocking were of sizes and dimensions that likely contained fewer soils, resulting in more consistent responses of macronutrients to stocking [15].

The stocking methods applied during this study also have some capacity to result in changes in organic matter and bulk density in soils of the US southern Great Plains (SGP). Research in Oklahoma, USA, [31] noted increased bulk density of soils of a perennial grassland in response to 10 years of different, increasing levels of stocking density applied by rotational grazing, compared to no grazing. Other research showed similar effects on bulk density after 27 years of different stocking methods [32]. However, bulk density and organic matter of soils of native grasslands in the SGP also have naturally variable distribution patterns, which appear to be related to the catenae-scale organization of plant–soil complexes that exist within landscapes of native prairies [32,33]. High levels of variance in both bulk density and organic matter were recorded within soil profiles at scales <20 m, both within and across a series of adjacent 1.6 ha experimental paddocks of tallgrass prairie managed by different stocking methods [33]. Such results indicated that soils in the area of the current study also have non-uniform distributions [17], and variable patterns of spatial distribution for soil properties beyond the effects related to the stocking method.

3.2. Mineral N

Main effects related to soil depth and time of growing season were significant for total mineral ($P \leq 0.02$) and NO_3^--N ($P \leq 0.05$). Higher amounts of total mineral and NO_3^--N were noted during the spring than summer (37.8 versus 27.1 µg N 10 cm^{-2} probe^{-1} 7.5 cm soil^{-1} 14 d^{-1} (Diff = 5.7 µg); 31.8 versus 22.3 µg NO_3^- 10 cm^{-2} probe 7.5 cm soil^{-1} 14 d^{-1} (Diff = 8.2 µg)). The highest available amount of total mineral and NO_3^- N within significant effects of soil depth was noted for the upper 7.5 cm compared to the 7.5 to 15 cm depth (47.0 versus 21.8 µg N 10 cm^{-2} probe^{-1} 7.5 cm soil^{-1} 14 d^{-1} (Diff = 5.7 µg); 39.2 versus 18.1 µg NO_3^- 10 cm^{-2} probe^{-1} 7.5 cm soil^{-1} 14 d^{-1} for (Diff = 8.2 µg)).

The stocking method × paddock location interaction was significant for both total mineral ($F_{9, 96}$ = 4.8; $P < 0.01$) and NO_3^--N ($F_{9, 96}$ = 3.3; $P < 0.01$); all other interactions among effects for both total and NO_3^--N were not significant ($0.15 < P < 0.84$). The highest flux in availability of total mineral (Table 4) and NO_3^--N within the interactions (Figure 1) were recorded at 0.7 PMP in paddocks receiving the HDRG-1× treatment. A group with the second-highest flux occurred at PMP of HDRG-1× paddocks, and 0.7 PMP in rotational-stocked paddocks. The group with the lowest flux recorded occurred at 0.7 PMP and PMP of HDRG-2×-stocked paddocks. Amounts of flux at all other locations within paddocks belonged to different intermediate, and usually multiple, groups of means. The amount of flux at 0.7 PMP of HDRG-1× paddocks was 510% greater overall than the amounts across all locations in continuous-stocked (control) paddocks, while the group with second-highest flux was 164% greater than those in the control paddocks. In contrast, the locations in HDRG-2×-stocked paddocks with the lowest amounts of recorded flux had amounts that were 49% to 52% lower than fluxes in NO_3^- and total N recorded in continuous-stocked paddocks.

Table 4. Stocking method × paddock location effects on flux of total mineral N in soils; LSD was least significant difference and numbers with the same letter were not different at $P = 0.10$.

Stocking Method [‡]	Location[†]			
	Water	0.25 PMP	0.7 PMP	PMP
	—(µg 10 cm^{-2} probe 7.5 cm soil^{-1} 14 d^{-1})—			
Continuous	31.7 cde	27.2 def	23.1 efg	25.0 ef
Rotational	19.1 fg	32.4 cde	61.0 b	42.5 c
HDRG-1×	42.3 c	22.2 efg	141.1 a	61.2 b
HDRG-2×	37.5 cd	33.9 cde	11.9 g	12.2 g
Diff		11.9		

[†] PMP, pasture mid-point; [‡] HDRG-1× and HDRG-2× were high-density rapid-rotation grazing at one and two-times normal stocking density (62 and 123 AU ha^{-1} d^{-1}, respectively).

Distribution of flux in plant-available NO_3^--N and total mineral N were more uniform among sampled locations in paddocks that were continuous-stocked than in paddocks managed under the other stocking systems (Figure 1, Table 4). This was a surprising result given that these paddocks were grazed year-long by cattle (as a control treatment), and the sampled transects covered substantially larger areas than in either the rotational- or HDRG-stocked paddocks. There was an expectation of the occurrence of specific hotspots of mineral N (or higher overall amounts of mineral N) in soils within the continuous-stocked paddocks. This premise was based on the greater opportunity (longer occupancy time) for cattle to congregate in certain areas (e.g., water sources) of paddocks, resulting in less-even distributions [10]. However, the response of mineral N to the form of stocking can be variable. A study of the effects of 2- and 12-day rotations with continuous stocking on Bermudagrass (*Cynodon dactylon*) paddocks reported no differences in paddock-scale NO_3^--N, but greater accumulations closer to water and shade [15]. The same results were noted in Bahiagrass (*Paspalum notatum*) paddocks managed by similar stocking methods [34]. In comparison, the amount of NO_3^--N flux in the rotational- and HDRG-grazed paddocks of this study showed definable hotspots, and greater ranges in amounts of flux among locations within paddocks, despite the shorter residence times of cattle

This result runs counter to one of the hypothesized effects for systems of rotational stocking, i.e., that subdivision into smaller sub-paddocks would result in more uniform use of paddock areas by cattle, and hence, provide more uniform distribution of excreta by grazing animals [16,22]. However, the inherent behavior and preferences of cattle for certain features of landscapes may prevent the achievement of uniform distribution of grazing in production-scale paddocks, regardless of stocking method [23]. For example, cattle wearing global positioning system (GPS) collars on rangeland in northeastern Colorado spent ~27% of their total time on paddocks in locations (water sources and corners) that represented just 2.5% of the total area of 65 to 130 ha paddocks [13]. Cattle redistributed 49% of all N in consumed biomass to these areas [13]. Similar results have been reported for different types of perennial warm-season grasses managed under both continuous and rotational stocking [15,34]. In comparison, the current study on native tallgrass prairie showed a series of high-N areas at more distant locations from water under rotational and HDRG-1× stocking, but responses were not consistent across all forms of rotational stocking. Such results indicate that the application of more management-intensive methods of stocking cattle may not achieve uniform use of the entire area of paddocks, and hence, uniform distribution of NO_3^- or mineral N. Application of animal densities, that do not exceed carrying capacity of the plant community (as in the current study) to continuously-stocked paddocks may be as effective at preventing hotspots of labile NO_3^--N as rotational methods.

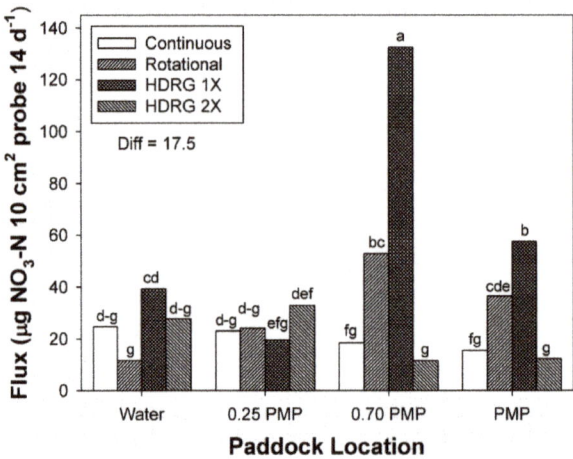

Figure 1. Grazing treatment × location interactions in flux of NO_3^--N in soils; DIFF was significant difference used in means test. Columns with the same letter were not different at $P = 0.10$

3.3. Phosphorus

There were significant differences in main effects related to soil depth ($F_{1, 96} = 29.3; P < 0.01$) for flux of available phosphorus (P), with greater fluxes within the upper 7.5 cm than the 7.5–15 cm increment of soil profiles (4.0 versus 2.4 µg P 10 cm^{-2} probe 7.5 cm $soil^{-1}$ 14 d^{-1}). Studies reporting the effects of rotational and continuous stocking on soil P in other perennial grasslands have noted similar effects related to soil depth [15,35] The stocking method × date ($F_{3, 96} = 10.3; P < 0.01$), paddock location × date ($F_{9, 96} = 2.6; P = 0.01$), and date × paddock location interactions ($F_{3, 96} = 4.9; P < 0.01$) were also significant, while all other interactions were not ($0.47 < P < 0.93$). Within stocking method × time of season interactions (Table 5), the largest P flux occurred under rotational stocking during spring, and the second-largest under continuous stocking during summer. All other levels of P flux within the interaction were low and belonged to the same means group. In spring, rotationally-stocked paddocks had 109% greater amounts of P flux than in continuously-stocked (control) paddocks. In comparison, paddocks managed under HDRG had 36% lower P flux than continuous-stocked paddocks.

Within the paddock location × time of season interaction (Table 5), the greatest flux in P occurred during spring at 0.25 PMP. This amount of P flux was 97% greater than the flux recorded at other locations in spring. During summer, the 0.7 PMP locations generated the second-largest flux in available P, with 33% greater flux than other locations within paddocks. All other fluxes in available P within the interaction were at similar low amounts. Within the stocking method × paddock location interaction (Table 6), the highest flux of available P in soils was noted at 0.25 PMP and PMP locations under continuous stocking. A group with the second-highest amounts of P flux was noted at 0.7 PMP in rotational-stocked paddocks and at water sources in continuous-stocked paddocks. Amounts of P flux in the elements of the interaction occurred at low and similar levels (1.3 to 2.1 µg). The highest amount of flux (0.25 PMP in continuous-stocked paddocks) was 34% greater than rotational-stocked paddocks at the same location, and 162% and 358% greater respectively, than P flux in HDRG-1× and 2×-stocked paddocks.

Table 5. Stocking method × time of year, and paddock location × time of year, interactions in P flux (µg 10 cm^{-2} probe 7.5 cm soil depth^{-1} 14 d^{-1}) within soils. Diff were significant differences in means tests; numbers with the same letter were not different ($P < 0.10$).

		Time of Year	
Main Effect	**Level of Effect**	Spring	Summer
Stocking Method [†]	Continuous	3.2 bc	5.1 ab
	Rotational	6.7 a	2.5 c
	HDRG-1×[†]	2.4 c	2.4 c
	HDRG-2×[†]	2.3 c	2.3 c
	Diff	2.5	
Location [‡]	Water	3.2 b	2.5 b
	0.25 PMP	6.3 a	3.3 b
	0.7 PMP	2.3 b	3.6 ab
	PMP	2.5 b	2.3 b
	Diff	2.7	

[†] HDRG-1× and HDRG-2× were high-density rapid-rotation stocking at 62 and 124 AUD ha^{-1} 24 h^{-1}, respectively.
[‡] PMP, pasture mid-point.

While significantly higher amounts of P flux were noted in all three interactions (Tables 5 and 6), these amounts were not large, and may have limited biological significance. The entire range in flux of available P in the interactions was 4.4 (grazing regime × time of year), 4.0 (paddock location × time of year), and 4.3 (grazing regime × paddock location) µg per 10 cm^2 surface area of probes 7.5 cm soil^{-1} 14 d^{-1}. The majority of the means of P flux noted in interactions also fell within narrow ranges (0.9 to 1.3 µg), when the statistically highest responses were excluded. Such results indicate that the amounts of plant-available P in soils in response to stocking method, time of year, and paddock locations was relatively small. Also, the higher amounts recorded (continuous and rotational stocking) were not consistent in terms of paddock locations or times of year, so variability was present in the responses. It is difficult to assess what these low fluxes in plant-available P represent, as it is only a portion of the entire P pool. Other research on P distributions within grass paddocks under rotational and continuous stocking methods noted differences in amounts related to lengths of applied grazing periods, and higher amounts close to water or shade (78–130 [15], 8–17 [34], and 15–22 [35] mg P kg^{-1}) in different soils.

The primary source of plant-available P within perennial grasslands is largely derived through recycling of amounts in soil organic matter consumed by soil microbes, and through inputs from decomposition of livestock feces [3]. Therefore, amounts of plant-available P in grassland soils are generally low, compared to other macronutrients like N, K, Ca, and Mg [3,36]. Another likely driver for the lower amounts of P flux within paddocks receiving the HDRG forms of stocking may be the short periods that cattle were assigned to these units. Though large numbers of cattle were present on these small areas, they spent only 24 h on these paddocks each year, which limited the potential input of feces, compared to the longer residence times of cattle in continuous- and rotational-stocked paddocks. Other studies have reported more random forms of feces distribution under short-term (24 to 48 h) rotational stocking on small paddocks, instead of patterns related to water or shade [34]. Research on paddocks of introduced warm-season perennial grasses in Florida also reported greater amounts of soil P in closer proximity to water across longer rotational (10–14-day grazing periods) and continuous stocking systems [15,34], which was a different result from the current study.

Table 6. Effect of stocking method × paddock location interactions on P flux. Diff was significant difference used in means test; numbers with the same letter were not different ($P < 0.10$).

Stocking Method ‡	Location †			
	Water	0.25 PMP	0.7 PMP	PMP
	(μg P 10 cm^{-2} probe 7.5 cm soil depth^{-1} 14 d^{-1})			
Continuous	3.2 cd	5.5 a	1.6 d	1.3 d
Rotational	1.5 d	4.1 ab	3.5 bc	1.7 d
HDRG-1×	1.3 d	2.1 c	1.1 d	1.4 d
HDRG-2×	1.4 d	1.2 d	1.5 d	1.5 d
Diff	1.9			

† PMP, pasture mid-point. ‡ HDRG-1× and HDRG-2× were high-density rapid-rotation grazing at one and two-times stocking density (62 and 123 AU ha^{-1} d^{-1}, respectively).

3.4. Potassium

There were significant differences in main effects related to soil depth ($F_{1, 96} = 22.8$; $P < 0.01$) for plant-available potassium (K), with greater flux within upper 7.5 cm depth of profiles than the 7.5 to 15 cm increment (335 versus 137 μg 10 cm^{-2} probe area 14 d^{-1} in the 0–7.5 and 7.5–15 cm increments, respectively). Interactions between stocking method and date were also significant ($F_{3, 96} = 2.7$; $P = 0.05$), but the main effects of paddock location ($P = 0.66$) and all other interactions were not ($0.23 < P < 0.93$).

Within the stocking method × date interaction in flux of available K (Figure 2A), the greatest and second greatest amounts occurred, respectively, under rotational and continuous stocking during spring. The lowest flux in available K included a group of responses under HDRG-2× stocking during spring and summer, and continuous and rotational stocking in summer. The flux in available K in soils of rotational-stocked paddocks was 149% larger than under continuous stocking during the spring, while K flux under HDRG-1× and HDRG-2× stocking were 30% and 44% lower than the control. In contrast, there was little difference in amounts of flux in available K among stocking methods during summer.

The location of lounging areas of cattle within paddocks has been defined as a potential key to change in distribution of K within landscapes of paddocks [3]. Movement of K in excreta from the larger area of paddocks to high traffic zones can translate to higher amounts of K (and N) in plant biomass produced near urine patches, relative to amounts actually retained within the body mass of cattle [3,36]. Higher amounts of K were noted in lounging areas of Bermudagrass and Bahiagrass paddocks in Florida, USA [15,34], likely due to greater inputs of urine in such areas. In contrast, the current study found no notable redistribution of plant-available K within any of the sampled paddock locations related to stocking method. While the two sampled locations near water within paddocks represented extremes for positions near water, these locations may have missed lounging areas near water [34]. More intense sampling regimes within paddocks may be required to identify areas with high amounts of plant-available K. The only definable effects noted on flux of available K were related to conditions during different times of the year, with greater amounts of flux during spring (80% and 150% greater under continuous and rotational stocking), and 2.4 times less flux in the deeper section of sampled soil.

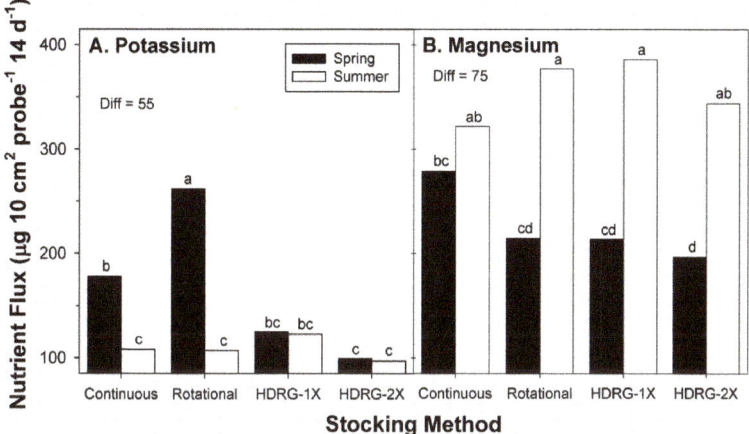

Figure 2. Stocking method × time of year interactions in flux of (**A**) K and (**B**) Mg in soils. Diff were significant differences used in means tests. Columns in panels with the same letter were not different at $P = 0.10$.

3.5. Magnesium

Main effects of soil depth were significant ($F_{1, 96} = 9.4; P < 0.01$), with the upper 7.5 cm increment having greater magnesium (Mg) flux than the 7.5 to 15 cm increment (342 versus 316 µg 10 cm^{-2} 7.5 cm soil depth^{-1} 14 d^{-1}; Diff = 23 µg). Similar patterns in amount of Mg in different sections of the soil profile were noted in other grazed grasslands [34,35]. Stocking method × time of growing season ($F_{3, 96} = 4.3; P = 0.01$), and stocking method × paddock location ($F_{9, 96} = 2.4; P = 0.02$) interactions were also significant. All other interactions among main effects were not significant ($0.19 < P < 0.78$). Within the stocking method × time of year interaction, the greatest amount of Mg flux occurred in response to rotational and HDRG-1× stocking during summer (Figure 2B), and slightly lower levels of Mg flux were noted in response to continuous and HDRG-2× stocking. In contrast, the lowest amounts of flux were noted during spring under HDRG-2× stocking, while responses to the other stocking methods during spring belonged to means groups with intermediate and lower responses.

Rotational and HDRG stocking resulted in 23% to 29% lower flux in Mg during spring than continuous stocking, but 7% to 20% greater amounts of flux during summer (Figure 2B). In contrast, the amount of flux present under continuous stocking was more consistent during different times of year than responses to the other stocking methods. Magnesium flux between spring and summer under continuous stocking increased 15%, while rotational and HDRG stocking resulted in larger increases. Higher fluxes in available Mg occurred under all stocking methods during summer, with an average difference of 61%.

Factors that generated the stocking method × time of year interaction in Mg flux were not immediately evident. The longer residence time of cattle on the continuous stocked paddocks was thought to result in increased amounts of Mg (and other nutrients) within paddocks [34,36]. Sandier soils, such as those in the continuous-stocked paddocks also tend to have lower amounts of available Mg [3,34]. The difference in timing of sampling between spring (delayed seven months) and summer grazing might have provided the higher amounts of Mg during summer. However, the primary excretal source of Mg is from feces (>75% of total animal inputs) and has slower rates of movement than N or K into the soil [3]. The summer sampling event occurred within seven days of termination of grazing, so there was not enough time for Mg enrichment prior to summer sampling. A more likely agent for the interaction would be leaching of Mg from soil profiles during the delay between grazing and spring sampling [34–36].

Among stocking methods, both rotational and HDRG-1× stocking resulted in greater amounts of Mg during summer, despite being grazed for shorter periods than the continuous paddocks, and the drier conditions that were prevalent during summer. In comparison, research on Bahiagrass paddocks reported no differences in Mg concentrations within soils of paddocks in Florida after three years of different forms of rotational stocking were applied during summers [34]. Alternatively, research on tall fescue (*Lolium arundinaceum*) paddocks grazed during April through September reported differences in amounts of Mg in soils in response to a series of rotational stocking systems, and consistent declines in amounts across all systems after five years [35].

Within the stocking method × paddock location interaction, the highest and lowest flux in available Mg occurred respectively, in proximity to water tanks and the 0.25 PMP locations of rotational- and continuous-stocked paddocks (Figure 3A). The second-highest Mg flux occurred for a group of mean responses for locations near water tanks, 0.7 PMP and PMP locations in continuous-stocked paddocks, and PMP locations in HDRG-2×-stocked paddocks. A difference in distribution patterns of Mg flux was the driver of the interaction between stocking methods and paddock locations. Both continuous and rotational stocking had higher levels of Mg flux at locations closer to water tanks, while HDRG-1× stocking generated more uniform levels across locations, and HDRG-2× stocking resulted in increased flux in available Mg with increasing distance from water. This change in distribution pattern generated by stocking methods resulted in a range of responses belonging to the means groups with intermediate amounts of flux in available Mg. Other research on paddocks of warm-season grasses managed under continuous and rotational stocking reported increasing amounts of Mg in areas closer to water or shade [15,34]. A similar effect was noted in the current study, with the continuous and rotational paddock, though the effects was not consistent across or within all stocking methods.

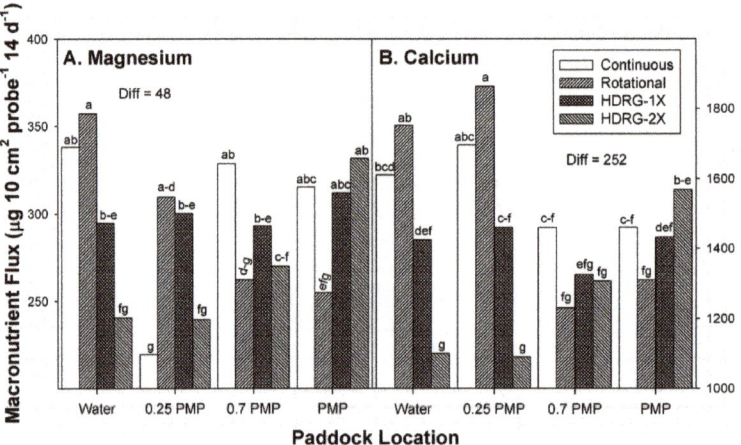

Figure 3. Stocking method × location interactions in (**A**) Mg and (**B**) Ca flux in soils; Diff were significant differences used in means tests. Columns in panels with the same letter were not different at $P = 0.10$.

3.6. Calcium

Main effects related to soil depth ($F_{1, 96} = 13.7$; $P < 0.01$) and time of growing season ($F_{1, 96.1} = 102$; $P < 0.01$) on calcium (Ca) flux were significant, as was the stocking method × paddock location interaction ($F_{9, 96} = 1.7$; $P = 0.09$). All other interactions were not significant ($0.11 < P < 0.84$). Among main effects related to soil depth, the upper 7.5 cm had greater flux in available Ca than the 7.5 to 15 cm increment (1779 versus 1551 µg 10 cm^{-2} probe area^{-1} 7.5 cm soil^{-1} 14 d^{-1}; Diff = 59 µg). Among times of growing season, lower flux in available Ca was noted at the time of spring sampling compared to summer (1512 versus 1817 µg 10 cm^{-2} probe area^{-1} 7.5 cm soil^{-1} 14 d^{-1}; Diff = 59 µg).

Within the stocking method × paddock location interaction, the highest flux in Ca occurred under rotational stocking at 0.25 PMP locations (Figure 3B). A group with the second-highest Ca flux included locations near water tanks in continuous- and rotational-stocked paddocks, and 0.25 PMP locations in continuous-stocked paddocks. In contrast, the lowest recorded Ca flux occurred near water in paddocks under HDRG-2× stocking. The interaction between stocking method and paddock locations was due to differences in the pattern of amounts of flux among forms of grazing. Both continuous and rotational stocking had higher levels of Ca flux at locations closer to water tanks. In contrast, HDRG-1× stocking generated more uniform levels of flux across locations, and HDRG-2× stocking resulted in increasing amounts of flux with distance from water tanks.

Calcium in animal excreta is primarily in feces; < 3% of animal inputs to soil are reported to be in urine [3]. Therefore, fecal sources have been considered a significant factor for flux in plant-available Ca, compared to decomposition of plant residues. However, amounts of Ca input to grassland soils via cattle excreta varies widely, and movement rates of Ca from dung to soils is not well-defined. Evidence suggests an increase in Ca flux with greater amounts of precipitation [36,37]. In comparison, the current study showed lower amounts of flux during spring, when greater amounts of soil moisture were present, compared to the amounts recorded during summer after a prolonged dry period. The delay between grazing and sampling for macronutrients may have contributed to this result. However, heavier rainfalls than those which were encountered during this study may be required. Movement of Ca from dung into soil without such events occurs at slower rates, similar to the change in amounts of organic matter in dung [38].

3.7. Sulfur

The effects of distance to water as a covariate in defining flux of plant-available sulfur (S) in soils were significant ($F_{1,\ 12.9} = 3.7$; $P = 0.08$), indicating distance between sampled locations and water sources were important to the amounts of flux recorded in plant-available S. However, the related effect of sample location of these distances (a measure of paddock organization) as a main effect ($F_{3,\ 96} = 1.2$; $P = 0.30$), and interaction of sample location with other factors, were not significant ($0.12 < P < 0.72$). Main effects of stocking method ($F_{3,\ 93.1} = 8.8$; $P < 0.01$) and soil depth ($F_{1,\ 96} = 5.4$; $P = 0.02$) were the only significant factors that affected flux in plant-available S in soils.

Among stocking methods, continuous and rotational stocking had the highest and second-highest flux (58.7 and 43.3 µg S 10 cm^{-2} 7.5 cm probe^{-1} soil depth^{-1} 14 d^{-1}; Diff = 4.4 µg), while HDRG-2× and HDRG-1× stocking had significantly lower and similar amounts (30.2 and 27.3 µg S, respectively). The range from greatest to least amounts of flux (31.4 µg) was 78% of the mean amount of flux across all stocking methods. In comparison, research on S in soils of other grasslands reported greater amounts in soils managed under rotational stocking with long residence times and continuous stocking, compared to rotations with short residence times on paddocks [15]. Among soil depths, the surface 7.5 cm increment had significantly more S flux than the 7.5 to 15 cm increment (47.1 versus 30.7 µg S 10 cm^{-2} probe^{-1} 7.5 cm soil depth^{-1} 14 d^{-1}; Diff = 8.6).

Sulfur is present in soils in both organic and inorganic forms, and in variable amounts [3]. The amounts of S present in soils also depends on such factors as amounts and types of organic matter, mineral composition, and pH of soils [3,39]. The majority (>90%) of S in soils of temperate grasslands was found in the upper 10 cm of the profile, in organic forms derived from animal excreta and plant residues [35]. Large proportions of this S are labile forms, including water-soluble sulfates derived from excreta and residues [39]. In comparison, there was 53% more S flux in the upper 7.5 cm increment of soil in the current study, which corresponded to findings noted in earlier research [15].

Inputs to soils from excreta of grazing animals, primarily in urine, represent an important source of S flux in graze grasslands [15,39], and stocking method has been reported to effect distributions. Sources of plant-available S from grazing animals is related to both retention and throughput of forage consumed. Roughly 25% of S in the biomass of forage consumed by cattle is retained within animals, with the remainder excreted in urine and feces [3,15]. Amounts of S in both forms of excreta varies with type of grassland, location within landscape where deposited, and form of management.

In the current study, higher amounts of flux in available S appeared to occur in response to the length of time animals were assigned to paddocks, with a sequential decrease from the longest (continuous) to shortest (HDRG) periods of applied grazing. Research in Australia and New Zealand noted that 50% to 70% of S excreted by cattle was in urine, primarily in labile forms of sulfate [36,39]. In contrast, S concentration in dung tends to be small (~0.3% of dung dry matter) and generally in more stable organic forms that mineralize slowly, at rates similar to those noted in mineralization of soil organic matter [39]. Therefore, the longer residence times of cattle on the continuous and rotational paddocks was an important contributor to the greater amounts of plant-available S that were recorded.

4. Conclusions

Responses during this study provided a mixed result for defining how, or if, seven years of applied stocking methods affected levels and distributions of plant-available macronutrients at contrasting locations within production-scale paddocks. All stocking methods tested, from continuous through to different forms of rotational stocking, appeared to generate varied responses in different plant-available macronutrients among paddock locations and time of growing season, though amount of sampling was limited. These variable responses were likely driven by interactions of climate and local soil–plant assemblages (catenae) with paddock management and time of year [1,6,11,22], which were noted in a range of other studies [11,13,15,16,31,34–36]. Though the number of locations within paddocks, number of paddocks, and amount of sampling in the study were limited, such responses indicate that stocking methods may not prevent high levels of flux in plant-available macronutrients within paddocks, which is similar to results reported for other ecosystems [15,34,35]. Production-scale paddocks of tallgrass prairie are generally large in size, and variable in terms of landscape features and soils that are included within their boundaries. Such variability makes identifying changes in soil attributes to management a difficult and complex process. Though limited in number of locations within paddocks, number of paddocks sampled, and amount of sampling, the current study does provide a preliminary example of whether stocking methods influence distribution of macronutrients within paddocks after seven years of application. Longer and/or more detailed studies of macronutrient distributions across a broader range of landscape positions are required to better describe responses within production-scale paddocks.

Author Contributions: Conceptualization, P.J.S. and K.E.T.; methodology and experimental design, B.K.N., P.J.S., and K.E.T.; statistical analysis, B.K.N.; resource management and project administration, P.J.S., K.E.T. and B.K.N.; data curation, P.J.S. and B.K.N.; writing—original draft preparation, B.K.N.; review and editing, B.K.N., K.E.T. and P.J.S.; funding acquisition, P.J.S.

Funding: This research was partially supported by funds from the Agriculture and Food Research Initiative Competitive Projects 2012-02355 and 2013-69002-23146, USDA National Institute of Food and Agriculture.

Acknowledgments: This research is a contribution from the Long Term Agroecosystem Research (LTAR) network. The authors wish to recognize ARS technicians Kory Bollinger and Jeff Weik for their assistance in managing the experiment. Mention of trademarks, proprietary products, or vendors does not constitute guarantee or warranty of products by USDA and does not imply its approval to the exclusion of other products that may be suitable. All programs and services of the USDA are offered on a nondiscriminatory basis, without regard to race, color, national origin, religion, sex, age, marital status, or handicap.

Conflicts of Interest: The authors declare no conflict of interest.

References

1. Archer, S.; Smeins, F.E. Ecosystem-level processes. In *Grazing Management: An Ecological Perspective*; Heitschmidt, R.K., Stuth, J.W., Eds.; Timber Press: Portland OR, USA, 1991; pp. 109–139.
2. Jenny, H. *Factors of Soil Formation: A System for Quantitative Pedology*; Dover Publications: Mineola, NY, USA, 1994.

3. Whitehead, D.C. *Nutrient Elements in Grassland: Soil-Plant-Animal Relationships*; CABI Publishing: New York, NY, USA, 2000.
4. Oades, J.M. The role of biology in the formation, stabilization, and degradation of soil structure. *Geoderma* **1993**, *56*, 377–400. [CrossRef]
5. Burke, I.; Lauenroth, W.; Vinton, M.; Hook, P.; Kelly, R.; Epstein, H.; Agular, M.; Robles, M.; Aguilera, M.; Murphy, M.; et al. Plant-soil interactions in temperate grasslands. *Biogeochemistry* **1998**, *42*, 121–143. [CrossRef]
6. Hook, P.; Burke, I.C. Biogeochemistry in a shortgrass landscape: Control by topography, soil texture and microclimate. *Ecology* **2000**, *81*, 2686–2703. [CrossRef]
7. Norman, M.J.T.; Green, J.O. The local influence of cattle dung and urine upon yield and botanical composition of permanent pasture. *J. British Grassl. Soc.* **1958**, *13*, 39–45. [CrossRef]
8. Sheldrick, W.; Syers, J.K.; Lindgard, J. Contribution of livestock excreta to nutrient balances. *Nutr. Cycl. Agroecosyst.* **2003**, *66*, 119–131. [CrossRef]
9. Orwin, K.H.; Bertram, J.E.; Clough, T.J.; Condron, L.M.; Sherlock, R.R.; O'Callaghan, M. Short-term consequences of spatial heterogeneity in soil nitrogen concentrations caused by urine patches of different sizes. *Appl. Soil Ecol.* **2009**, *42*, 271–278. [CrossRef]
10. Bailey, D.W.; Gross, J.E.; Laca, E.A.; Rittenhouse, L.R.; Coughenor, M.B.; Swift, D.M.; Sims, P.L. Mechanisms that result in large herbivore grazing patterns. *J. Range Manag.* **1996**, *49*, 386–400. [CrossRef]
11. Ganskopp, D.C.; Bohnert, D.W. Landscape nutritional patterns and cattle distribution in rangeland pastures. *Appl. Anim. Behav. Sci.* **2009**, *116*, 110–119. [CrossRef]
12. Schnyder, H.; Locher, F.; Auerswald, K. Nutrient redistribution by grazing cattle drives patterns of topsoil N and P stocks in a low-input pasture ecosystem. *Nutr. Cycl. Agroecosyst.* **2010**, *88*, 183–195. [CrossRef]
13. Augustine, D.J.; Milchunas, D.G.; Derner, J.D. Spatial redistribution of nitrogen by cattle in semiarid rangeland. *Rangeland Ecol. Manag.* **2013**, *66*, 56–62. [CrossRef]
14. Saunders, W.H.M. Effects of cow urine and its major constituents on pasture properties. *N. Z. J. Agric. Res.* **1982**, *25*, 61–68. [CrossRef]
15. Mathews, B.W.; Sollenberger, L.E.; Nair, V.D.; Staples, C.R. Impact of grazing management on soil nitrogen, phosphorus, potassium, and sulfur distribution. *J. Environ. Qual.* **1994**, *23*, 1006–1013. [CrossRef]
16. Barnes, M.K.; Norton, B.E.; Maeno, M.; Malachek, J.C. Paddock size and stocking density affect spatial heterogeneity of grazing. *Rangeland Ecol. Manag.* **2008**, *61*, 380–388. [CrossRef]
17. USDA-NRCS. *Soil Survey of Canadian County, Oklahoma. Supplement Manuscript*; USDA-NRCS and Okla. Agric. Exp. Stn.: Stillwater, OK, USA, 1999.
18. Northup, B.K.; Phillips, W.A.; Daniel, J.A.; Mayeux, H.S. Managing southern tallgrass prairie: Case studies on grazing and climatic effects. In Proceedings of the 2nd National Conference on Grazing Lands, Nashville, TN, USA, 8–10 December 2003; Theurer, M., Peterson, J., Golla, M., Eds.; Omnipress Inc.: Madison, WI, USA, 2003; pp. 834–890.
19. Phillips, W.A.; Coleman, S.A. Productivity and economic return of three warm-season grass stocker systems of the southern Great Plains. *J. Prod. Agric.* **1995**, *8*, 334–339. [CrossRef]
20. Phillips, W.A.; Northup, B.K.; Mayeux, H.S.; Daniel, J.A. Performance and economic returns of stocker cattle on tallgrass prairie under different grazing management strategies. *Prof. Anim. Sci.* **2003**, *19*, 416–423. [CrossRef]
21. Goodman, J.M. Physical environments of Oklahoma. In *Geography of Oklahoma*; Morris, J.W., Ed.; Oklahoma Historical Society: Oklahoma City, OK, USA, 1977; pp. 9–25.
22. Briske, D.D.; Derner, J.D.; Brown, J.R.; Fuhlendorf, S.D.; Teague, W.R.; Havstad, K.M.; Gillen, R.L.; Ash, A.J.; Willms, W.D. Rotational grazing on rangelands: Reconciliation of perception and experimental evidence. *Rangeland Ecol. Manag.* **2008**, *61*, 3–17. [CrossRef]
23. Arnold, G.W.; Dudzinski, M.L. *Ethology of Free-Ranging Domestic Animals*; Elsevier Scientific Publ.: New York, NY, USA, 1979.
24. Grossman, R.B.; Reinsch, T.G. Bulk density and linear extensibility. In *Methods of Soil Analysis. Part. 4. Physical Methods*; Dane, J.H., Topp, G.C., Eds.; SSSA: Madison, WI, USA, 2002; pp. 201–228.
25. Day, P.R. Particle fractionation and particle size analysis. In *Methods of Soil Analysis, Part. I*; Black, C.A., Evans, D.D., White, J.L., Ensminger, L.E., Clark, F.E., Eds.; Agronomy Monograph 9; ASA and SSSA: Madison, WI, USA, 1965; pp. 545–567.

26. Nelson, D.W.; Sommers, L.E. Total carbon, organic carbon, and organic matter. In *Methods of Soil Analysis: Part. 3. Chemical Methods*; Sparks, D.L., Page, A.L., Helmke, P.A., Loeppert, R.H., Eds.; SSSA: Madison, WI, USA, 1996; pp. 961–1010.
27. Steel, R.G.D.; Torrie, J.H. *Principles and Procedures of Statistics: A Biometrical Approach*, 2nd ed.; McGraw-Hill: New York, NY, USA, 1980.
28. Littel, R.C.; Milliken, G.A.; Stroup, W.W.; Wolfinger, R.D. *SAS Systems for Mixed Models*; SAS Institute Inc.: Cary, NC, USA, 1996.
29. Patetta, M. *Longitudinal Data Analysis with Discrete and Continuous Responses: Course Notes for Instructor-Based Training*; SAS Institute Inc.: Cary, NC, USA, 2005.
30. Buol, S.W.; Hole, F.D.; McCracken, R.J. *Soil Genesis and Classification*, 2nd ed.; Iowa State University Press: Ames, IA, USA, 1980.
31. Daniel, J.A.; Potter, K.N.; Altom, W.; Aljoe, H.; Stevens, R. Long-term grazing density impacts on soil compaction. *Trans. ASAE.* **2002**, *45*, 1911–1915. [CrossRef]
32. Northup, B.K.; Daniel, J.A. Distribution of soil bulk density and organic matter along an elevation gradient in central Oklahoma. *Trans. ASABE.* **2010**, *53*, 1749–1757. [CrossRef]
33. Williams, R.D.; Ahuja, L.R.; Naney, J.W.; Ross, J.D.; Barnes, B.B. Spatial trends and variability of soil properties and crop yield in a small watershed. *Trans. ASAE.* **1987**, *30*, 1653–1660. [CrossRef]
34. Dubeux, J.C.B., Jr.; Sollenberger, L.E.; Vendramini, J.M.B.; Interrante, S.M.; Lira, M.A., Jr. Stocking methods, animal behavior, and soil nutrient redistribution: How are they linked? *Crop. Sci.* **2014**, *54*, 2341–2350. [CrossRef]
35. Jones, G.B.; Tracy, B.J. Pasture soil and herbage nutrient dynamics through five years of rotational stocking. *Crop. Sci.* **2014**, *54*, 2351–2361. [CrossRef]
36. Haynes, R.J.; Williams, P.H. Nutrient cycling and soil fertility in the grazed pasture ecosystem. *Adv. Agron.* **1993**, *49*, 119–199.
37. Dickinson, C.H.; Craig, J. Effects of water on the decomposition and release of nutrients from cow pats. *New Phytol.* **1990**, *115*, 139–147. [CrossRef]
38. Underhay, V.H.S.; Dickinson, C.H. Water, mineral and energy fluctuations in decomposing cow pats. *J. Br. Grassl. Soc.* **1978**, *33*, 189–196. [CrossRef]
39. Nguyen, M.L.; Goh, K.M. Sulphur cycling and its implications on sulphur fertilizer requirements of grazed grassland ecosystems. *Agric. Ecosys. Environ.* **1994**, *49*, 173–206. [CrossRef]

© 2019 by the authors. Licensee MDPI, Basel, Switzerland. This article is an open access article distributed under the terms and conditions of the Creative Commons Attribution (CC BY) license (http://creativecommons.org/licenses/by/4.0/).

Article

Feed Value of Barn-Dried Hays from Permanent Grassland: A Comparison with Fresh Forage

Donato Andueza [1],*, Fabienne Picard [1], Philippe Pradel [2] and Katerina Theodoridou [3]

1   Université Clermont Auvergne, INRA, VetAgro Sup, UMR Herbivores,
    F-63122 Saint-Genès-Champanelle, France; fabienne.picard@inra.fr
2   INRA-Herbipôle, Unité 1414, F-63122 Saint-Genès Champanelle, France; philippe.pradel@inra.fr
3   Institute for Global Food Security, Queen's University Belfast, 19 Chlorine Gardens, Belfast BT9 5DL, UK; k.theodoridou@qub.ac.uk
*   Correspondence: donato.andueza@inra.fr; Tel.: +33-4736-24071

Received: 29 March 2019; Accepted: 28 May 2019; Published: 30 May 2019

Abstract: In mountain areas, hays are the main forage in winter diets for livestock. Barn-dried hays can be an alternative to traditional hays, which are generally characterized by a low feed value. The aim of this study was to compare the feed value of barn-dried hays with that of the fresh forage from a permanent meadow. The study was carried out over three periods during the first growth cycle of the meadow's vegetation (from 30 May to 3 June, from 13 to 17 June, and from 27 June to 1 July). Fresh forage and barn-dried hays of the same fresh forages were tested for dry matter digestibility (DMD), organic matter digestibility (OMD), and voluntary intake (VI). Both types of forage obtained each period were tested with an interval of 15 days. Chemical composition and OMD of forages did not change ($p > 0.05$) according to the feeding method. However, the DMD values for barn-dried hays were higher ($p < 0.05$) than for fresh forages at the end of the cycle. VI and digestible organic matter intake of barn-dried hays were higher ($p < 0.05$) than that of fresh forages. In conclusion, barn-dried hays obtained from permanent grasslands presented a higher feed value than fresh forages.

Keywords: permanent grassland; fresh forage; barn-dried hay; chemical composition; digestibility; voluntary intake

1. Introduction

In many regions of Europe and particularly in mountain areas, semi-natural grasslands are the predominant land types, and milk and meat production are based on forages they supply. The grassland systems of these areas seek to maximize the use of grass for grazing purposes, but preserved forages are also needed for feeding during the winter or to offset shortages during summer droughts. However, poor knowledge of their feed value and the reduced performance obtained when they were compared to seeded forages, such as that provided by *Lolium perenne* L. grasslands [1], make farmers reluctant to use forages obtained from semi-natural grasslands. The results obtained by Bruinenberg et al. (2002) [1], partially disagree to those obtained by Andueza et al. (2010) [2] and Andueza et al. (2013) [3] which reported a broad variability in digestibility and voluntary intake (VI) of different permanent grasslands at different maturity stages.

Haymaking is the most popular forage preservation method in mountain areas. However, the nutritive quality of hays is not optimal as it is often conditioned by weather, particularly in spring. According to Rotz and Muck (1994) [4] average dry matter (DM) losses in hay making are estimated between 24% and 28% of the original forage. To overcome this problem, farmers increasingly use barn-dried hays. However, their relative nutritive value is often not well known. French National Institute for Agricultural Research (INRA) nutritive value estimates for permanent grasslands are not exhaustive, while nutritive value of barn-dried hays is unknown, and the few relevant references in the literature mostly concern seeded forages. In general, in this preservation system, changes in the quality of forages are mainly related to plant respiration of grass after cutting. However changes in the feed value of artificially dried forages are often conflicting in the literature: Pasha et al. (2004) [5], reported higher digestibility and intake values for forced-air-dried hays than for frozen grass, but Archimède et al. (1999) [6] found higher organic matter digestibility (OMD) and VI for fresh *Digitaria decumbens* Stent than the equivalent forage dried for 20 h at 60 °C. Similar results were found in sheep by Dulphy and Rouel (1987) [7] for VI. Finally, Demarquilly (1970) [8] and Delaby and Peccatte (2008) [9] reported similar results between fresh forage and barn dried hay or those dehydrated at low temperatures, but Andueza et al. (2009) [10] found higher OMD values for lucerne hays than the same dehydrated forages, although no differences between preservation methods were found for VI. From these results, we hypothesized that the feed value of barn-dried permanent grassland hays might be similar to that of fresh forage.

Our aim in this work was to assess the effects of the barn-dried preservation method as a management alternative option to fresh forage feeding, for a permanent grassland across its first growth cycle on the chemical composition and feed value of the forage.

2. Materials and Methods

The study was conducted indoors at the French National Institute for Agricultural Research of the Marcenat experimental farm (INRA-Herbipôle) in France. The animals were handled by specialized personnel who applied animal care and welfare in accordance with European Union Directive no. 609/1986, under agreement no. A63 565. The experimental procedures complied with French regulations for the use of experimental animals (statutory order no. 87-848, guideline of 19 April 1988).

2.1. Forages and Climatic Conditions

One permanent grassland, located at Marcenat (Cantal, France 2°49′ E, 45°18′ N) at 1060 m above sea level, was used. It received 33 kg of N/ha in March 2005. The experimental period started on 20 May and lasted 8 weeks including an 8-day period of adaptation to the experiment (between 20 to 29 May). The sward was used three times during the first growth cycle of 2005 for fresh forage or barn-dried forage (from 30 May to 3 June, from 13 to 17 June and from 27 June to 1 July). Every Wednesday, (1, 15, and 29 June) an area of the 0.5 ha of this plot was cut, and forage was harvested for drying in an experimental barn. The rest of the Wednesdays between 18 May and 29 June, forages were barn-dried but they were not used in the current study. Briefly, air was heated to a maximum of 40 °C and then blown into the bottom of a box (2 × 2 × 2 m^3) containing the forage harvested using a fan and an electric heating element. The 3 kW centrifuge fan produced an air flow of 4000 m^3/h. The ambient air at the inlet of the ventilator was heated by 6 °C using a 9-kW electric heating element. Two boxes were used. The drying procedure lasted between 4 and 7 days according to the dry matter content of the fresh forage. Forage was manually turned once a day during each drying period to ensure consistent dehydration. Finally, 300 kg of dried forage was obtained each week.

Total rainfall during the growth cycle (1 April–30 July) was 300 mm. Total rainfall during the period of the study was 31 mm (Figure 1). Average temperatures ranged between 14 °C and 18 °C. Previous management of the grassland consisted in general on cutting the plot for hays at the end of June and then grazed in August and October. Plots received usually 30 kg of N/ha early in spring.

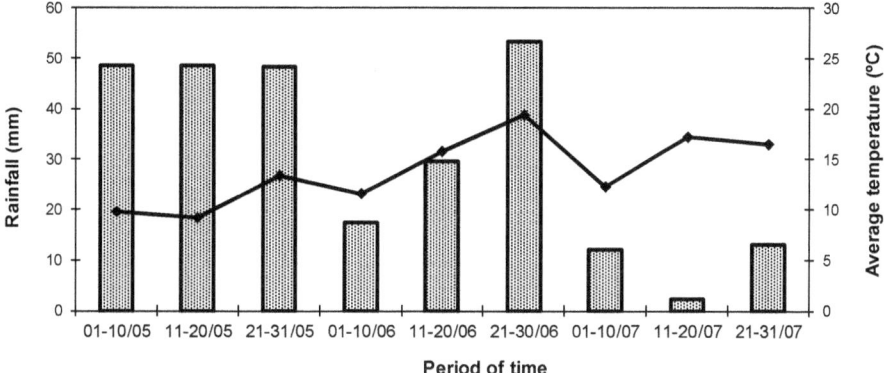

Figure 1. Gaussen ombrothermic diagram of the study site (Marcenat, France) from 1 May until 31 July in the period of 10-days including the experimental period when forage tested was grown (from 30 May to 1 July). Bars indicate rainfall. Average temperatures are line joined.

2.2. In Vivo Digestibility Trials

Twelve 2-year old Texel wethers (mean live weight: 63 kg) were used. Each forage type (fresh or barn dried forage) was randomly assigned to six wethers. The in vivo digestibility trials were run in each of the three measurement periods over 15 days, in which the first 10 days were devoted to adaptation to diet and the last 5 days (dates showed previously) to data collection (daily offering, refusals and feces). The offered diet consisted of fresh unpreserved forage cut daily from the grassland, or barn-dried hay, chopped to a length of 5–7 cm and offered ad libitum twice a day, at 8:00 a.m. and 4:00 p.m. [11]. The amount of forage offered was adjusted daily on the basis of the previous day's intake. A refusal of 0.1 of the offered quantity was allowed. Throughout the experimental period, the animals had free access to water and vitamin–mineral blocks. Barn-dried hays were tested 15 days after testing the same forage cut daily and offered fresh.

2.3. Samples

During the data collection of each period in the digestibility trials, three samples of about 200 g were randomly collected per day and per forage before chopping of herbage (total = 15 samples). These samples were stored at −20 °C and were used for the determination of the botanical composition and the phenological stage. Botanical composition was determined by hand separation and weighing of samples of different plant species on five samples per forage. Senescent material (Sm) was considered globally for all plants. Phenological stage was determined on three samples per forage using 50 random grass tillers according to an adaptation of the method proposed by Moore et al. (1991) [12]. The mean plant stage by weight (MPW) for each sample was calculated by

$$MPW = \Sigma\ (C_i\ D_i)/D \tag{1}$$

where C_i, is the code of stage i as defined in Table 1, D_i the total dry weight for tillers in stage i, and D the total dry weight for all tillers.

Furthermore, during digestibility trials, forage offered refusals and the feces of each animal were sampled daily. At the end of each period, three samples of forage and one sample of refusals and feces were made up and they were used for chemical analyses.

Table 1. Definition of the plant stages (in grasses) used to characterize the medium plant stage of the grasslands (adapted from [12]).

Stage	Code	Description
Vegetative		
Vegetative leaf development	1.5	
Stem Elongation		
Elongation (beginning)	2	First node palpable/visible
Elongation	2.5	Nodes palpable/visible
Elongation (end)	3	Boot stage
Reproductive/Floral Development		
Inflorescence emergence	3.1	First spikelet visible
Inflorescence	3.3	Spikelets fully emerged/peduncle not emerged
Inflorescence emerged	3.5	Inflorescence emerged/peduncle fully elongated
Anther emergence/anthesis	3.8	
Seed Development and Ripening		
Caryopsis visible	4	
Milk	4.1	
Dough	4.4	
Endosperm hard/physiological maturity	4.7	
Endosperm dry/seed ripe	4.9	

2.4. Analyses

All the samples (hand-separated plant species, forage offered, refusal, and feces) were oven-dried at 60 °C for 72 h. Forage offered, refusal and feces samples were ground through a 0.8 mm screen. Samples of forage offered and refusals were analyzed for crude ash (CA) according to AOAC, (1990) [13]. Forages offered were analyzed for nitrogen (N) [13], water soluble carbohydrates (WSC) [14] neutral detergent fiber (NDF) according to the method described by Van Soest et al. (1991) [15] and for acid detergent fiber (ADF) and acid detergent lignin (ADL) according to Van Soest and Robertson (1980) [16]. Neutral detergent fiber and ADF analyses were performed using the Ankom apparatus (Ankom® Tech. Co., Fairport, NY, USA). Analyses of CA were determined in feces samples.

2.5. Calculations

Crude protein (CP) was obtained by multiplying the N concentrations by 6.25. The relative proportions of grasses, legumes, and forbs were calculated on a DM basis from botanical composition data. The in vivo results and DM and CA data were used to calculate dry-matter digestibility (DMD), OMD, and digestible organic matter intake (DOMI). Voluntary intake (calculated as the average of a 5-day period of daily forage offered minus refusal) and DOMI were expressed in g DM/kg of metabolic body weight ($BW^{0.75}$).

2.6. Statistical Analysis

Data on chemical composition of the samples for the two types of forages in the experimental periods underwent to repeated-measures ANOVA according to the model

$$Y_{ijk} = \mu + F_i + P_j + (F \times P)_{ij} + R(F \times P)_{ijk} + \varepsilon_{ijk}, \qquad (2)$$

where Y is the dependent variable, μ is the overall mean, F is type of forage (1 df), P is the period (2 df), F × P is the interaction between type of forage and period (2 df), R is the replicate effect (2 df), and ε is the experimental error. Replicate was considered as a random effect, and period as a repeated measure.

For ANOVA of in vivo digestibility data (DMD, OMD, and DOMI), we used the model

$$Y_{ijk} = \mu + F_i + P_j + (F \times P)_{ij} + A(F)_{ik} + \varepsilon_{ijk}, \qquad (3)$$

where Y is the result for a single animal, µ is the overall mean, F is type of forage (1 df), P is the effect of period (2 df), A is the effect of animal (5 df), F × P is the interaction between type of forage and period (2 df), and ε is the experimental error. Each animal was considered as a block factor and as a random variable. Single degree-of-freedom orthogonal polynomial contrasts were used to detect linear and quadratic effects of time on chemical composition, digestibility coefficients, daily feed intake, and DOMI. Because forage samples were taken at 665, 822, and 1059 growing degree days (GDD), polynomial contrasts were adjusted for unequally spaced time effects (i.e., 665, 822, and 1059 GDD). The relationships between temperature accumulation from 1 February using 0 °C as minimum base temperature and 18 °C as maximum base temperature [17], phenological stage determined by MPW, chemical composition variables, in vivo digestibility, and intake data were evaluated by Pearson correlation coefficients. All analyses were performed using the mixed procedure of the SAS statistical package [18].

3. Results

3.1. Botanical Composition and Phenological Stage

The average botanical composition of the permanent meadow during the three periods of the experiment is reported in Table 2. Permanent meadow was characterized by higher proportions of grasses than that of forbs. The ratio between the two groups increased with the advance of the phenological cycle. At young stages, the proportion of grasses was 0.76 and at more mature stages, the proportion of grasses increased to 0.83. Proportion of legumes was always less than 0.01. Initially, grassland species >0.10 were *Agrostis capillaris* L., *Dactylis glomerata* L., *L. perenne*, and *Taraxacum officinale* F.H. Wigg., whereas at more advanced phenological stages dominated species were *A. capillaris*, *L. perenne*, and *Festuca rubra* L. Mean plant weight at P1 (665 GDD) was 2.65 (boot stage, Table 1) and increased until P2 (822 GDD) before finally levelled off between P2 and P3 (1059 GDD) (Table 2).

Table 2. Botanical composition (percentage of dry matter) and mean plant weight (MPW) (dimensionless) phenological stage of the permanent grassland in the three periods of the study.

	P1 [1]	P2 [2]	P3 [3]
Grasses			
Agrostis capillaris L.	20.23	12.96	29.87
Trisetum flavescens L.	4.41	4.01	7.02
Bromus mollis L.	1.03	0.55	0.00
Dactylis glomerata L.	15.39	4.92	0.84
Festuca rubra L.	3.14	2.02	9.97
Anthoxanthum odoratum L.	0.19	0.00	1.61
Poa pratensis L.	6.06	7.37	8.25
Holcus lanatus L.	1.44	0.19	6.34
Lolium perenne L.	18.81	56.46	16.80
Phleum pratense L.	5.64	0.27	2.62
Forbs			
Cerastium fontanum Baumg.	9.57	1.54	5.41
Taraxacum officinale P.H. Wigg.	11.64	2.54	3.16
Urtica dioica L.	0.00	1.01	0.02
Sum of grasses+forbs	97.55	93.84	91.91
Others			
Senescent material	0.72	4.70	6.36
MPW	2.65	3.79	3.82

[1] P1 = first period (from 30 May to 3 June; 665 GDD); [2] P2 = second period (from 13 to 17 June; 822 GDD); [3] P3 = third period (from 27 June to 1 July; 1059 GDD). Only those species that presented a percentage greater than 1 percent are shown.

3.2. Chemical Composition

There were no differences between types of forages ($p > 0.05$) for any determinations of chemical composition (Table 3). Concerning the period or the cumulation of temperature, for the CA content, there were no differences ($p > 0.05$) between values found among periods. The CP content decrease in a quadratic pattern along the first cycle of growth with a more pronounced decrease at the beginning (between the first and second period) and a weaker one at the end of the phenological cycle (between the second and third period). Neutral detergent fiber, ADF and ADL concentrations increased linearly over the beginning (first period) and the end of the experiment third period. Finally, for WSC concentration there were no significant ($p > 0.05$) differences between forages obtained across the growth cycle (in the three periods of the study); this concentration was approximately 100 g/kg DM. For all determinations, the interaction between type of forage and period (F × P) was non-significant ($p > 0.05$).

Table 3. Crude ash (CA) (g/kg dry matter (DM)), crude protein (CP) (g/kg DM), neutral detergent fiber (NDF) (g/kg DM), acid detergent fiber (ADF) (g/kg DM), acid detergent lignin (ADL) (g/kg DM), and water-soluble carbohydrates (g/kg DM) of the permanent grasslands throughout the first growth cycle.

	Forage		Period				Significance			
	FF [1]	BDF [2]	P1 [3]	P2 [4]	P3 [5]	SEM [6]	Period	Contrast	Forage	Period × Forage
CA	80	79	82	71	84	10.2	ns [10]	ns	ns	ns
CP	11.9	12.1	15.6	10.6	9.9	0.60	***	L [8],* Q [9],*	ns	ns
NDF	598	592	545	590	652	16.8	***	L *	ns	ns
ADF	320	322	287	321	355	6.3	***	L *	ns	ns
ADL	45	40	33	42	53	5.9	*	L *	ns	ns
WSC	104	92	99	105	90	11.1	ns	ns	ns	ns

[1] FF = fresh forage; [2] BDF = barn-dried forage; [3] P1 = first period (from 30 May to 3 June; 665 GDD); [4] P2 = second period (from 13 to 17 June; 822 GDD); [5] P3 = third period (from 27 June to 1 July; 1059 GDD). [6] SEM = standard error of the mean; [7] P= Probability; [8] L = linear; [9] Q = quadratic; [10] ns: $p > 0.05$; *, $p < 0.05$; ***, $p < 0.001$.

3.3. Digestibility and Intake

Dry matter digestibility and OMD declined with the increase in accumulated temperature ($p < 0.001$) (Table 4). For the DMD, this decline did not varied with the type of forage ($p < 0.10$) but a significant ($p < 0.001$) interaction between type of forage and GDD at which forages were cut. For DMD of fresh forages, linear effects ($p < 0.05$) were observed, whereas the DMD of preserved forages decreased, showing linear and quadratic effects ($p < 0.05$). The main differences in DMD between barn-dried hays and fresh forages were obtained at 1059 GDD (P3) (0.09 points between the two types of forage ($p < 0.001$). No significant differences between forages ($p > 0.05$) were found for OMD, and this declined linearly ($p < 0.05$) along the development cycle.

Barn-dried hays showed higher mean VI values ($p < 0.05$) and higher mean DOMI values ($p < 0.01$) than fresh forage (65.01 vs. 55.70 g/kg BW$^{0.75}$ respectively for VI and 36.11 vs. 28.46 g/kg BW$^{0.75}$ respectively for DOMI) (Table 4). Voluntary Intake and DOMI for both types of forage decreased quadratically ($p < 0.001$) along the cycle of growth. For these determinations, the interaction type of forage by period was non-significant ($p > 0.05$).

Temperature accumulation (GDD) was closely correlated with digestibility coefficients values and chemical composition, particularly with cell wall content and cell wall partitioning (Table 5). A close correlation was also observed with VI of barn-dried hays than with VI of fresh forages. Furthermore, phenological stage code was more closely correlated with CP than with cell wall contents. On the other hand, it was also less closely correlated compared to temperature accumulation with digestibility coefficients of both fresh forages and barn-dried hays.

Table 4. Average values of the grassland in the three cuts, and significance levels obtained in the analysis of variance developed for dry matter digestibility (DMD) (g/g), organic matter digestibility (OMD) (g/g), voluntary intake (VI) (g/kg BW$^{0.75}$), and digestible organic matter intake (DOMI) (g/kg BW$^{0.75}$).

	Fresh Forage					Barn-Dried Forage					Significance		
	P1 [1]	P2 [2]	P3 [3]	SEM [4]	Contrast	P1	P2	P3	SEM	Contrast	F [5]	P [6]	F × P
DMD	0.64	0.56	0.41	0.564	L [7],*	0.64	0.55	0.50	0.013	L * Q [8],*	†	***	***
OMD	0.66	0.60	0.50	0.012	L *	0.66	0.59	0.52	0.012	L *	ns [9]	***	ns
VI	67.07	50.27	49.75	4.408	L * Q *	80.87	60.96	53.20	4.408	L * Q *	*	***	ns
DOMI	35.94	26.95	22.48	2.665	L * Q *	49.54	33.60	25.18	2.665	L * Q *	**	***	ns

[1] P1 = first period (from 30 May to 3 June; 665 GDD); [2] P2 = second period (from 13 to 17 June; 822 GDD); [3] P3 = third period (from 27 June to 1; 1059 GDD); [4] SEM = standard error of the mean; [5] F = forage; [6] P = period; [7] L = linear; [8] Q = quadratic; [9] ns = $p > 0.05$; †, $p < 0.10$; *, $p < 0.05$; **, $p < 0.01$; ***, $p < 0.001$.

Table 5. Correlations between sum of temperatures, phenological stage, chemical composition, botanical composition, digestibility coefficients, voluntary intake and digestible organic matter intake of fresh forage and barn dried forage.

	ST [1]	MPW	Ash	CP	NDF	ADF	ADL	WSC	DMDff	OMDff	VIff	DOMIff	DMDbdf	OMDbdf	VIbdf	DOMIbdf	Gram	Leg	for
MPW [2]	0.66																		
Ash	0.31	−0.51																	
CP [3]	−0.87	−0.94	0.19																
NDF [4]	0.99 *	0.67	0.29	−0.88															
ADF [5]	0.99 †	0.74	0.20	−0.92	0.99 †														
ADL [6]	0.99 *	0.70	0.26	−0.90	0.99 *	0.99 *													
WSC [7]	−0.39	0.44	−0.99 †	−0.11	−0.36	−0.28	−0.33												
DMDff [8]	−0.99 *	−0.60	−0.38	0.83	−0.99 †	−0.98	−0.99 †	0.45											
OMDff [9]	−0.99 *	−0.63	−0.34	0.86	−0.99 *	−0.99 †	−0.99 †	−0.41	0.99 *										
VIff [10]	−0.81	−0.97	0.30	0.99 †	−0.82	−0.87	−0.84	−0.23	0.76	0.79									
DOMIff [11]	−0.95	−0.86	−0.01	0.98	−0.96	−0.98	−0.97	0.08	0.92	0.94	0.95								
DMDbdf [12]	−0.96	−0.84	−0.04	0.97	−0.97	−0.98	−0.98	0.12	0.94	0.95	0.94	0.99 †							
OMDbdf [13]	−0.99 †	−0.74	−0.20	0.92	−0.99 †	−0.99 *	−0.99 *	−0.28	0.98	0.99 †	0.87	0.98	0.99 †						
VIbdf [14]	−0.93	−0.88	0.04	0.99 †	−0.94	−0.97	−0.95	−0.03	0.91	0.92	0.97	0.99 *	0.99 *	0.97					
DOMIbdf [15]	−0.96	−0.84	−0.03	0.98	−0.96	−0.98	−0.97	0.10	0.93	0.95	0.94	0.99 *	0.99 *	0.98	0.99 *				
Gram [16]	0.45	0.97	−0.71	−0.83	0.47	0.55	0.50	0.65	−0.38	−0.42	−0.89	−0.70	−0.68	−0.55	−0.74	−0.69			
leg [17]	−0.05	0.72	−0.96	−0.45	−0.02	0.07	0.01	0.94	0.12	0.08	−0.55	−0.27	−0.23	−0.07	−0.31	−0.24	0.87		
for [18]	−0.62	−0.99 *	0.54	0.93	−0.64	−0.71	−0.67	−0.48	0.56	0.60	0.96	0.84	0.81	0.71	0.86	0.82	−0.98	−0.75	
sm [19]	0.94	0.87	−0.03	−0.98	0.95	0.97	0.96	−0.05	−0.91	−0.93	−0.96	−0.99 †	−0.99 *	−0.97	−0.99 *	−0.99 *	0.73	0.30	−0.85

[1] ST = sum of temperatures; [2] MPW = phenological stage mean plant weight; [3] CP = Crude protein; [4] NDF = neutral detergent fibre; [5] ADF = acid detergent fibre; [6] ADL = acid detergent lignin; [7] WSC = water soluble carbohydrates; [8] DMDff = dry matter digestibility for fresh forages; [9] OMDff = organic matter digestibility for fresh forages; [10] VIff = voluntary intake for fresh forages; [11] DOMIff = digestible organic matter intake for fresh forages; [12] DMDbdf = dry matter digestibility for barn-dried hays; [13] OMDbdf = organic matter digestibility for barn-dried hays; [14] VIbdf = voluntary intake for barn-dried hays; [15] DOMIbdf = digestible organic matter intake for barn-dried hays; [16] gram = grasses; [17] leg = legumes; [18] for = forbs; [19] sm = senescent material. †, $p < 0.1$; *, $p < 0.05$.

4. Discussion

The grassland studied was composed mainly of L. perenne, A. capillaris, and T. officinale, although its composition varied throughout the growth cycle as reported also by Andueza et al. (2016) [19]. At young stages, competitive species, as L. perenne and D. glomerata characterized according to Gross et al. (2007) by fast growth rate and high specific leaf area, formed the largest group, whereas at late stages, conservative species as A. capillaris, Trisetum flavescens L., and F. rubra characterized by slow growth rate and low specific leaf area Gross et al. 2007 were dominant. Other authors [20] recorded changes in the proportion of A. capillaris from 0.19 to 0.69 of the total biomass in different periods of the growing season in permanent grassland. The sward used in our study can be considered representative of many situations where forage from seeded temperate grasslands provides a major share of ruminant diets [1].

The present study covers the period between grazing and the time when most permanent grasslands were cut for hays. Despite the study duration (665–1059 GDD) the MPW ranged only between 2.65 (elongation) and 3.82 (anthesis). The presence of different species characterized by a different phenology and the heterogeneity of phenological stages within each one, a typical characteristic of permanent grasslands, [1] can explain this short evolution of maturity stages observed in the permanent grassland used in the present study.

The absence of any significant differences ($p > 0.05$) in chemical composition between fresh forages and barn-dried hays is difficult to explain. In the literature, DM losses due to respiration are well documented, and have been estimated at 2–5% [21–23]. According to these authors, WSC and CP were the main constituents involved in respiratory losses. However, Rees (1982) [21], in his review, point out that some researchers have produced evidence that crops can sometimes increase their weight of DM after cutting. This may be due to continuing of photosynthesis and the weight of nutrients assimilated being greater than that used for respiration [21]. In the present study, the balance between the nutrient assimilated in the photosynthesis after cutting and that used for respiration might explain the lack of differences between fresh forages and the barn-dried forages despite the different conditions of plant species proportions, maturity stage, temperature, and moisture of forages processed.

An interesting result of the present study was the closer relationship between CP and MPW than between MPW and structural carbohydrates or lignin. The evolution of cell wall content and the partitioning components remain more closely correlated to temperature accumulation in agreement with other reported results [24]. In general, factors such as age and maturity stage are confounded, but Buxton (1996) [24] states that under most circumstances, maturity stage rather than age can be more closely related to quality. The results found in the present study for CP agree with this statement, but not those obtained for structural carbohydrates, which are more closely related to the digestibility values. The coexistence in the permanent grassland of different species and different phenological stages within each one at a given date could explain these results.

Differences between forages were not significant ($p > 0.05$) for OMD, in agreement with chemical composition results. However, a surprising result of the present study is the differences between DMD and OMD values for the two types of forage and specifically differences for the last period. These results suggest that the digestibility of minerals could be greatly diminished in P3 for fresh forages. There is no obvious explanation for this lower DMD for fresh forages at P3. The possible presence of antinutritive compounds (e.g., condensed tannins) [25] in fresh forages which could be inactivated by the drying process might explain these results.

The presence of condensed tannins is influenced by the environmental conditions under which the plant is grown. It is likely that longer photoperiod and higher temperatures lead to an increased biosynthesis of tannins, as observed by Lees et al. (1994) [26] on lotus. In our study, the average temperatures were higher in the last period (18 °C) than in the first two (14 °C). In addition, it has been found that CT concentration tends to increase with phenological stage [27], so we would expect the concentration to be higher at P3 than in the other periods in our study. In line with this finding, Frutos et al. (2004) [28] state that condensed tannins are chelating agents that can reduce the availability of certain metal ions. By contrast, Scharenberg et al. (2007) [29] report a lower digestibility of Ca, P, Mg, and Na of dried and ensiled sainfoin without polyethylene glycol than of dried and ensiled sainfoin with added polyethylene glycol. This suggests that complexes of condensed tannins formed with the minerals were the most likely reason for the low mineral digestibility of sainfoin without condensed tannins.

Another interesting result of this study is the higher VI of barn-dried hays than of fresh forage. Fresh forages are usually consumed in spring, summer, and autumn, and hays in winter when fresh forages do not grow. As seasons influence the VI of animals [30], comparison of fresh forages and hays is difficult to carry out. In some studies, fresh forage is stored frozen until the comparison trial is performed [5]. In the present study, digestibility and VI trials for fresh forages were run only 15 days before trials for the same forages preserved as barn-dried hays, thus minimizing the influence of season on the VI of animals. In the literature, the reported effects of artificial drying on forage intake are conflicting. Whereas Archimède et al. (1999) [6] reported higher intake of *Digitaria decumbens* when it was ingested as fresh forage than as barn-dried hay, Demarquilly (1970) [8] found no differences between the VI of fresh forages and those dehydrated at low temperature. However, Estrada et al. (2004) [31] reported higher VI for lactating cows when ryegrass was offered partially dried than when it was offered after cutting. The results of Archimède et al. (1999) [6] and those of Demarquilly (1970) [8] could be explained by possible losses associated with the barn drying process, mainly respiratory losses, [23]. By contrast, Estrada et al. (2004) [31] explain their results by the influence of water content of fresh grassland forage, which could limit their intake. However, in the present study, this effect still does not fully explain the differences between VI of barn-dried hays and fresh forages. In the first period (proportion of DM in the fresh forage of 0.18) the difference between VI of barn-dried hays and fresh forages was 13.8 g DM/kg $BW^{0.75}$, whereas in P3 the difference between VI of barn-dried hays and fresh forages, (proportion of DM 0.29), was 3.45 g DM/kg $BW^{0.75}$. Other factors that might explain these results are the possible presence of antinutritive compounds in fresh forage, which could depress the VI of animals, and which could be inactivated during the barn drying, or also possible differences in factors associated with ruminal fermentation, which could influence forage intake [5].

5. Conclusions

Barn-drying hay is a very good method of preserving forages from permanent grasslands. The results of the present study fail to support the hypothesis that the feed value of barn-dried hays is similar to that of fresh forages. We conclude that feed value of barn-dried hays of permanent grasslands is higher than that of fresh forage. This higher feed value of barn-dried hays of permanent grasslands is mainly a consequence of their higher VI in relation to that of fresh grass. More research is now needed to find a cogent explanation for this higher VI of hays in relation to fresh forages, and particularly to determine the influence of botanical composition of permanent grassland in this effect.

Author Contributions: D.A. carried out the experimental design, data interpretation, manuscript writing, and editing. He was also involved in the data recovery. F.P. was involved in the experimental design, data recovery, data analysis and manuscript revision. P.P.: contributed in the experimental design and data recovery. K.T. contributed in the interpretation of results, manuscript writing, and manuscript revision. All authors read and approved the final manuscript.

Funding: This work was partially performed thanks to the program "Investissement d'Avenir" (16-IDEX-0001 CAP 20-25) funded by the Agence Nationale de la Recherche of the French government.

Acknowledgments: The authors thank the Herbipôle team of the Marcenat experimental farm for their technical help.

Conflicts of Interest: The authors declare no conflict of interest.

References

1. Bruinenberg, M.H.; Valk, H.; Korevaar, H.; Struik, P.C. Factors affecting digestibility of temperate forages from seminatural grasslands: A review. *Grass Forage Sci.* **2002**, *57*, 292–301. [CrossRef]
2. Andueza, D.; Cruz, P.; Farruggia, A.; Baumont, R.; Picard, F.; Michalet-Doreau, B. Nutritive value of two meadows and relationships with some vegetation traits. *Grass Forage Sci.* **2010**, *65*, 325–334. [CrossRef]
3. Andueza, D.; Picard, F.; Jestin, M.; Aufrere, J. The effect of feeding animals ad libitum vs. at maintenance level on the in vivo digestibility of mown herbage from two permanent grasslands of different botanical composition. *Grass Forage Sci.* **2013**, *68*, 418–426. [CrossRef]
4. Rotz, C.A.; Muck, R.E. Changes in forage quality during harvest and storage. In *Forage Quality, Evaluation, and Utilization*; Fahey, G.C., Collins, M., Mertens, D.R., Moser, L.E., Eds.; American Society of Agronomy, Inc.; Crop Science Society of America, Inc.; Soil Science Society of America, Inc.: Madison, WI, USA, 1994; pp. 828–868.
5. Pasha, T.N.; Prigge, E.C.; Russell, R.W.; Bryan, W.B. Influence of moisture-content of forage diets on intake and digestion by sheep. *J. Anim. Sci.* **1994**, *72*, 2455–2463. [CrossRef]
6. Archimede, H.; Poncet, C.; Boval, M.; Nipeau, F.; Philibert, L.; Xande, A.; Aumont, G. Comparison of fresh and dried *Digitaria decumbens* grass intake and digestion by Black-belly rams. *J. Agric. Sci.* **1999**, *133*, 235–240. [CrossRef]
7. Dulphy, J.P.; Rouel, J. Effect of wilting on changes in the voluntary feed-intake in cattle compared to sheep. *Ann. Zootech.* **1988**, *37*, 31–41. [CrossRef]
8. Demarquilly, C. Effect of low temperature-dehydration on forage feed-value. *Ann. Zootech.* **1970**, *19*, 45–51. [CrossRef]
9. Delaby, L.; Peccatte, J.R. Feeding value of ventilated hay from multi-specific pastures. *Fourrages* **2008**, *195*, 354–356.
10. Andueza, D.; Delgado, I.; Muñoz, F. Effect of lucerne preservation method on the feed value of forage. *J. Sci. Food Agric.* **2009**, *89*, 1991–1996. [CrossRef]
11. Demarquilly, C.; Chenost, M.; Giger, S. Pertes fécales et digestibilité des aliments et des rations. In *Nutrition des Ruminants Domestiques. Ingestion et Digestion*; Jarrige, R., Ruckebusch, Y., Demarquilly, C., Farce, M.H., Journet, M., Eds.; INRA Éditions: Paris, France, 1995; pp. 601–647.
12. Moore, K.J.; Moser, L.E.; Vogel, K.P.; Waller, S.S.; Johnson, B.E.; Pedersen, J.F. Describing and quantifying growth-stages of perennial forage grasses. *Agron. J.* **1991**, *83*, 1073–1077. [CrossRef]
13. AOAC. *Official Methods of Analysis*, 15th ed.; Association of Official Analytical Chemists: Arlington, VA, USA, 1990; p. 1298.
14. Somogyi, M. Notes on sugar determination. *J. Biol. Chem.* **1952**, *195*, 19–23.
15. Van Soest, P.J.; Robertson, J.B.; Lewis, B.A. Methods for dietary fiber neutral detergent fiber, and nonstarch poysaccharides in relation to animal nutrition. *J. Dairy Sci.* **1991**, *74*, 3583–3597. [CrossRef]
16. Van Soest, P.J.; Robertson, J.B. *Systems of Analysis for Evaluating Fibrous Feeds*; Pidgen, W.J., Balch, C.C., Graham, M., Eds.; IDRC No 134; International Development Research Centre: Ottawa, ON, USA, 1980; pp. 49–60.
17. Otto, S.; Masin, R.; Chiste, G.; Zanin, G. Modelling the correlation between plant phenology and weed emergence for improving weed control. *Weed Res.* **2007**, *47*, 488–498. [CrossRef]
18. SAS. *SAS/STAT UsersGuide, Version 6.12*; Statistical Analysis System Institute: Cary, NC, USA, 1998.
19. Andueza, D.; Rodrigues, A.M.; Picard, F.; Rossignol, N.; Baumont, R.; Cecato, U.; Farruggia, A. Relationships between botanical composition, yield and forage quality of permanent grasslands over the first growth cycle. *Grass Forage Sci.* **2016**, *71*, 366–378. [CrossRef]
20. Todorova, P.A.; Kirilov, A.P. Changes in the permanent grassland composition and feeding value during the growing season. *Grassl. Sci. Eur.* **2002**, *7*, 170–171.
21. Rees, D.V.H. A discussion of sources of dry-matter loss during the process of haymaking. *J. Agric. Eng. Res.* **1982**, *27*, 469–479. [CrossRef]
22. Dulphy, J.P. Fenaison: Pertes en cours de récolte et de conservation. In *Les Fourrages Secs: Récolte, Traitement, Utilisation*; Demarquilly, C., Ed.; INRA: Paris, France, 1987; pp. 103–124.
23. McGechan, M.B. A review of losses arising during conservation of grass forage. 1. Field losses. *J. Agric. Eng. Res.* **1989**, *44*, 1–21. [CrossRef]

24. Buxton, D.R. Quality-related characteristics of forages as influenced by plant environment and agronomic factors. *Anim. Feed Sci. Technol.* **1996**, *59*, 37–49. [CrossRef]
25. Montossi, F.; Liu, F.; Hodgson, J.; Morris, S.; Barry, T.; Risso, D. Influence of low-level condensed tannins concentrations in temperate forages on sheep performance. In Proceedings of the XVIIIth International Grassland Congress, Saskatoon, SK, Canada, 8–19 June 1997; pp. 8.1–8.2.
26. Lees, G.L.; Hinks, C.F.; Suttill, N.H. Effect of high-temperature on condensed tannin accumulation in leaf tissues of big trefoil (Lotus Uliginosus Schkuhr). *J. Sci. Food Agric.* **1994**, *65*, 415–421. [CrossRef]
27. Theodoridou, K.; Aufrere, J.; Andueza, D.; Le Morvan, A.; Picard, F.; Stringano, E.; Pourrat, J.; Mueller-Harvey, I.; Baumont, R. Effect of plant development during first and second growth cycle on chemical composition, condensed tannins and nutritive value of three sainfoin (*Onobrychis viciifolia*) varieties and lucerne. *Grass Forage Sci.* **2011**, *66*, 402–414. [CrossRef]
28. Frutos, P.; Hervas, G.; Giraldez, F.J.; Mantecon, A.R. Review. Tannins and ruminant nutrition. *Span. J. Agric. Res.* **2004**, *2*, 191–202. [CrossRef]
29. Scharenberg, A.; Arrigo, Y.; Gutzwiller, A.; Wyss, U.; Hess, H.D.; Kreuzer, M.; Dohme, F. Effect of feeding dehydrated and ensiled tanniniferous sainfoin (*Onobrychis viciifolia*) on nitrogen and mineral digestion and metabolism of lambs. *Arch. Anim. Nutr.* **2007**, *61*, 390–405. [CrossRef]
30. Michalet Doreau, B.; Gatel, F. Annual variations in voluntary food-intake of wethers. *Ann. Zootech.* **1988**, *37*, 151–158. [CrossRef]
31. Estrada, J.I.C.; Delagarde, R.; Faverdin, P.; Peyraud, J.L. Dry matter intake and eating rate of grass by dairy cows is restricted by internal, but not external water. *Anim. Feed Sci. Technol.* **2004**, *114*, 59–74. [CrossRef]

© 2019 by the authors. Licensee MDPI, Basel, Switzerland. This article is an open access article distributed under the terms and conditions of the Creative Commons Attribution (CC BY) license (http://creativecommons.org/licenses/by/4.0/).

Article

Evaluating the Impacts of Continuous and Rotational Grazing on Tallgrass Prairie Landscape Using High-Spatial-Resolution Imagery

Shengfang Ma [1,2], Yuting Zhou [3,4,*], Prasanna H. Gowda [5], Liangfu Chen [1], Patrick J. Starks [5], Jean L. Steiner [5] and James P. S. Neel [5]

[1] State Key Laboratory of Remote Sensing Science, Institute of Remote Sensing and Digital Earth, Chinese Academy of Sciences, Beijing 100101, China; shengfangma@gmail.com (S.M.); chenlf@radi.ac.cn (L.C.)
[2] University of Chinese Academy of Sciences, Beijing 100049, China
[3] Department of Plant and Soil Sciences, Oklahoma State University, Stillwater, OK 74078, USA
[4] Department of Geography, Oklahoma State University, Stillwater, OK 74078, USA
[5] USDA-ARS Grazinglands Research Laboratory, El Reno, OK 73036, USA; prasanna.gowda@ars.usda.gov (P.H.G.); patrick.starks@ars.usda.gov (P.J.S.); jean.steiner@hughes.net (J.L.S.); jim.neel@ars.usda.gov (J.P.S.N.)
* Correspondence: yuting.zhou@okstate.edu; Tel.: +1-405-845-2102

Received: 1 April 2019; Accepted: 7 May 2019; Published: 9 May 2019

Abstract: This study evaluated the impacts of different grazing treatments (continuous (C) and rotational (R) grazing) on tallgrass prairie landscape, using high-spatial-resolution aerial imagery (1-m at RGB and near-infrared bands) of experimental C and R pastures within two replicates (Rep A and Rep B) in the southern Great Plains (SGP) of the United States. The imagery was acquired by the National Agriculture Imagery Program (NAIP) during the agricultural growing season of selected years (2010, 2013, 2015, and 2017) in the continental United States. Land cover maps were generated by combining visual interpolation, a support vector machine, and a decision tree classifier. Landscape metrics (class area, patch number, percentage of landscape, and fragmentation indices) were calculated from the FRAGSTATS (a computer software program designed to compute a wide variety of landscape metrics for categorical map patterns) based on land cover results. Both the metrics and land cover results were used to analyze landscape dynamics in the experiment pastures. Results showed that both grass and shrubs of different pastures differed largely in the same year and had significant annual dynamics controlled by climate. High stocking intensity delayed grass growth. A large proportion of bare soil occurred in sub-paddocks of rotational grazing that were just grazed or under grazing. Rep A experienced rapid shrub encroachment, with a large proportion of shrub at the beginning of the experiment. Shrub may occupy 41% of C and 15% of R in Rep A by 2030, as revealed by the linear regression analysis of shrub encroachment. In contrast, shrub encroachment was not significant in Rep B, which only had a small number of shrub patches at the beginning of the experiment. This result indicates that the shrub encroachment is mainly controlled by the initial status of the pastures instead of grazing management. However, the low temporal resolution of the NAIP imagery (one snapshot in two or three years) limits our comparison of the continuous and rotational grazing at the annual scale. Future studies need to combine NAIP imagery with other higher temporal resolution imagery (e.g., WorldView), in order to better evaluate the interannual variabilities of grass productivity and shrub encroachment.

Keywords: tallgrass prairie landscape; grazing management; classification; shrub encroachment; Normalized Difference Vegetation Index (NDVI)

1. Introduction

Tallgrass prairie provides various ecological (e.g., carbon sequestration and biodiversity) and economic (e.g., forage) benefits for people in the Great Plains of the United States. However, the distribution and condition of the tallgrass prairie is threatened by agricultural land use (e.g., conversion to crops) and land management (e.g., different grazing management systems). The tallgrass prairie provides nutritious forage for cattle grazing [1] in the Great Plains, which is a major revenue stream for farmers in the region [2]. Cattle grazing is an important human disturbance to the tallgrass prairie landscape patterns (e.g., reducing landscape heterogeneity) and can affect its ecosystem productivity and health [3,4]. Continuous grazing with a moderate stocking rate (around six acres per animal unit for a tallgrass prairie, with an average production of 2,885 kg per acre in the region [5]) and rotational grazing with a relatively high stocking rate are two main grazing management systems nowadays. Rotational grazing has been recommended since the mid-20th century, as an effective way to maintain pasture productivity and sustainability [6–10]. Theoretically, multi-paddock rotational grazing has high stocking in each paddock, which may lead to more even spatial distribution of livestock grazing [11] and reduce bush encroachment [12]. Rotational grazing also allows for the rest of paddocks during different periods of the year, which may help improve landscape diversity or productivity [13–15]. In continuously grazed pastures, cattle may preferentially graze specific locations the majority of the time, which can cause landscape fragmentation and uneven grass coverage. However, there is still debate on whether rotational grazing clearly outperforms continuous grazing or not in terms of plant or livestock productivity [13,15–27].

As a landscape dominated by grasses, the grass coverage of pastures is an important indicator of grassland landscape and productivity in tallgrass prairie. The presence of shrubs in the tallgrass prairie as a result of invasion is becoming an important component in many of tallgrass prairie ecosystems. Woody shrub invasion in the Great Plains [28,29] can cause a decrease in species richness and diminish cattle grazing capacity [12]. Many researchers have found that shrub invasion may be mitigated by grazing [12,30]. Thus, investigating the impacts of grazing management on grass condition and shrub invasion is of great importance.

Since different grazing management systems could exert different impacts on tallgrass prairie landscape patterns. Evaluating the impacts and further identifying an efficient and sustainable grazing management system for the tallgrass pastures is essential to maintain the health of prairie ecosystem [31–33]. However, evaluation of the impacts of grazing on tallgrass prairie landscape is challenging, mostly due to the potential mismatch between pasture size and spatial resolution of free and commonly used satellite datasets (e.g., MODIS and Landsat, with spatial resolutions of 500 m and 30 m, respectively). The National Agriculture Imagery Program (NAIP) acquires high spatial resolution (1 m) multi-spectral (red, green, blue, and near-infrared) aerial imagery during the agricultural growing season in the continental United States [34]. The high spatial resolution of NAIP imagery provides us an opportunity to study landscape dynamics within paddocks not of sufficient size to be studied using medium- to low-spatial-resolution data sets.

To compare and contrast differences in continuously and rotationally grazed pastures with regard to variations in grass condition and woody shrub, experimental pastures composed of two treatments (continuous grazing (C) versus rotational grazing (R)), with two repetitions for each treatment (Rep A and Rep B) located in El Reno, Oklahoma, United States were used in this study. This study evaluated the impacts of different grazing treatments (continuous versus rotational grazing) on tallgrass prairie landscape, using NAIP imagery of the experimental pastures. Visual interpretation and two supervised classification methods were combined to get accurate classification results. Landscape metrics are used to quantitatively analyze the response of landscape patterns to continuous and rotational grazing. The findings of this study can help develop better grazing management systems for forage production and shrub encroachment control.

2. Materials and Methods

2.1. Study Area and Experiment Design

The study area is located at the United States Department of Agriculture Agricultural Research Service (USDA-ARS) Grazinglands Research Laboratory (GRL) in El Reno, OK, United States. The study area is typical tallgrass prairie in the Southern Great Plain (SGP). The main grass species in the study area are big bluestem (*Andropogon gerardii Vitman*), little bluestem (*Schizachyrium scoparium* (Michs.) Nash), indiangrass (*Sorghastrum nutans* (L.) Nash), and switchgrass (*Panicum virgatum* L.). The main woody shrub in the area is buckbrush (*Symphoricarpos orbiculatus* Moench), and is not edible by cattle. Photos of main grass and shrub species were shown in Figure S1. Long term annual average air temperature is 15.14 °C, and total precipitation is 91.3 cm. Dominant soils consist mainly of Norge silt loams (Fine–silty, mixed, active thermic Udic Paleustolls), Renfrow (fine, mixed super-active, thermic Udertic Paleustolls), and Kirkland silt loams (Fine–silty, mixed super-active, thermic Cumulic Hapulstolls) [35].

The study was comprised of two grazing treatments (continuous (C) and rotational (R)), with two replicates (Rep A and Rep B) for each treatment (Figure 1). The areas of the pastures were 58.6, 62.7, 78.6, and 82.7 ha for C in Rep A (Ca), C in Rep B (Cb), R in Rep A (Ra), and R in Rep B (Rb), respectively. The R pastures were further sub-divided into 10 approximately equally-sized cells. Grazing in Ca and Ra began in 2009, while Cb and Rb started grazing in 2011. The number of grazing cattle varied across years: 10 to 20 cow–calf pairs in the C treatments, and 20 to 27 cow–calf pairs in the R treatments. Cattle herds stayed in the C treatments year-round, while they rotated within the 10 cells in 7- to 10-day grazing bouts, depending on the vegetation conditions. More detail on the grazing experiment can be found in Steiner et al. [36].

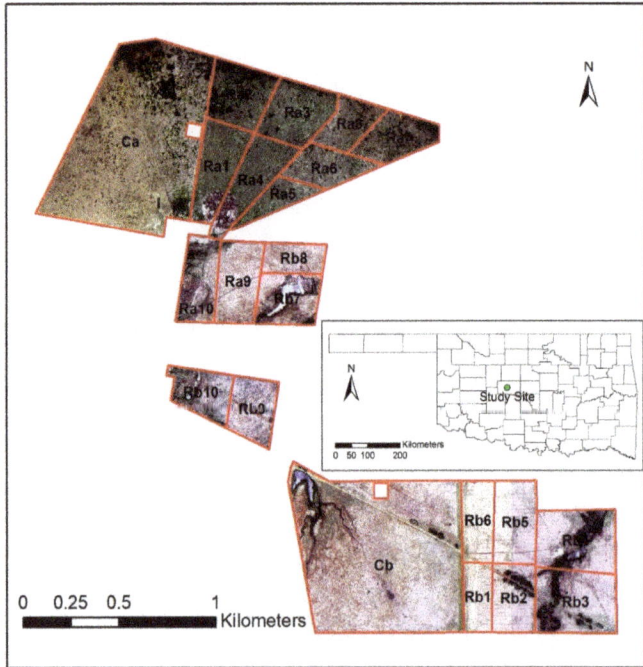

Figure 1. The study area with the National Agriculture Imagery Program (NAIP) mosaic true color image as background. The location of the study site within the state of Oklahoma is shown in the inset.

2.2. National Agriculture Imagery Program Imagery and the Normalized Difference Vegetation Index

The NAIP acquires multi-spectral aerial imagery (natural color (red, green, and blue, or RGB) and near-infrared (NIR)) during the agricultural growing seasons. NAIP imagery is acquired at a one-meter ground sample distance, with a horizontal accuracy that matches within six meters of photo-identifiable ground control points that are used during image inspection. Every NAIP imagery has no more than 10% cloud cover, weather conditions permitting. Horizontal accuracy and tonal quality are inspected for all imagery. Thus, these images are analysis-ready. We downloaded the four-band (RGB and NIR) NAIP product of the experimental pastures for 5 May 2010, 2 June 2013, 6 June 2015, and 21 May 2017 using the Google Earth Engine. The Normalized Difference Vegetation Index (NDVI) was calculated from the red and near-infrared bands (Equation 1) [37], and the index was then used to estimate grass coverage.

$$NDVI = \frac{\rho_{nir} - \rho_{red}}{\rho_{nir} + \rho_{red}} \tag{1}$$

where ρ_{red}, and ρ_{nir} represent surface reflectance from red and near infrared band (841–876 nm), respectively.

2.3. Image Classification

According to our field survey of land cover types, the main land cover classes in the study area are grass, shrub, tree, water body, and miscellaneous (e.g., paths and shades for cattle). The grass category can be further subdivided into four levels of coverage: high, medium, low cover, and bare soil. Bare soil is temporary in paddocks, especially in R pastures that are currently under grazing. Therefore, it is also included in the grass category. In order to achieve a high classification accuracy, we combined visual interpretation and two supervised classification methods (support vector machine (SVM) and a decision tree classifier). The work flow is shown in Figure 2.

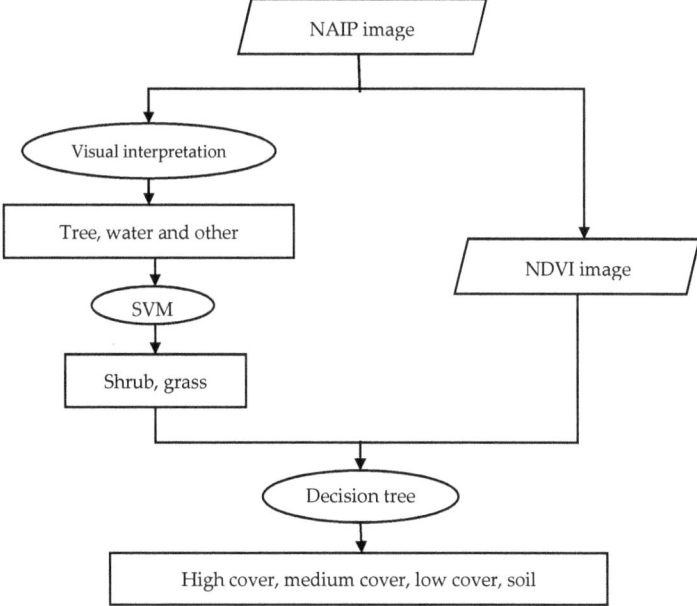

Figure 2. Flow of the land cover classification.

2.3.1. Visual Interpretation

Visual interpretation is commonly used, when good prior knowledge of the study area exists, and one or more cover types are easily distinguished based on color, size, shape, and tonal qualities, especially for high spatial aerial imagery. In this study, trees, man-made features, and water bodies were easily identified, and were manually classified by this method. As trees occupied a small area and distributed in a fixed location, with different sizes in different years depending on growth conditions, it is easier to extract them by interpretation.

2.3.2. Support Vector Machine

Although shrub and grass have some overlapping spectral features, they have different visual textural features. We initially evaluated the performances of the support vector machine (SVM), maximum likelihood (ML), and artificial neural network (ANN) in identifying shrub and the various levels of grass cover after the manually-identified tree, water body, and man-made feature classes were masked from the images. Before the classification, a training dataset and validation dataset representative of shrub and grass were acquired by combining a field survey, Google Earth images, and the region of interest (ROI) tool of NAIP imagery. The initial analysis revealed that the SVM achieves a higher level of classification accuracy than the other two methods (Table 1); thus, SVM was adopted in classification throughout the study years.

Table 1. Accuracy assessment of the support vector machine (SVM), maximum likelihood (ML), and artificial neural network (ANN).

Classifier	Overall Accuracy(%)	Kappa Coefficient
SVM	93.58	0.87
ANN	92.6	0.85
ML	87.15	0.75

The SVM classifier is based on statistical learning theory and was initially designed for two-class problems [38,39]. The SVM has been widely reported as an outstanding supervised classification method. With given training datasets, SVM outputs an optimal hyperplane that separates categorizes. Kernel function selection and its parameter (gamma, pyramid levels, etc.) determinations affect SVM classifier performance. SVM classification in this study was executed in radial basis function kernel types by using the default value. The default parameter value of gamma is 0.25, the penalty parameter is 100, the pyramid level is 0, and the classification probability threshold is 0.

2.3.3. Decision Tree

It was noticed during the field survey that the grass coverage was not homogeneous. Thus, it is necessary to divide them into different categories to reflect the impacts of grazing management on grass coverage. The NDVI has been shown to have a good relationship with green vegetation coverage [40–42]. In order to distinguish different proportions of grass coverage, a decision tree based on the NDVI was used to categorize grass into high-, medium-, and low-cover, as well as bare soil. Using one of the most recognized algorithms [43], we have calculated the green vegetation fraction from the NDVI. The thresholds for low-, medium-, and high-coverage grass were 41%, 63%, and 76%, respectively.

To build the decision tree, about 6000 pixel samples of each class were selected in every pasture of different years via the region of interest (ROI) tool in ENVI. The samples were used to develop statistical relationships between the NDVI and different classes. Figure 3 shows that the potential of misclassifying the tree, shrub, and high-grass-cover categories due to overlap in the NDVI value range (Table S1); this demonstrates the necessity of classifying trees (through visual interpretation) and shrubs (through the SVM) separately before using the NDVI-based decision tree (Figure 2). From Figure 3, it is observed that the mean and range of the NDVI for the non-tree and non-shrub classes should be sufficient to distinguish high, medium, and low grass cover and soil with the NDVI-based decision tree. The average values of the NDVI increased from bare soil to high cover grass. NDVI cutoff values were the mean of the upper lower quartile of nearby two-class coverage. The decision tree used in this study is given in Figure 4.

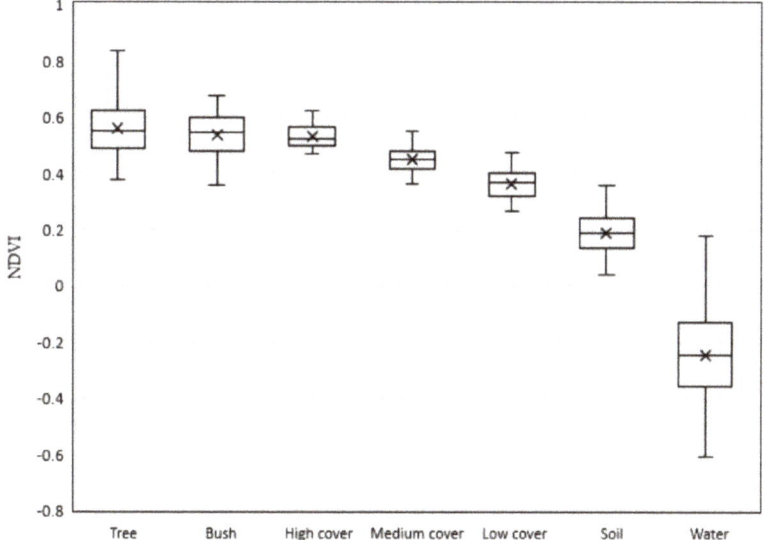

Figure 3. Normalized Difference Vegetation Index (NDVI) box-plot diagram for each category.

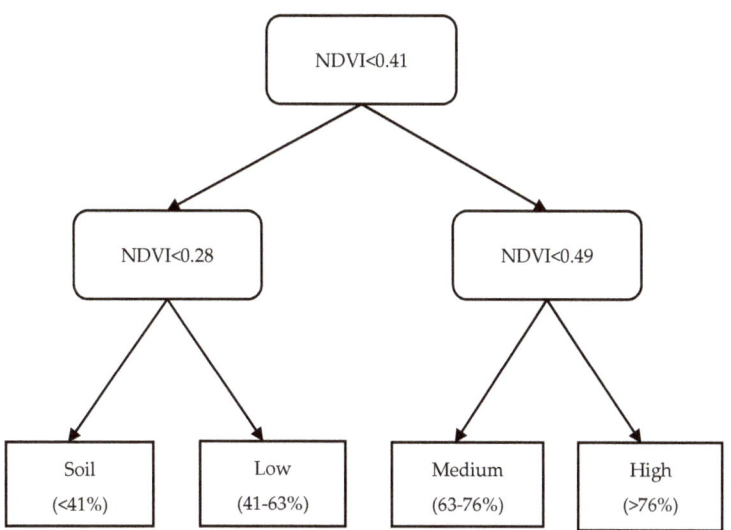

Figure 4. NDVI-based decision tree. The percentage in the lower box represents the green vegetation fraction for corresponding grass class.

2.4. Landscape Metrics and Statistical Analysis

Landscape metrics are a group of quantitative indices that can highly concentrate the information of landscape patterns and reflect the structural composition and spatial configuration of the landscape [44,45]. Landscape metrics can be used to show the status of classes in pastures, while interannual comparison of landscape metrics can show the changes caused by disturbances [46]. The land cover classification results were used to calculate the following landscape metrics for each year of NAIP data: class area (CA), number of patches (NP), class percentage of landscape area (PLAND), and fragmentation (FN) using the FRAGSTATS 4.2 [47]. The landscape metrics used in this study are shown in Table 2.

Table 2. Landscape metrics used in this study. TA, CA, NP, PLAND, and FN represent total area, class area, number of patches, proportion of the landscape, and degree of fragmentation. The meaning of indices are as follows: i is class I, j is patch j in class I, m is the total class number, and n is the total patch number in class I.

Landscape Index	Description	Formula
TA	Area of total landscape	$TA = \sum_{i=1}^{m} A_i$
CA	Area of class I	$CA = A_i = \sum_{i=1}^{n} A_{ij}$
NP	Number of patches in class I	$NP = N_i = n$
PLAND	Proportion of the landscape occupied by patch type (class) I	$PLAND = A_i/A$
FN	The degree of fragmentation of the class I	$FN = N_i/A_i$

For the impacts of grazing treatments on grass, we focused on the grass coverage and NDVI value. Percentage of landscape (PLAND) for the four levels of grass cover (high cover, medium cover, low cover, and bare soil) and average NDVI of the grass category were compared between C and R in different years. For the impacts of grazing treatments on shrub distribution, we compared the shrub class metrics (area (CA), number of patches (NP), percentage of landscape (PLAND), and fragmentation (FN)) of the same pasture in different years. A paired-samples t-test was conducted to compare proportions of bare soil as well as low-, medium-, and high-coverage grass in continuous and rotational grazing treatments.

3. Results

3.1. Land Cover Classification

Land cover classification of the study area is shown in Figure 5. It is observed from Figure 5 that grass was clearly the dominant class in the landscape. Grass coverage of different pastures also largely differed within the same year, and grass coverage within pasture varied among years. Overall, grass in continuous pastures was more homogeneous (less mixture of different grass coverage classes) than rotational pastures, which had some bare soil in some of the paddocks (e.g., Figure 5b,d). The years 2010 and 2015 had better grass cover than other years, whereas grass coverage in 2013 was the poorest. Shrubs mainly occurred in Rep A, with scattered distribution, with a smaller number of shrubs in Rep B. Comparison of NAIP images from 2010 and 2017 indicate that shrubs increased in terms of number and area of patches in Ca and Ra while there were no visible changes in Cb and Rb. The dynamics of grass condition in different pastures is shown in Section 3.2, and shrub encroachment during these years is introduced in Section 3.3. Trees are generally concentrated along water bodies, with some scattered in the pastures and along old road beds. The proportion of trees showed little difference between years. The proportion of land area occupied by the water bodies varied slightly due to variation in rainfall.

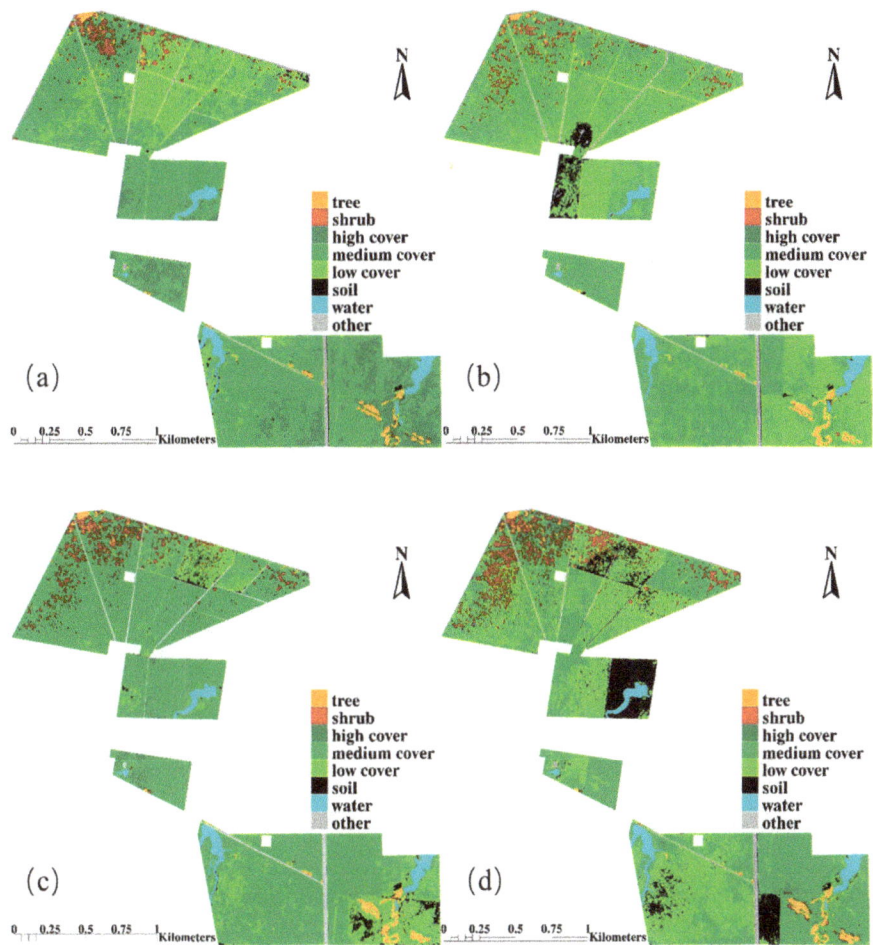

Figure 5. Land classification results in 2010 (**a**), 2013 (**b**), 2015 (**c**), and 2017 (**d**).

3.2. Grass Dynamics

Fractional percent for four levels of grass coverage (high, medium, low, and bare soil) and the NDVI for both the continuous (C) and rotational (R) replicates (Rep A and Rep B) for 2010, 2013, 2015, and 2017 (Figure 6) were used to quantitatively investigate grass dynamics (refer to Tables S2–S5 for absolute values).

On a year-by-year basis, it is observed from Figure 6 that in 2010, Ca had more grassland that was classified as medium coverage (60%) compared to the low coverage (32%) or high coverage (5%) categories. In contrast to the Ca, the Ra replication had more area classified as low coverage (59%), which mainly occurred in sub-paddocks Ra1 to Ra4 (Figure 5a), and less in the medium coverage (35%) category. However, land area occupied by high grass cover was about the same as in Ca. The NDVI value was higher in Ca (0.43) than Ra (0.4). Cb had a lower NDVI value (0.45) than Rb (0.49). Cb exhibited less grass in the high category (20%) than Rb (48%), and more in the medium coverage category (60%) than in Rb (47%). Grass in Rep B was in better condition than in Rep A, shown by higher grass coverage and NDVI value in the former.

In 2013, the proportion of grass class classified as low grass coverage was highest in all four pastures among the four study years. The C pastures had higher percentages of grass classified as medium coverage (37–44%) than observed for the R pastures (18–26%). Little to no land area was classified as high coverage grass for any of the four pastures in this year. Comparatively, the R pastures had more land area classified as low coverage grass (~70%) than observed in the C pastures (55–63%). The C pastures showed a higher NDVI value (0.4) than the R pastures (0.36–0.37). It is also observed that for the R pastures during this year, more land was classified as bare soil than was observed in 2010. Pasture Ra exhibited much higher bare soil areas than Rb. Ra 10 and a small portion of Ra1, and Ra4 (Figure 5b) had large bare soil area.

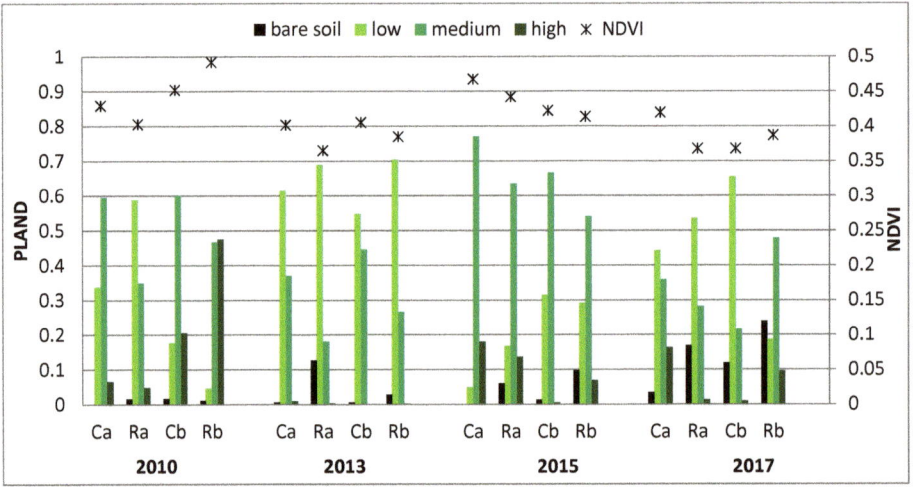

Figure 6. NDVI and fractional percent of grass coverage (bare soil, low coverage, medium coverage, and high coverage) for 2010, 2013, 2015, and 2017.

In 2015, the C and R treatments all showed improved grass coverage over the prior years. The C pastures had larger percentages of grass classified as medium cover (66–78%) than observed in the R pastures (53–63%). Pastures Ca, Ra, and Rb had grass classified as high cover, but Cb had very little land area in this category. Cb and Rb had ~30% low cover grass, while Ca and Ra only had 5% and 16%, respectively. Bare soil was presented in both Ra (Ra3) and Rb (Rb2 and Rb3) (Figure 5c).

In 2017, the low-coverage grass category was highest in all replicates except Rb. Considering the medium- and high-coverage categories together, Ca had more area in these classes than Ra (51% versus 30%, respectively), while Cb had less area in these classes than Rb (20% versus 58%, respectively). The same trend was also found in the NDVI comparison between Ca and Ra (0.42 versus 0.37, respectively), and Cb and Rb (0.37 versus 0.39, respectively). Bare soil occurred in Ra2, Ra3, and Ra6. In the case of the Rb replicate, three sub-paddocks of Rb (Rb1, Rb7, and Rb8) contributed most to the total bare soil value (Figure 5d). A noticeable portion (12%) of pasture Cb was bare soil.

Tables 3 and 4 summarize the paired t-test results for four grass coverage categories in Reps A and B, respectively. There were significant differences (95% confidence interval) in the scores for Ca and Ra in low- and medium-grass-coverage categories (Table 3). However, there was no significant difference between Cb and Rb in any grass coverage category. These results suggest that the differences in grass coverage between rotational and continuous grazing were relatively small.

Table 3. Paired *t*-test results for four grass coverage categories in Rep A. The parameters t, df, and p represent t-statistic, degree of freedom, and the probability value for the t-statistic.

	t	df	p	Mean of the Differences
Bare soil	−2.95	3	0.06	−0.08
Low grass cover	−3.38	3	0.04	−0.13
Medium grass cover	4.45	3	0.02	0.16
High grass cover	1.66	3	0.20	0.05

Table 4. Paired *t*-test results for four grass coverage categories in Rep B. Refer to Table 3. For the meaning of the parameters.

	t	df	p	Mean of the Differences
Bare soil	−1.95	3	0.15	−0.06
Low grass cover	0.89	3	0.44	0.12
Medium grass cover	0.43	3	0.69	0.04
High grass cover	−1.83	3	0.17	−0.11

3.3. Shrub Encroachment

Table 5 includes landscape matrices for shrub in two treatments among the study years. Using the data presented in Table 5, the trend of shrub encroachment is analyzed with linear regressions between year and *PLAND* (Figure 7).

Table 5. Landscape metrics for shrub in the four pastures in the years 2010, 2013, 2015, and 2017.

ID	Year	TA (ha)	CA (ha)	PLAND (%)	NP	Patch Fragmentation
Ca	2010		5.49	9.37	1727	314.68
	2013	58.6	7.40	12.62	2054	277.69
	2015		9.15	15.61	2230	243.76
	2017		12.44	21.23	3034	243.87
Ra	2010		3.02	3.85	762	252.03
	2013	78.6	3.46	4.40	1266	365.79
	2015		4.35	5.54	1353	310.88
	2017		6.26	7.96	1438	229.88
Cb	2010		0.28	0.45	238	851.83
	2013	62.7	0.08	0.12	129	1710.88
	2015		0.06	0.10	93	1492.78
	2017		0.05	0.09	90	1679.10
Rb	2010		0.48	0.58	477	988.19
	2013	82.7	0.33	0.40	419	1268.54
	2015		0.39	0.47	331	842.88
	2017		0.63	0.77	346	545.40

From Figure 7 it is readily observed that replicates Ca and Ra show increasing land area being encroached by shrubs, whereas the increase in shrub area is minimal in Rb and slightly decreased in Cb. The shrub encroachment was consistent among years in Ca, while there was a turning point at year 2013 in Ra, which indicates that shrub area was increasing slowly before 2013 and had a larger increasing rate after that. In 2010, the shrub area was 9% of the land area in Ca and 4% in Ra. By 2017, the shrub area had increased 12% in Ca and 4% in Ra. From 2010 to 2017, the number of patches and the area occupied by shrub (*NP* and *CA* in Table 2) increased in both Ca and Ra. However, patch fragmentation decreased, indicating that the patches were growing in size and beginning to cluster together. The linear regressions (r^2 = 0.94 and 0.85 for Ca and Ra, respectively) reveal that by the year

2030, if shrub encroachment is not addressed, potentially 24 ha (41%) of Ca may be occupied by shrubs, and that shrubs may account for 11.8 ha (15%) of the land area of Ra.

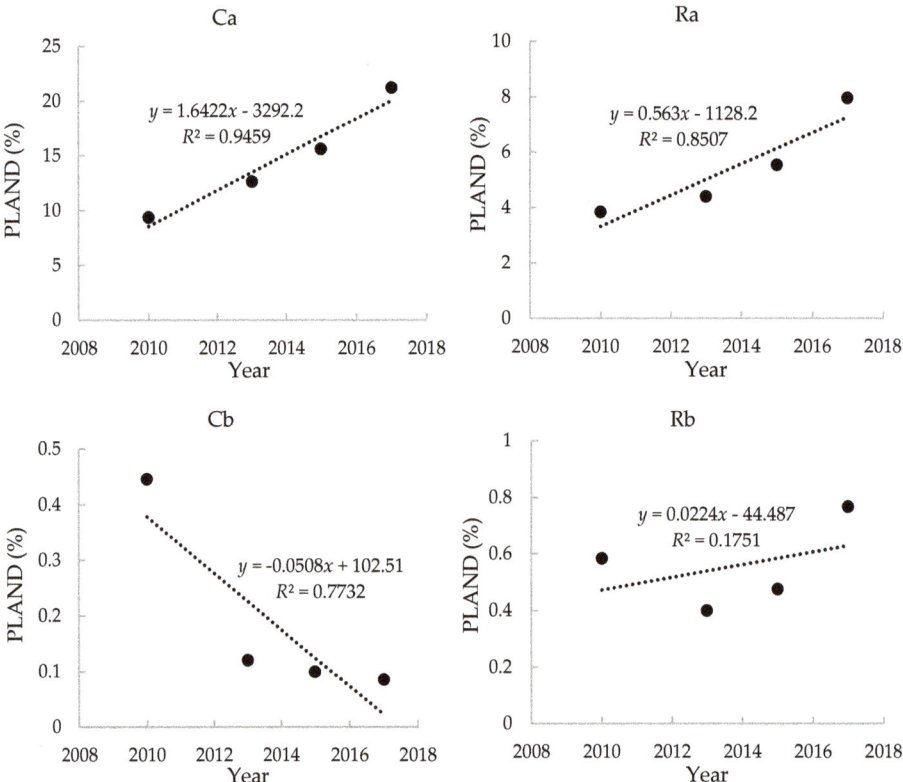

Figure 7. Shrub *PLAND* by year for each replicate (Ca, Ra, Cb and Rb).

As mentioned earlier, both the initial proportion (less than 1%) and dynamics of shrubs were small in Cb and Rb. Shrubs only occupied a minimal portion of Cb, and can almost be ignored after 2015. A turning point appeared in 2013, which means the shrub area changing rate decreased after 2013. In 2010, shrub *PLAND* was only ~0.5%, and this value was less than 0.1% by 2017. Patch fragmentation increased during 2010 to 2017. The linear regressions (r^2 = 0.77) implies that shrub will disappear in Cb after 2018. In pasture Rb, the shrub area decreased from 2010 to 2013, then increased after 2013. On the contrary, patch fragmentation increased first, then decreased after 2013. The increasing shrub was mainly contributed by newly growing small woody trees distributed around existing trees (Figure 5).

4. Discussion

Grass interannual dynamics were jointly controlled by climate conditions and grazing management. In 2013, worse grass conditions were caused by severe drought in 2011 and 2012, although the region had good temperature (T) and precipitation (P) (Figure S2). However, Zhou et al. [48] indicate that grass productivity in 2013 was high, which means grass grew well after the imaging date. With good T and P in May and June 2015, grass coverage and NDVI of all pastures improved over 2013, and Rep A had the best grass condition during the study period. However, grass condition of Rb in 2015 was not better than in 2010, with less high-coverage area and a lower NDVI; this is probably because Rep B was not grazed in 2010. The climate was similar in years 2017 and 2015, but the imaging date was a half-month earlier in 2017. Thus, there was higher percentage of low-coverage grass in 2017 than in 2015, due to a shorter growing period until image acquisition date in the former. Rb had highest medium coverage area, because of most sub-paddocks were rested.

Shrub encroachment rates between years indicated that climate was a key factor controlling shrub encroachment. In harsh climate conditions, shrub might die. Because of previous consecutive drought in 2011 and 2012, the shrub encroachment rate in Rep A was the slowest from 2010 to 2013; meanwhile, shrub area decreased in Rep B. After that, the climate conditions were less harsh, and the shrub encroachment was faster in Rep A. The shrub distribution in Rb also increased after 2013. In contrast, the shrub area continued to decrease in Cb at a much lower rate, probably because the area was too small and isolated distribution made it easy to be destroyed by cattle. Initial shrub area and distribution in pastures was another key factor in determining shrub encroachment. Rep A had large shrub encroachment, with a high proportion of shrub area and small patch fragmentation at the beginning of the experiment. Rep B, which had a small number of shrub area and high patch fragmentation at the beginning of the experiment, showed little change in shrub distribution.

Our results showed that grazing management affected grass coverage and is not a significant factor in controlling shrub encroachment. However, since NAIP images were snapshots in the early growing season, they cannot represent the grass condition for the whole grazing season. Thus, it is hard to evaluate how the grazing management affected grass production between years, because of low temporal resolution of NAIP images. The high stocking intensity created bare soil in some of paddocks in R pastures, which indicates that high grazing intensity in rotational grazing can cause the late growth of grasses. However, vegetation growth might be better later in the year and compensate for the damage caused by intensive grazing when grazing pressure is removed. As we found, Ca had better grass condition than Ra. Zhou et al. [48] studied grass productivity and did not show an advantage in Ca. More frequent observation is needed to better evaluate grazing management at annual scale. Soil type is also an important factor in affecting grass condition. Before grazing started, Rb had better grass condition than Cb. However, Cb had better grass condition than Rb in 2013 and 2015 after grazing, but also had the grass condition among pastures in 2017. Grass condition getting worse in Cb might be because of poor soil fertility, which cannot support recovery from grazing.

5. Conclusions

The comparison between paddocks in the same year showed that high grazing intensity in rotational grazing can cause the late growth of grasses. Ca always had better overall vegetation growth than Ra. Cb had a better vegetation growth than Rb at the beginning of the experiment (2013). However, their respective growth was similar in 2015, and Rb outperformed Cb in 2017, which indicates that soil fertility in Cb might not have been enough to support the grass recovery at current grazing pressure. The grass condition had large variation and no trend within study years. In the years with good T and P, grass recovered faster. To sum up, grass coverage was jointly affected by grazing management (stocking intensity) and climate.

Climate and initial status of shrub distribution (total area, percentage of landscape, fragmentation) instead of grazing management were the controlling factors of shrub encroachment. Higher initial proportion and better climate facilitated shrub encroachment. Shrub encroachment at a rapid speed in pasture Ca may cover 40% of the land area in 2030, and 15% of the land area in Ra. Scattered distribution of low-percentage shrub might even get suppressed. It is hard to conclude how grazing management affects shrub encroachment. Both high spatial and temporal resolution images are required to better monitor the tallgrass prairie landscapes.

Supplementary Materials: The following are available online at http://www.mdpi.com/2073-4395/9/5/238/s1. Figure S1: main plants in study area: (**a**) trees and shrubs, (**b**) shrubs, and (**c–f**) grasses; Figure S2: Monthly normal temperature and rainfall values from year 2009 to 2017: (**a**) temperature, (**b**) rainfall. This figure is an excerpt from Zhou et al. [1]; Table S1: NDVI value range for different classes; Table S2: PLAND and NDVI values of four grass categories in 201; Table S3: PLAND and NDVI values of four grass categories in 2013; Table S4: PLAND and NDVI values of four grass categories in 2015; Table S5: PLAND and NDVI values of four grass categories in 2017.

Author Contributions: Conceptualization: S.M., Y.Z., and P.H.G; methodology: S.M., Y.Z., P.H.G, and L.C.; formal analysis: S.M., Y.Z., and L.C.; investigation: S.M., Y.Z., P.J.S., and J.L.S.; data curation: S.M. and J.P.S.N.; writing—original draft preparation: S.M., L.C., and J.P.S.N.; writing—review and editing: Y.Z., P.H.G., J.L.S., P.J.S., and J.P.S.N.; visualization: S.M., Y.Z., P.H.G, P.J.S., and J.P.S.N.; funding acquisition: P.H.G, L.C., and J.L.S.

Funding: This study was supported in part by a research grant (Project No. 2013-69002) through the USDA-NIFA's (United States Department of Agriculture National Institute of Food and Agriculture) Agriculture and Food Research Initiative (AFRI).

Conflicts of Interest: The authors declare no conflict of interest.

References

1. Samson, F.B.; Knopf, F.L.; Ostlie, W.R. Great Plains ecosystems: past, present, and future. *Wildl. Soc.* **2004**, *32*, 6–15. [CrossRef]
2. Cunfer, G. *On the Great Plains: Agriculture and Environment*; Texas A&M University Press: College Station, TX, USA, 2005.
3. Fuhlendorf, S.D.; Engle, D.M. Restoring Heterogeneity on Rangelands: Ecosystem Management Based on Evolutionary Grazing Patterns: We propose a paradigm that enhances heterogeneity instead of homogeneity to promote biological diversity and wildlife habitat on rangelands grazed by livestock. *BioScience* **2001**, *51*, 625–632.
4. Westoby, M.; Walker, B.; Noy-Meir, I. Opportunistic Management for Rangelands Not at Equilibrium. *J. Range Manag.* **1989**, *42*, 266. [CrossRef]
5. Bidwell, T.; Elmore, D.; Hickman, K. Stocking Rate Determination on Native Rangeland. Available online: https://www.cattlemen.bc.ca/docs/factsheet_stocking_rate_determination.pdf (accessed on 25 March 2019).
6. Hubbard, W.A. Rotational Grazing Studies in Western Canada. *J. Range Manag.* **1951**, *4*, 25. [CrossRef]
7. Hyder, D.N.; Sawyer, W.A. Rotation-Deferred Grazing as Compared to Season-Long Grazing on Sagebrush-Bunchgrass Ranges in Oregon. *J. Range Manag.* **1951**, *4*, 30. [CrossRef]
8. McIlvain, E.H.; Savage, D.A. Eight-Year Comparisons of Continuous and Rotational Grazing on the Southern Plains Experimental Range. *J. Range Manag.* **1951**, *4*, 42. [CrossRef]
9. Rogler, G.A. A Twenty-Five Year Comparison of Continuous and Rotation Grazing in the Northern Plains. *J. Range Manag.* **1951**, *4*, 35. [CrossRef]
10. Sampson, A.W. A Symposium on Rotation Grazing in North America. *J. Range Manag.* **1951**, *4*, 19. [CrossRef]
11. Norton, B.E.; Barnes, M.; Teague, R. Grazing Management Can Improve Livestock Distribution. *Rangelands* **2013**, *35*, 45–51. [CrossRef]
12. Eldridge, D.J.; Soliveres, S.; Bowker, M.A.; Val, J. Grazing dampens the positive effects of shrub encroachment on ecosystem functions in a semi-arid woodland. *J. Appl. Ecol.* **2013**, *50*, 1028–1038. [CrossRef]
13. Heady, H.F. Continuous vs. Specialized Grazing Systems: A Review and Application to the California Annual Type. *J. Range Manag.* **1961**, *14*, 182. [CrossRef]
14. Teague, W.; Dowhower, S.; Baker, S.; Haile, N.; Delaune, P.; Conover, D. Grazing management impacts on vegetation, soil biota and soil chemical, physical and hydrological properties in tall grass prairie. *Agric. Ecosyst.* **2011**, *141*, 310–322. [CrossRef]
15. Teague, R.; Provenza, F.; Kreuter, U.; Steffens, T.; Barnes, M. Multi-paddock grazing on rangelands: Why the perceptual dichotomy between research results and rancher experience? *J. Environ. Manag.* **2013**, *128*, 699–717. [CrossRef] [PubMed]
16. Bryant, F.C.; Dahl, B.E.; Pettit, R.D.; Britton, C.M. Does short-duration grazing work in arid and semiarid regions? *J. Soil Water Conserv.* **1989**, *44*, 290–296.
17. Gillen, R.L.; Mccollum, F.T.; Tate, K.W.; Hodges, M.E. Tallgrass Prairie Response to Grazing System and Stocking Rate. *J. Range Manag.* **1998**, *51*, 139. [CrossRef]
18. Briske, D.D.; Derner, J.D.; Brown, J.R.; Fuhlendorf, S.D.; Teague, W.R.; Havstad, K.M.; Gillen, R.L.; Ash, A.J.; Willms, W.D. Rotational Grazing on Rangelands: Reconciliation of Perception and Experimental Evidence. *Rangel. Ecol. Manag.* **2008**, *61*, 3–17. [CrossRef]
19. Budd, B.; Thorpe, J. Benefits of Managed Grazing: A Manager's Perspective. *Rangelands* **2009**, *31*, 11–14. [CrossRef]
20. Briske, D.D.; Sayre, N.F.; Huntsinger, L.; Fernandez-Gimenez, M.; Budd, B.; Derner, J.D. Origin, Persistence, and Resolution of the Rotational Grazing Debate: Integrating Human Dimensions Into Rangeland Research. *Rangel. Ecol. Manag.* **2011**, *64*, 325–334. [CrossRef]

21. Becker, W.; Kreuter, U.; Atkinson, S.; Teague, R. Whole-Ranch Unit Analysis of Multipaddock Grazing on Rangeland Sustainability in North Central Texas. *Rangel. Ecol. Manag.* **2017**, *70*, 448–455. [CrossRef]
22. Danvir, R.; Simonds, G.; Sant, E.; Thacker, E.; Larsen, R.; Svejcar, T.; Ramsey, D.; Provenza, F.; Boyd, C. Upland Bare Ground and Riparian Vegetative Cover Under Strategic Grazing Management, Continuous Stocking, and Multiyear Rest in New Mexico Mid-grass Prairie. *Rangelands* **2018**, *40*, 1–8. [CrossRef]
23. Sampson, A.W. *Range Improvement by Deferred and Rotation Grazing*; Bulletin of the U.S. Department of Agriculture No. 34: Washington, DC, USA, 1913.
24. Leo, B.M. A Variation of Deferred Rotation Grazing for Use under Southwest Range Conditions. *J. Range Manag.* **1954**, *7*, 152–154.
25. White, M.R.; Pieper, R.D.; Donart, G.B.; Trifaro, L.W. Vegetational Response to Short-Duration and Continuous Grazing in Southcentral New Mexico. *J. Range Manag.* **1991**, *44*, 399. [CrossRef]
26. Jacobo, E.J.; Rodriguez, A.M.; Bartoloni, N.; Deregibus, V.A. Rotational Grazing Effects on Rangeland Vegetation at a Farm Scale. *Rangelands* **2006**, *59*, 249–257. [CrossRef]
27. Wang, T.; Teague, W.R.; Park, S.C. Evaluation of Continuous and Multipaddock Grazing on Vegetation and Livestock Performance—a Modeling Approach. *Rangel. Ecol. Manag.* **2016**, *69*, 457–464. [CrossRef]
28. Van Auken, O.W. SHRUB INVASIONS OF NORTH AMERICAN SEMIARID GRASSLANDS. *Annu. Ecol. Syst.* **2000**, *31*, 197–215. [CrossRef]
29. Knapp, A.K.; Briggs, J.M.; Collins, S.L.; Archer, S.R.; Ewers, B.E.; Peters, D.P.; Young, D.R.; Shaver, G.R.; Pendall, E.; Cleary, M.B.; et al. Shrub encroachment in North American grasslands: shifts in growth form dominance rapidly alters control of ecosystem carbon inputs. *Chang. Boil.* **2008**, *14*, 615–623. [CrossRef]
30. Angell, D.L.; McClaran, M.P. Long-term influences of livestock management and a non-native grass on grass dynamics in the Desert Grassland. *J. Environ.* **2001**, *49*, 507–520. [CrossRef]
31. Batáry, P.; Orci, K.M.; Báldi, A.; Kleijn, D.; Kisbenedek, T.; Erdős, S. Effects of local and landscape scale and cattle grazing intensity on Orthoptera assemblages of the Hungarian Great Plain. *Basic Appl. Ecol.* **2007**, *8*, 280–290. [CrossRef]
32. Schönbach, P.; Wan, H.; Gierus, M.; Bai, Y.F.; Müller, K.; Lin, L.J.; Susenbeth, A.; Taube, F. Grassland responses to grazing: Effects of grazing intensity and management system in an Inner Mongolian steppe ecosystem. *Plant Soil* **2011**, *340*, 103–115. [CrossRef]
33. Baldi, A.; Batary, P.; Kleijn, D. Effects of grazing and biogeographic regions on grassland biodiversity in Hungary—Analysing assemblages of 1200 species. *Agric. Ecosyst.* **2013**, *166*, 28–34. [CrossRef]
34. NAIP Imagery. Available online: https://www.fsa.usda.gov/programs-and-services/aerial-photography/imagery-programs/naip-imagery (accessed on 20 March 2018).
35. U.S. Department of Agriculture, Natural Resources Conservation Service. National soil survey handbook, title 430-VI. Available online: http://www.nrcs.usda.gov/wps/portal/nrcs/detail/soils/ref/?cid=nrcs142p2_054242 (accessed on 20 March 2019).
36. Steiner, J.L.S.; Neel, P.J.; Northup, B.K.; Gowda, P.H.; Brown, M.A.; Coleman, S. Managing, Tallgrass Prairies for Productivity and Ecological Function: A Long Term Grazing Experiment in the Southern Great Plains, USA. *Agronomy* **2019**. In press.
37. Tucker, C.J. Red and photographic infrared linear combinations for monitoring vegetation. *Remote. Sens. Environ.* **1979**, *8*, 127–150. [CrossRef]
38. Cortes, C.; Vapnik, V. Support-vector networks. *Mach. Learn.* **1995**, *20*, 273–297. [CrossRef]
39. Pal, M.; Mather, P.M. Support vector machines for classification in remote sensing. *Int. J. Sens.* **2005**, *26*, 1007–1011. [CrossRef]
40. DeFries, R.S.; Townshend, J.R.G. NDVI-derived land cover classifications at a global scale. *Int. J. Sens.* **1994**, *15*, 3567–3586. [CrossRef]
41. Carlson, T.N.; Ripley, D.A. On the relation between NDVI, fractional vegetation cover, and leaf area index. *Remote. Sens. Environ.* **1997**, *62*, 241–252. [CrossRef]
42. Geerken, R.; Zaitchik, B.; Evans, J.P. Classifying rangeland vegetation type and coverage from NDVI time series using Fourier Filtered Cycle Similarity. *Int. J. Sens.* **2005**, *26*, 5535–5554. [CrossRef]
43. Gutman, G.; Ignatov, A. The derivation of the green vegetation fraction from NOAA/AVHRR data for use in numerical weather prediction models. *Int. J. Sens.* **1998**, *19*, 1533–1543. [CrossRef]
44. Wagner, H.H.; Fortin, M.-J. Spatial Analysis of Landscapes: Concepts and Statistics. *Ecology* **2005**, *86*, 1975–1987. [CrossRef]

45. McGarigal, K. Landscape Pattern Metrics. Available online: https://www.umass.edu/landeco/pubs/mcgarigal.2002.pdf (accessed on 12 October 2018).
46. Fichera, C.R. Land Cover classification and change-detection analysis using multi-temporal remote sensed imagery and landscape metrics. *Eur. J. Sens.* **2012**, *45*, 1–18. [CrossRef]
47. McGarigal, K.; Marks, B.J. *Fragstats: Spatial Pattern Analysis Program for Quantifying Landscape Structure*; U.S. Department of Agriculture, Forest Service, Pacific Northwest Research Station: Corvallis, OR, USA, 1995; Volume 351, p. 122.
48. Zhou, Y.; Gowda, P.H.; Wagle, P.; Ma, S.; Neel, J.P.S.; Kakani, V.G.; Steiner, J.L. Climate Effects on Tallgrass Prairie Responses to Continuous and Rotational Grazing. *Agronomy* **2019**, *9*, 219. [CrossRef]

© 2019 by the authors. Licensee MDPI, Basel, Switzerland. This article is an open access article distributed under the terms and conditions of the Creative Commons Attribution (CC BY) license (http://creativecommons.org/licenses/by/4.0/).

Article

Assessment of the Standardized Precipitation and Evaporation Index (SPEI) as a Potential Management Tool for Grasslands

Patrick J. Starks *, Jean L. Steiner, James P. S. Neel, Kenneth E. Turner, Brian K. Northup, Prasanna H. Gowda and Michael A. Brown

United States Department of Agriculture-Agricultural Research Service, El Reno, OK 73036, USA; jean.steiner@hughes.net (J.L.S.); Jim.neel@ars.usda.gov (J.P.S.N.); Ken.turner@ars.usda.gov (K.E.T.); Brian.northup@ars.usda.gov (B.K.N.); Prasanna.gowda@ars.usda.gov (P.H.G.); michaelbrown@atlinkwifi.com (M.A.B.)
* Correspondence: Patrick.starks@ars.usda.gov; Tel.: +01-405-262-5291

Received: 15 March 2019; Accepted: 5 May 2019; Published: 9 May 2019

Abstract: Early warning of detrimental weather and climate (particularly drought) on forage production would allow for tactical decision-making for the management of pastures, supplemental feed/forage resources, and livestock. The standardized precipitation and evaporation index (SPEI) has been shown to be correlated with production of various cereal and vegetable crops, and with above-ground tree mass. Its correlation with above-ground grassland or forage mass (AGFM) is less clear. To investigate the utility of SPEI for assessing future biomass status, we used biomass data from a site on the Konza Prairie (KP; for years 1984–1991) and from a site at the United States Department of Agriculture-Agricultural Research Service's (USDA-ARS) Grazinglands Research Laboratory (GRL; for years 2009–2015), and a publicly-available SPEI product. Using discriminant analysis and artificial neural networks (ANN), we analyzed the monthly timescale SPEI to categorize AGFM into above average, average, and below average conditions for selected months in the grazing season. Assessment of the confusion matrices from the analyses suggested that the ANN better predicted class membership from the SPEI than did the discriminant analysis. Within-site cross validation of the ANNs revealed classification errors ranging from 0 to 50%, depending upon month of class prediction and study site. Across-site ANN validation indicated that the GRL ANN algorithm better predicted KP AGFM class membership than did the KP ANN prediction of GRL AGFM class membership; however, misclassification rates were ≥25% in all months. The ANN developed from the combined datasets exhibited cross-validation misclassification rates of ≤20% for three of the five months being predicted, with the remaining two months having misclassification rates of 33%. Redefinition of the AGFM classes to identify truly adequate AGFM (i.e., average to above average forage availability) improved prediction accuracy. In this regard, results suggest that the SPEI has potential for use as a predictive tool for classifying AGFM, and, thus, for grassland and livestock management. However, a more comprehensive investigation that includes a larger dataset, or combinations of datasets representing other areas, and inclusion of a bi-weekly SPEI may provide additional insights into the usefulness of the SPEI as an indicator for biomass production.

Keywords: forage management; above ground biomass; standardized precipitation and evaporation index (SPEI)

1. Introduction

Forage-based agricultural systems in the US Southern Plains (SP) states of Kansas, Oklahoma, and Texas are vulnerable to climate extremes, especially drought [1]. According to References [2–4], the beef cattle industry in these states accounted for ~$26 billion US in total agricultural sales in 2012. With respect to US cattle production, Texas, Kansas, and Oklahoma ranked 1, 3, and 5, respectively, in production of all cattle in 2017 [5]. Combined, these states accounted for 21.3 million head of cattle and contained 55 million ha of grass and range lands in 2012 [2–4]. As with most other agricultural systems, extremes in weather conditions can adversely impact production. For example, the drought of 2011 caused ~$7.6 US billion in agricultural losses in Texas alone [6]. Early warning of potential detrimental effects of weather and climate (particularly drought) on forage production would allow for tactical decision-making for the management of pastures and/or livestock (e.g., timely implementation of pasture rotations, adjustment of stocking density, advanced notice of possible needs of earlier supplemental feeding, or acquisition of additional forage resources).

Precipitation, soil water content, and evapotranspiration (ET) are key variables affecting crop yield and biomass production. The amount and timing of rainfall in relation to plant development clearly has an impact on crop yields and biomass production [7–9]. However, not all precipitation successfully infiltrates the soil, and that which does may percolate to depths beyond the reach of plant root systems. Many investigators have also shown that the amount of soil water greatly affects crop yield [10–12] and other grain crops and grasslands [13], but such measurements are not routinely collected and recorded at most locations.

Over the last several decades, numerous indices and indicators of drought have been developed to assess the onset/end, extent, and severity of drought. These indices or indicators may be classified as meteorological (e.g., the Palmer Drought Severity Index [14] and the Standardized Precipitation Index (SPI) [15]); may be designed to reflect soil moisture conditions (e.g., Soil Moisture Anomaly [16]) or the status of surface hydrology (e.g., the Standardized Streamflow Index [17]); or may be based on remotely sensed data (e.g., the Evaporative Stress Index [18]). A more complete listing of indicators may be found in References [19]. The standardized precipitation and evaporation index (SPEI, [20]), which builds upon the methodology used to develop the SPI, is a water-balance drought index based on the difference between precipitation (water supply) and potential evapotranspiration (ETp, water demand). It has been suggested that the SPEI may be functional for assessment of agricultural drought and, hence, biomass production [21,22]. Values of the SPEI may be positive (wet conditions) or negative (dry conditions), and are interpreted as the number of standard deviations away from the mean conditions. The SPEI is scalable over time periods of 1 to 48 months (or longer). For example, a 3 month SPEI value calculated for June (i.e., 3SPEI6) would represent the number of standard deviations (+/−) from the mean for the April–June time period. This scalability allows assessment of the cumulative effects of local weather/climate for time periods of various lengths.

The 1 month SPEI has been shown to closely follow the decline in soil water content, gross primary productivity, and ET measured for maize grown near Mead, NE, USA [23], and 1, 3, and 6 month timescale SPEIs were found to correlate well with biomass production for several different climate/vegetation regimes [24]. Several drought indices were evaluated by Reference [25], and the revised SPEI with crop-specific potential ET was found to follow the trends and magnitudes of water demand for several irrigated summer crops grown in the Texas High Plains in the USA. A linear regression model, based on a time-series of SPEI (April through September) values, accounted for ≈59% of the variability in yield for several vegetable crops for a study site in the Czech Republic [26]. This study was followed by another assessment [27], and it was shown that the SPEI, computed at various time scales, was correlated with productivity for 11 different crops grown in the Czech Republic. These investigators [27] found that the correlations between the SPEI and crop productivity were different for each crop and for different SPEI timescales, with the greatest correlation (r) achieved for cereal crops ($r = 0.52$–0.60; for a monthly timescale SPEI for the months May and June). Above-ground deciduous tree biomass for Mediterranean forests correlated significantly ($r = 0.55$ and 0.67, respectively) with the September and December SPEI [28] for Mediterranean forests, and it was shown that the interannual variability of above-ground tree (evergreen) mass in the Swiss Alps was also correlated with SPEI [29]. In terms of grassland above-ground mass, correlation has been reported between the 3, 6, 12, and 24 month timescale SPEI and grassland production in China [30], and it was shown that the monthly scale SPEI for July through September was better related to biomass production for grasslands and shrublands than SPEIs at other timescales [31]. However, Reference [32] assessed the SPEI's sensitivity to drought for six grasslands in the central US, and found no statistically significant relationships. These investigators also indicated that legacy effects of the previous (calendar) year's rainfall contributed little to biomass production in the following year. This lack of impact of the previous year's rainfall on the following year's biomass production was also noted for grasslands in central Oklahoma, USA [33,34].

Given the inconsistent results noted in the literature, and the probability of non-significant legacy effects of the previous calendar year's rainfall on grassland biomass production, we further examined the relationship of SPEI with grassland production in the Southern Plains of the US. More specifically, our objectives were: (1) to determine if the monthly-scale SPEIs obtained in the early part of the calendar year can be used to classify above-ground forage biomass (AGFM) into above average, below average, and average conditions for months later in the grazing season, and (2) to determine if successful assessment at one location can be directly applied to another location with similar forage and climatic conditions. If successful, application of the SPEI in this way could provide an early warning of probable shortfalls in forage production for grassland and livestock managers, and may be useful as a grassland management tool.

2. Materials and Methods

2.1. Study Sites

The study was conducted using data collected at Konza Prairie (KP) Long-Term Ecological Research site near Manhattan, Kansas, USA (39°06′0.38″ N, 96°36′29.5″ W), and at the US Department of Agriculture's Agricultural Research Service Grazinglands Research Laboratory (GRL), located in El Reno, Oklahoma, USA (35°33′29″ N, 98°01′50″ W) (Figure 1). The two sites are similar in that they both belong to the Level 1 (Great Plains) and Level 2 (South Central Semi-Arid Prairie) ecoregions, as defined by the US Environmental Protection Agency (https://www.epa.gov/eco-research/ecoregions-north-america). Additionally, both sites fall within the Koeppen–Geiger (http://koeppen-geiger.vu-wien.ac.at/usa.htm) Cfa climate classification (warm temperate, fully humid, hot summer). The 30 year (1981–2010) annual average ("normal") rainfall at the KP is ~904 mm, and ~913 mm at the GRL. The normal maximum air temperatures at the KP and GRL in January and July are 5 and 33.3, and 9.0 and 34.1 °C, respectively (data available at https://www.ncdc.novv.gov/normalsPDFaccess/). Warm season perennial grasses dominate both sites, with the most abundant being big bluestem (*Andropogon gerardii* Vtiman), little bluestem (*Schizacyrium scoparium*, (Michx.) Nash), Indiangrass (*Sorghastrum nutans* (L.) Nash), and switchgrass (*Pancium virgatum*, L.). The terrain at the KP is hilly, but is gently rolling at the GRL.

Established as a research site in 1971, the KP is 3487 ha in size and is the largest area of unplowed prairie in North America. The KP has been divided into numerous watershed units which receive various management treatments (e.g., grazed vs. ungrazed; prescribed burn frequency, patch burning), and provides a rich data source (including AGFM) related to tallgrass prairies. (See http://lter.konza.ksu.edu/ for a more complete site description and for data access and collection protocols.)

The AGFM data from the GRL was collected during a grazing study initiated in 2009 on 346 ha of native tallgrass prairie grasses, divided into two control paddocks of ~63 ha each and two rotational units of ~83 ha each. The control paddocks were assigned to continuous year-round stocking. The rotational units were divided into 10 approximately equally-sized sub-paddocks. The cattle in these sub-units were moved on a weekly basis. (For specifics on experimental design, stocking rates, etc. see Reference [34].) From 1949 to 2008, the site was primarily subjected to year-round continuous stocking by beef cattle.

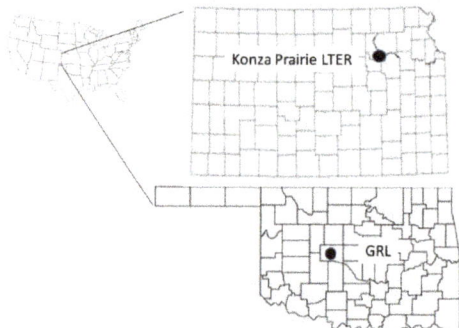

Figure 1. Study sites located at the Konza Prairie Long-Term Ecological Research (LTER) site near Manhattan, Kansas, and the United States Department of Agriculture-Agricultural Research Service's Grazinglands Research Laboratory (GRL) in El Reno, Oklahoma, USA.

2.2. Above-Ground Forage Mass

For the purposes of this study, and to minimize the confounding effects of grazing on AGFM classification, the AGFM data used herein reflect ungrazed conditions.

The AGFM data for the KP were acquired from their website (http://lter.konza.ksu.edu/data) from the dataset labeled PAB02 [35]. The AGFM data selected from PAB02 were collected on watershed 000a (KP000a), and represent bi-weekly harvests for the years 1984–1991 and for the months May through September. The samples were collected from five 0.1 m^2 areas (clipped to ground level) spaced 10 m apart along four transects (n = 20 per watershed per sampling date) and reflect ungrazed and unburned conditions. To ensure that areas were not re-sampled, subsequent sampling occurred "one step" away from and parallel to the previously sampled transect. The AGFM data used from the KP were provided on a g m^{-2} on a dry weight basis. The bi-weekly samples (n = 40) for a given month and year were averaged and converted to kg ha^{-1} to represent the AGFM for that month within that year (one value for each month for each year), yielding eight biomass observations per month.

The AGFM data for the GRL were collected from 2009 through 2015. Within the continuously stocked pastures, biomass samples were collected semi-monthly (May through September) at four random locations in areas visually identified as not having been grazed within the May through September timeframe of the current calendar year. Within the rotational units, biomass was collected (May through September) at two random locations within and prior to (or at the time of) the introduction of the cattle into the sub-units. All biomass samples were collected within a sampling area measuring 0.5 m^2 and were harvested to within 1 cm of the soil surface. The samples were weighed fresh, then dried in a forced-air oven at 65 °C for 48 h, then weighed dry to determine dry AGFM. The biomass measurements for a given month and year from both the continuously stocked and rotational sub-units were averaged to provide a monthly value for each year of study. This averaging resulted in seven biomass observations per month (one observation for each month for each of the seven years of study).

The monthly AGFM data from KP000a and GRL were tested for normality using the Shapiro–Wilk's W-test [36] (α = 0.05). Those data not meeting the normality test were further examined for outliers using box plots, and, if found, were removed from the dataset and re-analyzed for normality. Any remaining non-normal AGFM distributions were transformed using a Johnson Sb transformation [37]. Normally distributed monthly biomass values were assigned to the appropriate AGFM class based on the upper and lower 95% confidence intervals of the respective monthly means. Non-normally distributed monthly biomass values were first transformed using the Johnson Sb transformation, and then assigned to the appropriate AGFM class based on the upper and lower 95% confidence intervals of the mean of the transformed data. In all cases, the monthly values above the upper 95% confidence interval were designated as "above average" (abv), those biomass values below the lower 95% confidence interval were designated "below average" (blw), and all other values were designated "average" (avg).

Assuming that the month of September represents peak biomass, within-site annual variability in peak biomass was examined via plots of biomass and Tukey's Honestly Significant Difference (HSD, $\alpha = 0.05$) [38] means test. Monthly variation in biomass values is depicted using box plots.

2.3. Standardized Precipitation and Evaporation Index (SPEI)

The SPEI was developed by Reference [20], and is a standardized value of the difference (D_i) of precipitation (P_i) and potential evapotranspiration (ETp_i) for month i:

$$D_i = P_i - ETp_i \tag{1}$$

The D_i values can be aggregated over different timescales, however the D_i are first standardized from a three-parameter log–logistic distribution:

$$f(x) = \frac{\beta}{\alpha}\left(\frac{x-\gamma}{\alpha}\right)^{\beta-1}\left(1+\left(\frac{x-\gamma}{\alpha}\right)^{\beta}\right) \tag{2}$$

The parameters α, β, and γ are calculated from probability-weighted moments (PWM) and used to determine the log–logistic distribution, which is applied to the D_i dataset. The SPEI is then calculated as the probability of exceeding a given value of D_i (W = $-2 \times \ln(P)$):

$$SPEI = W - \frac{C0 + C1W + C2W^2}{1 + d1W + d2W^2 + d3W^3} \tag{3}$$

where C0, C1, C2, d1, d2, and d3 are constants. See References [19,35] for details.

For our study locations, we used the 0.5 degree gridded SPEI product, which was obtained from the Global SPEI database (https://climatedataguide.ucar.edu/climate-data/standardized-precipitation-evapotranspiration-index-spei). Latitude and longitude of the approximate centroid of each site were entered into the user interface, which returns the SPEI values for the grid point nearest to the input coordinates. In this dataset, SPEI is calculated using monthly precipitation and ET_p data provided by the Climatic Research Unit of the University of East Anglia, East Anglia, UK. For this product, ET_p is based on the FAO-56 Penman–Monteith equation [39] (http://spei.csic.es/database.html). Although the delivered product provides grid-specific SPEIs at timescales between 1 and 48 months, we only used the 1 month timescale.

Rather than calculate the SPEIs for each site, we chose to use this readily-available product for our study. A publicly-available SPEI represents the type of product that could be easily accessible and implemented for forecasting AGFM.

2.4. Statistical Analysis

We analyzed the ability of the 1 month SPEIs to be used as predictors for classifying AGFM into the abv, avg, and blw categories using discriminant analysis and artificial neural networks (ANN). Discriminant analysis is a multi-variate linear technique which seeks to separate observations into distinct classes based on a set of predictor variables (in this case, the SPEI values), and to assign new observations into previously defined groups [40]. All SPEI values for the months January through September, inclusive, were used to determine which months, if any, were most advantageous for AGFM classification. The objective functions used to evaluate the discriminant results included the overall misclassification rates and the confusion matrices. We used discriminant analysis and ANN in four ways: (1) we evaluated each site separately, by month, using the complete data; (2) we then separated the data for each site into calibration and validation datasets to test the calibrated discriminant function on data not used in the calibration phase; (3) we applied the site-specific discriminant and

ANN functions developed for a given site in (1) to the other site to examine the transferability of the discriminant and ANN classification algorithm across sites; and (4) we combined the two datasets, from which we created calibration and validation datasets to determine if the accuracy of the resulting prediction algorithms increased. In (2) and (4), the validation datasets were based upon a random selection of the dataset that was first stratified according to the AGFM classes. This approach yielded 75% of the observations assigned to the calibration datasets, and the remaining 25% were assigned to the validation datasets. Misclassification rates for a given month were calculated by dividing the number of misclassified observations by the total number of observations for that month.

Artificial neural networks are non-linear, non-statistical mathematical models that mimic the learning process of the human brain [41]. The ANN does not assume normality of the data, nor is a minimum amount of data specified to develop a predictive algorithm. However, the number of hidden nodes, which provides the mathematical connection between the input neurons (in this case the monthly SPEI values) and the output neurons (one neuron in this case, the AGFM class), is constrained either as a function of the number of records in the dataset, or the number of input and output neurons, or a combination of these. For this study, the number of input neurons equaled four and the number of output neurons equaled one for both datasets. The number of records per month varied from six to eight, depending upon site and month. We set the number of hidden neurons to two (half of the sum of number of inputs and outputs), in accordance with Reference [41]. The ANN predictive algorithm is developed through weighting the neurons in the hidden layer, which is located between the input neurons and the output neurons. At the outset of running the ANN, every input neuron is connected to every hidden neuron, and every hidden neuron is connected to every output neuron. As the data records are read and evaluated by the ANN, the weights of the hidden neurons are adjusted so that inputs are associated as strongly as possible with the intended output neuron. The reading and evaluating of the dataset is done iteratively until the prediction ability of the ANN is maximized. The ANN used in this study is described as a fully connected multi-layer perceptron. We used the hyperbolic tangent transformation function between the input neurons and the hidden neurons. All analyses were performed in JMP 14 Pro (SAS Institute, Cary, NC, USA).

3. Results

3.1. Above-Ground Forage Mass

No outliers were observed in the AGFM datasets. The Shapiro–Wilks W-test indicated that monthly biomass was normally distributed at KP000a; thus, the biomass data needed no transformation before assigning the AGFM values to the appropriate AGFM classes. However, the Shaprio–Wilks W-test indicated that the GRL biomass data for all months was not normally distributed. Therefore, these data were transformed before assignment to the appropriate AGFM class.

Assuming that September approximates peak biomass, it was observed (Figure 1) that AGFM varied by ~388 kg ha^{-1} at KP000a, which is about 11 times smaller than the range in values observed at GRL. The AGFM values at KP000a were low compared to what has been reported (1500 to 5000 kg ha^{-1}) for other locations on the KP [42]. The cause of this low biomass production is not precisely known, but may relate to its not having been burned or grazed for many years, thus reducing the nutrient cycling necessary for plant growth. At KP000a, the highest observed AGFM occurred in 1991 (1265 kg ha^{-1}) and the lowest in 1989 (886 kg ha^{-1}). At GRL, the AGFM was 7230 kg ha^{-1} in 2015 (Figure 2), which was statistically higher than all other years. The years 2009 and 2012 exhibited the lowest peak AGFM (~2800 kg ha^{-1}). It was also observed that considerable variation occurred in May at KP000a, while the variability was comparatively low and similar for the remaining months (Figure 3). AGFM at GRL was highly variable for all months.

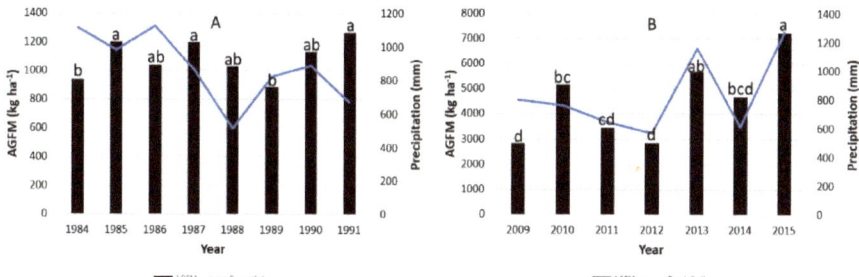

Figure 2. September (peak) above-ground forage mass by year for sites KP000a (**A**) and GRL (**B**). Above-ground grassland or forage mass (AGFM) levels within sites not connected by the same letter are statistically different. Annual rainfall is shown as the blue line.

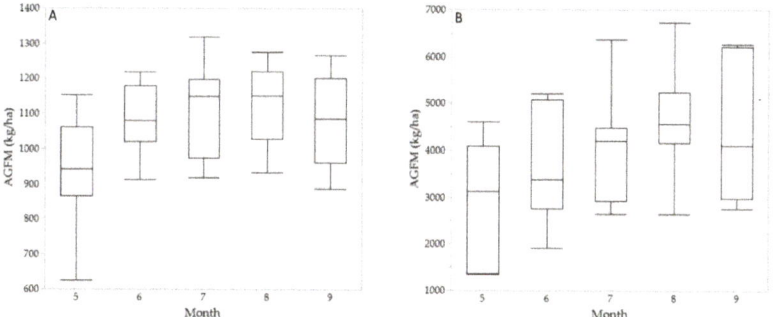

Figure 3. Box plots of AGFM for month of year for KP000a (**A**) and GRL (**B**). The number of years represented in the KP000a and GRL datasets is eight and seven, respectively.

3.2. Discriminant Analysis

Analysis of the entire site-specific datasets revealed that the 1 month timescale SPEIs representing months January through April (i.e., 1SPEI1–1SPEI4) were useful in differentiating AGFM classes at both sites (Table 1) for the May through September time periods. For this time period, only the avg class exhibited any misclassification error (17%) for the KP000a site, whereas only the abv category exhibited misclassification errors (33%) at the GRL site. Overall, adding SPEI values for months occurring after April did not substantially change the misclassification rate for site KP000a, but it did increase the misclassification rate at the GRL site, where the abv AGFM category was impacted more than the others. In Table 1, note that we are using the 1 month scale SPEIs up to, but not including, the month for which we are trying to predict AGFM classes. Thus, sequentially adding SPEI values for the months following April (i.e., 1SPEI4), produces a commensurate decrease in the number of months than can be evaluated. Given the results noted immediately above, we used only the 1SPEI1–1SPEI4 values in the following analysis to build and evaluate both the discriminant and ANN models.

At KP000a, it was observed (Table 2) that misclassifications ≥ 50% occurred for all months except September (0%). At the GRL site, the misclassification rates were 50% (Table 2) for all months except September, which exhibited no misclassifications. Application of the GRL discriminant function to KP000a resulted in misclassification errors ≥75% for all months (Table 3). Similarly, application of the KP000a discriminant function to the GRL site resulted in classification errors ranging from 43% in August to 86% in July. Results from combining the KP000a and GRL datasets led to calibration misclassification rates of 0% for May and June, and from 30 to 36% for the remaining months.

The misclassification rates for the validation dataset ranged from 40 to 100% for all months except September, which had a 0% misclassification rate.

Table 1. Misclassification rate results from the site-specific discriminant analysis of the 1 month time scale standardized precipitation and evaporation indices (SPEIs) and AGFM classifications for the Konza Prairie (KP000a) and GRL research sites. AGFM classifications are: abv = above average, avg = average, and blw = below average.

KP000a		Misclassification Rate [1]			GRL		Misclassification Rate		
SPEI Months	AGFM Month	abv	avg	blw	SPEI Months	AGFM Month	abv	avg	blw
1–4	May	0	17	0	1–4	May	0	0	0
	Jun	0	0	0		Jun	0	0	0
	Jul	0	0	0		Jul	0	-	0
	Aug	0	0	0		Aug	0	0	0
	Sep	0	0	0		Sep	33	0	0
1–5	Jun	0	0	0	1–5	Jun	0	0	0
	Jul	0	0	0		Jul	0	-	0
	Aug	0	0	0		Aug	33	0	0
	Sep	0	0	0		Sep	33	0	33
1–6	Jul	0	0	0	1–6	Jul	33	-	0
	Aug	0	0	0		Aug	33	50	0
	Sep	0	0	0		Sep	33	0	0
1–7	Aug	0	0	0	1–7	Aug	33	50	50
	Sep	0	0	0		Sep	33	0	0
1–8	Sep	0	25	0	1–8	Sep	66	0	0

[1] Percentage (%).

Table 2. Discriminant analysis within-site validation confusion matrices showing actual and predicted class membership for the months May through September for both sites. AGFM classes are above average (abv), average (avg), and below average (blw) above-ground forage biomass. The total misclassification rate (TMR) for each month is also shown, and is calculated as the number of misclassifications divided by the total number of observations.

		May			June			July			August			September		
	Actual	abv	avg	blw	abv	avg	blw	abv	avg	blw	abv	avg	blw	abv	avg	blw
Site = KP000a	abv	0	0	0	0	1	1	0	0	0	0	0	0	0	0	0
	avg	0	0	1	0	0	0	1	0	1	0	0	1	0	1	0
	blw	0	0	0	0	0	0	0	0	0	0	1	1	0	0	0
	TMR [1]	100			50			100			66			0		
Site = GRL	abv	0	1	1	0	0	0	0	0	0	0	1	1	3	0	0
	avg	0	0	0	1	0	0	0	0	0	0	0	0	0	1	0
	blw	0	0	0	0	0	1	1	0	1	0	0	0	0	0	3
	TMR	50			50			50			50			0		

[1] Percentage (%).

3.3. Neural Networks

The KP000a ANN calibrated well for all months except September (Table 4). Within-site cross-validation resulted in a 0% misclassification rate for May, July, and August, but a 33 and 50% misclassification rate for June and September, respectively (Table 4). The GRL ANN calibrated well for the months May, June, and August, but exhibited a 40 and 67% misclassification rate for July and September (Table 5). The GRL ANN cross-validation indicated a 33 and 50% misclassification rate for May and June, respectively, and 0% misclassification rates for the remaining months. Application of GRL's ANN algorithm to site KP000a resulted in misclassification rates ranging from 25% for May to 38% for June through August (Table 6). Application of KP000a's ANN algorithm to the GRL site produced misclassification rates ranging from 33% in May to 57% in June. The monthly average misclassification rate for GRL prediction of KP000a was ~34%, compared to ~44% where the KP000a ANN algorithm was applied to GRL. Results from combining the KP000a and GRL datasets led to calibration misclassification rates of 31% for July to 89% for June (Table 7). Although the ANN calibrated poorly, the validation misclassification rates were 0% for July and September, 20% for May, and 33% both June and August (Table 7).

Table 3. Discriminant analysis confusion matrices from cross-site prediction showing actual and predicted class membership for the months May through September for both sites. AGFM classes are above average (abv), average (avg), and below average (blw) above-ground forage biomass. The total misclassification rate (TMR) for each month is also shown, and is calculated as the number of misclassifications divided by the total number of observations. Site KP000a class membership predicted from GRL discriminant analysis algorithm, GRL class membership predicted from site KP000a discriminant analysis algorithm.

		May			June			July			August			September		
	Actual	abv	avg	blw	abv	avg	blw	abv	avg	blw	abv	avg	blw	abv	avg	blw
Site KP000a Predicted by GRL	abv	0	1	0	0	0	2	0	1	0	1	1	1	0	0	1
	avg	3	0	3	1	1	3	3	2	0	0	1	2	3	0	1
	blw	0	0	1	0	1	0	1	1	0	2	0	0	0	0	1
	TMR[1]		88			88			75			75			83	
Site GRL Predicted by KP000a	abv	1	1	1	1	1	1	1	2	1	2	0	1	0	3	0
	avg	0	1	0	0	0	1	0	0	0	1	1	0	0	0	1
	blw	0	2	0	1	2	1	2	1	0	0	1	1	0	1	2
	TMR		67			75			86			43			71	

[1] Percentage (%).

Table 4. Confusion matrices from artificial neural network (ANN) calibration and cross-validation, showing actual and predicted class membership for the months May through September for site KP000a. AGFM classes are above average (abv), average (avg), and below average (blw) above-ground forage biomass. The total misclassification rate (TMR) for each month is also shown, and is calculated as the number of misclassifications divided by the total number of observations.

		May			June			July			August			September		
	Actual	abv	avg	blw	abv	avg	blw	abv	avg	blw	abv	avg	blw	abv	avg	blw
ANN Calibration for Site = KP000a	abv	1	0	0	1	0	0	1	0	0	1	0	0	0	1	0
	avg	0	5	0	0	3	0	0	5	0	0	3	0	0	2	1
	blw	0	0	1	0	0	1	0	0	1	0	0	2	0	0	0
	TMR[1]		0			0			0			0			50	
Cross-Validation for Site = KP000a	abv	0	0	0	0	1	0	0	0	0	1	0	0	0	0	0
	avg	0	1	0	0	2	0	0	0	0	0	1	0	0	1	0
	blw	0	0	0	0	0	0	0	0	1	0	0	0	0	1	0
	TMR		0			33			0			0			50	

[1] Percentage (%).

Table 5. Confusion matrices from ANN calibration and cross-validation, showing actual and predicted class membership for the months May through September for the GRL site. AGFM classes are above average (abv), average (avg), and below average (blw) above-ground forage biomass. The total misclassification rate (TMR) for each month is also shown, and is calculated as the number of misclassifications divided by the total number of observations.

		May			June			July			August			September		
	Actual	abv	avg	blw	abv	avg	blw	abv	avg	blw	abv	avg	blw	abv	avg	blw
ANN Calibration for Site = GRL	abv	2	0	0	2	1	0	3	–	0	2	0	0	2	1	0
	avg	0	1	0	0	1	0	–	–	–	0	2	0	0	0	0
	blw	0	0	1	0	0	2	2	0	0	0	0	2	0	3	0
	TMR[1]		0			17			40			0			67	
Cross-Validation for Site = GRL	abv	2	0	0	0	0	0	0	–	1	1	0	0	0	0	0
	avg	0	0	0	0	0	0	–	–	–	0	0	0	0	1	0
	blw	0	1	0	0	0	1	0	0	1	0	0	0	0	0	0
	TMR		33			0			50			0			0	

[1] Percentage (%).

Table 6. Confusion matrices from ANN calibration at one site applied to the other site, showing actual and predicted class membership for the months May through September for the GRL site. AGFM classes are above average (abv), average (avg), and below average (blw) above-ground forage biomass. The total misclassification rate (TMR) for each month is also shown, and is calculated as the number of misclassifications divided by the total number of observations.

		May			June			July			August			September		
	Actual	abv	avg	blw	abv	avg	blw	abv	avg	blw	abv	avg	blw	abv	avg	blw
GRL ANN Prediction of Site KP000a	abv	1	0	0	2	0	0	1	0	1	1	1	0	0	1	0
	avg	1	5	0	2	3	0	1	4	0	0	4	0	0	4	0
	blw	1	0	0	0	1	0	1	1	0	0	2	0	0	1	0
	TMR [1]	25			38			38			38			33		
Site KP000a ANN Prediction of GRL	abv	3	0	0	3	0	0	4	0	0	3	0	0	2	1	0
	avg	0	1	0	1	0	0	0	0	0	1	1	0	0	0	1
	blw	1	1	0	2	1	0	3	0	0	1	1	0	1	0	2
	TMR	33			57			43			43			43		

[1] Percentage (%).

Table 7. Confusion matrices from ANN calibration and cross-validation, showing actual and predicted class membership for the months May through September for the GRL site. AGFM classes are above average (abv), average (avg), and below average (blw) above-ground forage biomass. The total misclassification rate (TMR) for each month is also shown, and is calculated as the number of misclassifications divided by the total number of observations.

		May			June			July			August			September		
	Actual	abv	avg	blw	abv	avg	blw	abv	avg	blw	abv	avg	blw	abv	avg	blw
ANN Calibration for Combined Site Datasets	abv	1	1	1	1	1	1	2	0	2	1	0	3	1	1	1
	avg	1	3	0	3	0	0	1	4	0	0	4	1	1	3	1
	blw	0	2	0	2	1	0	1	0	3	1	1	1	0	0	3
	TMR [1]	56			89			31			50			36		
ANN Cross-Validation for Combined Site Datasets	abv	1	0	0	2	0	0	1	0	0	0	1	0	1	0	0
	avg	0	3	0	0	2	1	0	0	0	0	1	0	0	0	0
	blw	0	1	0	1	0	1	0	0	1	0	0	1	0	0	1
	TMR	20			33			0			33			0		

[1] Percentage (%).

The SPEIs with the highest absolute weightings on the two hidden neurons (Hn1, Hn2) during calibration were tabulated (Table S1) with respect to the scenarios presented in Tables 4–7. Overall, and for Hn1, 1SPEI1 and 1SPEI3 each accounted for 36% of the highest weightings, followed by 1SPEI2 (16%), and 1SPEI4 (12%). For Hn2, SPEI3 accounted for 56% of the highest weightings, followed by 1SPEI2 (32%), 1SPEI4 (8%), and 1SPEI1 (4%). On a month-by-month basis for Hn1, 1SPEI2 accounted for 60% of the highest weightings in May. For July and September, 1SPEI3 accounted for 60 and 80% of the highest weightings, whereas 1SPEI1 accounted for 60% of the highest weightings in August. In June, both 1SPEI1 and 1SPEI4 accounted for 40% of the hieghest weightings. For Hn2, 1SPEI3 accounted for 60% of the highest weightings in June through and August, whereas 1SPEI2 accounted for 60% of the highest weightings in September. 1SPEI3 and 1SPEI4 each accounted for 40% of the highest weightings in May.

4. Discussion

4.1. Study Context

Estimating forage mass or amount of standing forage in pastures and grasslands is important to allow decisions for agronomic management and limitations to livestock stocking to be made. The determination/measurement of forage mass reflects the existing growing conditions. Reference [43] reported that scaling of spring precipitation from the long-term average precipitation provided useful information for predicting peak standing crop for destocking decisions. They also reported that grazing did not affect the relationship between peak standing crop prediction of mixed-grass prairie and precipitation. Reference [44] used soil water content at the beginning of the growing season to improve predictions of peak standing crop from the Great Plains Framework for Agricultural Resource

Management Range model in a mixed prairie. Reference [45] used annual precipitation and soil moisture to evaluate short- and long-term influences on above-ground biomass production in native grasslands, and Reference [46] used mean annual precipitation and drought to forecast reductions in above-ground biomass production. Reference [47] combined drought indices (1 and 3 month SPI) and remote sensing to estimate production of wheat and barley production in a semi-arid region of Spain. Prevailing patterns of biomass production in a tall-grass prairie site in the US Great Plains have also been estimated using the normalized difference vegetation index (NDVI) and integrated over time [47,48]. In the current study, the 1 month scale SPEI (a water-balance drought index) was assessed, via discriminant analysis and ANN, as a predictive tool for classes of AGFM in native grasslands of the southern Great Plains of the US.

4.2. Discriminant Analysis

The results in Table 1 suggest that when all the data for a given site were used, that the 1 month time scale SPEIs for the months January through April (i.e., 1SPEI1–1SPEI4) were useful in assigning AGFM classifications for the months of May through September. However, when the discriminant analysis was forced to cross-validate within site (Table 2), misclassification rates of 50% or more were observed for most months at both sites. Cross-site validation (i.e., the application of one site's predictive discriminant function using the other site's SPEI data) led to misclassification rates of ≥40% (Table 3). In both of these latter instances, these misclassification rates may be inflated, given the small n-size available at each site for cross-validation.

In discriminant analyses, classification can be problematic, since variance can cause classification groups to overlap [39]. Discriminant analysis is most effective when observations contain more variable information within the tails of the cumulative distribution function of the population, rather than within the upper and lower 95% confidence intervals of the mean. One potential driver of misclassification could be the effect of amount and timing of precipitation received during the growing season. There is variation in the timing and amount of precipitation from year to year, which has the capacity to affect the amount of biomass produced. Such issues are of particular relevance when considered in relation to amount and timing of growth by plant communities of native grasslands. One experiment near the GRL study site noted completely different rates of accumulation of above-ground biomass during two years that related to variance in precipitation received [34]. In that study, there was a shift in the rate of biomass accumulation beginning in early May, caused by drier conditions during late May through August of one year. This drought-affected period received one third less total accumulated precipitation during March through August of the growing season, compared to the second year, though amounts received in January through April were similar during both years. The result of this difference in accumulation of precipitation was a roughly 50% reduction in total biomass by the end of the more drought-affected growing season.

4.3. Artificial Neural Network

According to Reference [40], it is better for an ANN to validate and perform well than to calibrate well. For within-site cross-validation, the ANN for the KP000a site (Table 4) both calibrated and validated well. The GRL ANN site validated better than it calibrated (Table 5). Within the context of cross-site validation (Table 6) it was observed that the GRL ANN performed better when applied to site KP000a than when the site KP000a ANN was applied to the GRL site. Given that the sites are similar in terms of general climate characteristics, and that both are native warm-season tallgrass prairies (and dominated by the same species of grass), this may indicate that the relationship between SPEI and AGFM categories is different at the two sites. (This would also impact discriminant analysis.)

Comparison of the maximum and minimum values of 1SPEI1–1SPEI4 for both sites reveals significant differences between the two sites (Table 8). For example, at site KP000a, the maximum value of 1SPEI1 was 1.85, compared to a value of 0.80 at GRL. However, for this same SPEI variable site KP000a experienced a minimum value of −0.89, which is 35% larger than that observed at GRL.

The minimum 1SPEI2 value at site KP000a indicates much drier conditions than were experienced at GRL, but the maximum values are similar. The 1SPEI3 and 1SPEI4 values are somewhat similar for the two sites. These findings suggest it may be possible to develop a more accurate ANN prediction algorithm by combining datasets from similar climatic and forage types that represent a wider range of conditions.

Table 8. Maximum, minimum, and range of 1 month timescale SPEI values 1SPEI1 through 1SPEI4 for sites KP000a and GRL.

Statistic	1SPEI1	1SPEI2	1SPEI3	1SPEI4
		Site = KP000a		
Maximum	1.85	1.14	0.99	1.95
Minimum	−0.89	−1.69	−1.66	−1.20
Range	2.74	2.83	2.65	3.15
		Site = GRL		
Maximum	0.80	1.09	1.3	1.49
Minimum	−1.37	−0.75	−1.42	−1.35
Range	2.17	1.84	2.72	2.84

Examination of the importance (via the weightings) of the SPEIs on the hidden nodes of the ANNs revealed that, overall, 1SPEI1 and 1SPEI3 played leading roles in the development of the ANN predictive algorithms, whereas 1SPEI4 was of much less importance.

4.4. AGFM and AGFM Classes

The two study sites are similar in terms of climate and vegetation. However, despite the large differences in biomass production observed at the two sites (Figure 1), the seasonal evolution of this production is similar—increases steadily from May to about August/September (Figure 3). Variability in AGFM is more pronounced at the GRL site, but sufficient variability existed at both sites to assign AGFM values to AGFM classes. The tacit assumption in this approach is that, although biomass production may vary between two sites in absolute terms, both sites should trend similarly in response to precipitation and evapotranspiration (ET). That is, a given value of AGFM at a site will represent average, above average, or below average conditions in response to the variation in precipitation and ET, as represented by the SPEI. Given that the two sites are similar in terms of climate and vegetation, the application of the discriminant analysis and ANN predictive algorithms of AGFM class developed from the GRL data to the KP000a site, and vice versa, is appropriate.

As noted above, the discriminant analysis approach did not perform well. In some cases it may also be difficult for ANN algorithms to differentiate between categories that closely overlap. Assuming that the abv and avg categories indicate adequate forage availability, then "true adequate," "false adequate," "false below adequate," and "true inadequate" forage availability categories can be constructed from confusion matrices. False adequate conditions represent predictions of adequate forage, when in reality a below adequate forage availability condition occurred, and false below adequate would represent predictions of below average forage availability, but average or better conditions actually occurred. Using the results in Table 6 and the validation portion of Table 7, it can be seen that the GRL ANN prediction algorithm applied to site KP000a correctly predicted adequate forage for 79% of the cases (over all months), incorrectly classified 8% of the cases as having adequate forage (i.e., false adequate forage condition), while 13% of the cases were incorrectly classified as having below adequate forage (i.e., false inadequate condition). (True inadequate conditions did not occur for this scenario.) When the KP000a ANN algorithm was applied to the GRL dataset, it correctly predicted true adequate forage conditions for 59% of the cases, 3% of the cases fell within the false inadequate condition, 32% of the cases fell within the false adequate condition, and 3% of the cases were correctly placed in the true inadequate category. The ANN developed from the combined datasets correctly predicted true adequate conditions for 67% of the cases, 6% of the cases fell within the false inadequate category, 11% of the cases fell within the false adequate category, and true inadequate conditions were correctly predicted for 22% of the cases. This approach might be useful for a grassland or livestock manager who requires only a basic forecast of forage conditions.

4.5. SPEI

In this study, a gridded SPEI product was used which represented a 50 km × 50 km area, which may not be representative of the conditions experienced at the study sites. A SPEI calculated using local meteorological data, such as was done by Reference [25], might improve correlation with biomass production. Based on References [32,33], we assumed that previous calendar year precipitation had little to no effect on current year biomass production. The results reported in Table 1, and the ANN k-fold validation statistics reported in Tables 4 and 5 tend to support this assertion. However, as noted above, timing and amount of precipitation during the growing season can have a profound impact on biomass production. We speculate that a bi-weekly SPEI may provide the temporal resolution needed to improve prediction of AGFM. Moreover, relationships between SPEI and biomass production for the climatic area represented in this study will likely be different for areas having markedly different climate and forage (e.g., cool-season grasses) conditions.

5. Conclusions

This study investigated the potential use of early in the calendar year, 1 month scale SPEIs for the prediction of AGFM for the grazing months May through September of tallgrass prairies in the Southern Plains of the US. Initial results from the within-site discriminant analysis and ANN indicated that the January through April monthly timescale SPEIs could be used to predict AGFM class for the months May through September. However, within-site validation indicated that discriminant analysis led to high misclassification rates, whereas the ANN approach exhibited much lower misclassification rates. The predictive discriminant and ANN functions developed for one study site were not transferable to the other study site. This is likely due to two factors: (1) Neither site fully captured the variability in SPEI experienced by the other, and (2) The monthly timescale SPEI does not capture the timing of precipitation, which can have a profound impact on biomass production. A bi-weekly SPEI may capture this important aspect of biomass production. At present, the results presented herein suggest that the SPEI may be useful for predicting adequate AGFM (i.e., average to above average forage availability) conditions, but false positives are to be expected. A more comprehensive investigation that includes a longer time period of study, an increased number of study sites, and a bi-weekly SPEI would provide additional insight into the usefulness of the SPEI as an indicator for biomass production. If AGFM prediction can be based on the SPEI, then a prediction tool can be developed that could provide an early warning of probable shortfalls in forage production for grassland and livestock managers, and may be useful for improved grassland management.

Supplementary Materials: The following are available online at http://www.mdpi.com/2073-4395/9/5/235/s1, Table S1. Monthly scale SPEI exerting the most influence on the ANN hidden nodes 1 and 2 (Hn1, Hn2) by month. The SPEIs shown are for ANN calibrations for KP (upper portion of Table 4 in the main text), for GRL (upper portion of Table 5 in the main text), for GRL calibration and prediction of KP, for KP calibration and prediction of GRL (Table 6 in the main text), and for the calibration of the combined datasets (Table 7 in main text).

Author Contributions: P.J.S conceptualized the study, analyzed the data and wrote the manuscript; J.L.S. assisted in project design, was responsible for project administration, and edited the manuscript; J.P.S.N and K.E.T. assisted in data collection and edited the manuscript; B.K.N. and P.H.G. assisted in data analyses and statistical interpretation, and edited the manuscript; M.A.B. assisted in the experimental design and statistical interpretation, and edited the manuscript.

Funding: This research was partially supported by funds from the Agriculture and Food Research Initiative Competitive Projects 2012-02355 and 2013-69002-23146 from the USDA National Institute of Food and Agriculture.

Acknowledgments: This research is a contribution from the USDA-Agricultural Research Service's Long Term Agroecosystem Research (LTAR) network.

Conflicts of Interest: The authors declare no conflict of interest. The funder noted above had no role in the design of the study, in data collection, analyses, or interpretation, or in the writing or publication of the manuscript.

References

1. Steiner, J.L.; Schneider, J.M.; Pope, C.; Pope, S.; Ford, P.; Steele, R.F. *Southern Plains Assessment of Vulnerability and Preliminary Adaptation and Mitigation Strategies for Farmers, Ranchers, and Forest Land Owners*; Anderson, T., Ed.; United States Department of Agriculture: Washington, DC, USA, 2017; 61p.
2. National Agricultural Statistics Service. *2012 Census of Agriculture, Texas State and County Data*; Geographic Area Series; Part 43A, AC-12-A-43A; United States Department of Agriculture, National Agricultural Statistics Service: Washington, DC, USA, 2014; Volume 1.
3. National Agricultural Statistics Service. *2012 Census of Agriculture, Oklahoma State and County Data*; Geographic Area Series; Part 36, AC-12-A-36; United States Department of Agriculture, National Agricultural Statistics Service: Washington, DC, USA, 2014; Volume 1.
4. National Agricultural Statistics Service. *2012 Census of Agriculture, Oklahoma State and County Data*; Geographic Area Series; Part 16, AC-12-A-16; United States Department of Agriculture, National Agricultural Statistics Service: Washington, DC, USA, 2014; Volume 1.
5. Guerrero, B. The Impact of Agricultural Drought Losses on the Texas Economy. 2011. Available online: https://agecoext.tamu.edu/wp-content/uploads/2013/07/BriefingPaper09-01-11.pdf (accessed on 6 May 2019).
6. Ranking of States with The Most Cattle. Available online: http://beef2live.com/story-ranking-states-cattle-0-108182 (accessed on 6 April 2019).
7. Kunkel, K.E.; Stevens, L.E.; Stevens, S.E.; Sun, L.; Janssen, E.; Wuebbles, D.; Kruk, M.C.; Thomas, D.P.; Shulski, M.; Umphlett, N.; et al. *Regional Climate Trends and Scenarios for the U.S. National Climate Assessment Part 4. Climate of the U.S. Great Plains*; NOAA Technical Report; NESDIS: Washington, DC, USA, 2013; pp. 142–144.
8. Dukes, J.S.; Chiariello, N.R.; Cleland, E.E.; Moore, L.A.; Shaw, M.R.; Thayer, S.; Tobeck, T.; Mooney, H.A.; Field, C.B. Responses of Grassland Production to Single and Multiple Global Environmental Changes. *PLoS Biol.* **2005**, *3*, e319. [CrossRef] [PubMed]
9. Fay, P.A.; Carlisle, J.D.; Knapp, A.K.; Blair, J.M.; Collins, S.L. Productivity responses to altered rainfall patterns in a C4-dominated grassland. *Oecologia* **2003**, *37*, 245–251. [CrossRef]
10. Knapp, A.K.; Smith, M.D. Variation among biomes in temporal dynamics of aboveground primary production. *Science* **2001**, *291*, 481–484. [CrossRef] [PubMed]
11. Dale, R.F.; Shaw, R.H. Effect on corn yields of moisture stress and stand at two different fertility levels. *Agron. J.* **1965**, *57*, 475–479. [CrossRef]
12. Boonjung, H.; Fukai, S. Effects of soil water deficit at different growth stages on rice growth and yield under upland conditions. 2. Phenology, biomass production and yield. *Field Crops Res.* **1996**, *48*, 47–55. [CrossRef]
13. Earl, H.J.; Davis, R.F. Effect of drought stress on leaf and whole canopy maize radiation use efficiency and yield of maize. *Agron. J.* **2003**, *95*, 688–696. [CrossRef]
14. Flanagan, L.B.; Johnson, B.G. Interacting effects of temperature, soil moisture, and plant biomass production on ecosystem respiration in a northern temperate grassland. *Agric. For. Meteor.* **2005**, *130*, 237–253. [CrossRef]
15. Palmer, W.C. *Meteorological Drought*; Research Paper 45; US Weather Bureau, US Department of Commerce: Silver Spring, MD, USA, 1965; 58p.
16. McKee, T.B.; Doesken, N.J.; Kleist, J. The Relationship of Drought Frequency and Duration to Time Scales. In Proceedings of the 8th Conference on Applied Climatology, Anaheim, CA, USA, 17–22 January 1993; American Meteorological Society: Boston, MA, USA, 1993.
17. Bergman, K.H.; Sabol, P.; Miskus, D. Experimental indices for monitoring global drought conditions. In Proceedings of the 13th Annual Climate Diagnostics Workshop, Cambridge, MA, USA, 31 October–4 November 1988; US Department of Commerce: Washington, DC, USA.
18. Modares, R. Streamflow drought time series forecasting. *Stoch. Environ. Res. Risk Assess.* **2007**, *21*, 223–233. [CrossRef]
19. Anderson, M.C.; Hain, C.; Wardlow, B.; Pimstein, A.; Mecikalski, J.R.; Kustas, W.P. Evaluation of drought indices based on thermal remote sensing of evapotranspiration over the continental United States. *J. Clim.* **2011**, *24*, 2025–2044. [CrossRef]

20. Svoboda, M.; Fuchs, B.A. *Handbook of Drought Indicators and Indices, Integrated Drought Management Programme (IDMP), Integrated Drought Management Tools and Guidelines Series 2*; WMO-No. 2273; World Meteorological Organization (WMO): Geneva, Switzerland, 2016; 45p.
21. Vicente-Serrano, S.M.; Begueria, S.; Lopez-Moreno, J.I. A multi-scalar drought index sensitive to global warming: The standardized precipitation evapotranspiration index–SPEI. *J. Clim.* **2010**, *23*, 1696–1718. [CrossRef]
22. Wilhite, D.; Glantz, M. Understanding the drought phenomenon: The role of definitions. *Water Int.* **1985**, *10*, 111–120. [CrossRef]
23. Sivakumar, M.V.K.; Motha, R.P.; Wilhite, D.A.; Wood, D.A. Agricultural drought indices. In Proceedings of the A WMO/UNISDR Expert Group Meeting on Agricultural Drought Indices, Murcia, Spain, 1–4 June 2010. (AGM-11, WMO/TD No. 1572; WAOB-2011).
24. Hunt, E.D.; Svoboda, M.; Wardlow, B.; Hubbard, K.; Hayes, M.; Arkebauer, T. Monitoring the effects of rapid onset of drought on non-irrigated maize with agronomic data and climate-based drought indices. *Agric. For. Meteor.* **2014**, *191*, 1–11. [CrossRef]
25. Chen, T.; van der Werf, G.R.; de Ju, R.A.M.; Wang, G.; Dolman, A.J. A global analysis of the impact of drought on net primary productivity. *Hydrol. Earth Syst. Sci.* **2013**, *17*, 3885–3894. [CrossRef]
26. Moorehead, J.E.; Gowda, P.H.; Singh, V.P.; Porter, D.O.; Marek, T.H.; Howell, T.A.; Stewar, B.A. Identifying and evaluating a suitable index for agricultural drought monitoring in the Texas High Plains. *J. Am. Water Resour. Assoc.* **2015**, *51*, 807–820. [CrossRef]
27. Potop, V.; Mozny, M.; Soukup, J. Drought evolution at various time scales in the lowland regions and their impact on vegetable crops in the Czech Repubic. *Agric. For. Meteor.* **2012**, *156*, 121–133. [CrossRef]
28. Potopova, V.; Stepanek, P.; Mozny, M.; Turkott, L.; Soukup, J. Performance of the standardised precipitation evapotranspiration index at various lags for agricultural drought risk assessment in the Czech Republic. *Agric. For. Meteor.* **2015**, *202*, 26–38. [CrossRef]
29. Ogaya, R.; Barbeta, A.; Basnou, C.; Penuelas, J. Satellite data as indicators of tree biomass growth and forest dieback in a Mediterranean holm oak forest. *Ann. For. Sci.* **2015**, *72*, 135–144. [CrossRef]
30. Klesse, S.; Ettold, S.; Frank, D. Integrating tree-ring and inventory-based measurements of aboveground biomass growth: Research opportunities and carbon cycle consequences from a large snow breakage event in the Swiss Alps. *Eur. J. For. Res.* **2016**, *135*, 297–311. [CrossRef]
31. Liu, S.; Zhang, Y.; Cheng, F.; Hou, X.; Zhao, S. Response of grassland degradation to drought at different time-scales in Qinghai Province: Spatio-temporal characteristics, correlation, and implications. *Rem. Sen.* **2017**, *9*, 1329. [CrossRef]
32. Barnes, M.L.; Moran, M.S.; Scott, R.L.; Kolb, T.E.; Ponce-Campos, G.E.; Moore, D.J.P.; Ross, M.A.; Mitra, B.; Dore, S. Vegetation productivity responds to sub-annual climate conditions across semiarid biomes. *Ecosphere* **2016**, *7*, e0.119. [CrossRef]
33. Knapp, A.K.; Carroll, C.J.W.; Denton, E.M.; La Pierre, K.J.; Collins, S.L.; Smith, M.D. Differential sensitivity to regional-scale drought in six central US grasslands. *Oecologia* **2015**, *177*, 949–957. [CrossRef] [PubMed]
34. Northup, B.K.; Daniel, J.A. Impact of climate and management on species composition of southern tallgrass prairie in Oklahoma. In Proceedings of the 1st National Conference on Grazing Lands, Las Vega, NV, USA, 5–8 December 2000.
35. Norhtup, B.K.; Schneider, J.M.; Daniel, J.A. The effects of management and precipitation on forage composition of a southern tallgrass prairie. In Proceedings of the 15th Conference on Biometeorology and Aerobiology, Kansas City, MO, USA, 27 October–1 November 2002.
36. Knapp, A. PAB02 Biweekly Measurement of Aboveground Net Primary Productivity on an Unburned and Annually Burned Watershed. *Environ. Data Initiat.* **2018**. Available online: http://129.130.186.12/content/pab02-biweekly-measurement-aboveground-net-primary-productivity-unburned-and-annually-burned (accessed on 17 April 2019).
37. Shapiro, S.S.; Wilk, M.B. An analysis of variance test for normality (complete samples). *Biometrika* **1965**, *52*, 591–611. [CrossRef]
38. Slifker, J.F.; Shapiro, S.S. The Johnson system: Selection and parameter estimation. *Technometrics* **1980**, *22*, 239–246. [CrossRef]

39. Tukey, J.W. The Problem of Multiple Comparisons. In *Multiple Comparisons, 1948–1983*; Volume 8 of The Collected Works of John W. Tukey. Unpublished manuscript; Braun, H.I., Ed.; Chapman & Hall: London, UK, 1994; pp. 1–300.
40. Allen, R.G.; Pereira, L.S.; Raes, D.; Smith, M. *Crop Evapotranspiration–Guidelines for Computing Crop Water Requirements*; FAO Irrigation and Drainage Paper 56; Food and Agriculture Organization: Rome, Italy, 1998.
41. Johnson, R.A.; Wichern, D.W. *Applied Multivariate Statistical Analysis*, 2nd ed.; Prentice-Hall Inc.: Englewood Cliffs, NJ, USA, 1988.
42. Lawrence, J. *Introduction to Neural Networks: Design, Theory, and Applications*, 6th ed.; California Scientific Software: Nevada City, CA, USA, 1994.
43. Abrams, M.D.; Knapp, A.K.; Hulbert, L.C. A ten-year record of aboveground biomass in a Kansas tallgrass prairie: Effects of fire and topographic position. *Am. J. Bot.* **1986**, *73*, 1509–1515. [CrossRef]
44. Wiles, L.J.; Dunn, G.; Printz, J.; Patton, B.; Nyren, A. Spring precipitation as a predictor for peak standing crop of mixed-grass prairie. *Rangel. Ecol. Manag.* **2011**, *64*, 215–222. [CrossRef]
45. Andales, A.A.; Derner, J.D.; Ahuja, L.R.; Hart, R.H. Strategic and tactical prediction of forage production in northern mixed-grass prairie. *Rangel. Ecol. Manag.* **2006**, *59*, 576–584. [CrossRef]
46. Nippert, J.B.; Knapp, A.K.; Briggs, J.M. Intra-annual rainfall variability and grassland productivity: Can the past predict the future? *Plant Ecol.* **2006**, *184*, 65–74. [CrossRef]
47. Vincente-Serrano, S.M.; Cuadrat-Prats, J.M.; Romo, A. Early prediction of crop production using drough indices at different time-scales and remote sensing data: Application in the Ebro Valley (north-east Spain). *Int. J. Remote Sens.* **2006**, *27*, 511–518. [CrossRef]
48. Wang, J.; Rich, P.M.; Price, K.P.; Kettle, W.D. Relations between NDVI, grassland production, and crop yield in the central Great Plains. *Geocarto Int.* **2005**, *20*, 5–11. [CrossRef]

© 2019 by the authors. Licensee MDPI, Basel, Switzerland. This article is an open access article distributed under the terms and conditions of the Creative Commons Attribution (CC BY) license (http://creativecommons.org/licenses/by/4.0/).

20. Svoboda, M.; Fuchs, B.A. *Handbook of Drought Indicators and Indices, Integrated Drought Management Programme (IDMP), Integrated Drought Management Tools and Guidelines Series 2*; WMO-No. 2273; World Meteorological Organization (WMO): Geneva, Switzerland, 2016; 45p.
21. Vicente-Serrano, S.M.; Begueria, S.; Lopez-Moreno, J.I. A multi-scalar drought index sensitive to global warming: The standardized precipitation evapotranspiration index–SPEI. *J. Clim.* **2010**, *23*, 1696–1718. [CrossRef]
22. Wilhite, D.; Glantz, M. Understanding the drought phenomenon: The role of definitions. *Water Int.* **1985**, *10*, 111–120. [CrossRef]
23. Sivakumar, M.V.K.; Motha, R.P.; Wilhite, D.A.; Wood, D.A. Agricultural drought indices. In Proceedings of the A WMO/UNISDR Expert Group Meeting on Agricultural Drought Indices, Murcia, Spain, 1–4 June 2010. (AGM-11, WMO/TD No. 1572; WAOB-2011).
24. Hunt, E.D.; Svoboda, M.; Wardlow, B.; Hubbard, K.; Hayes, M.; Arkebauer, T. Monitoring the effects of rapid onset of drought on non-irrigated maize with agronomic data and climate-based drought indices. *Agric. For. Meteor.* **2014**, *191*, 1–11. [CrossRef]
25. Chen, T.; van der Werf, G.R.; de Ju, R.A.M.; Wang, G.; Dolman, A.J. A global analysis of the impact of drought on net primary productivity. *Hydrol. Earth Syst. Sci.* **2013**, *17*, 3885–3894. [CrossRef]
26. Moorehead, J.E.; Gowda, P.H.; Singh, V.P.; Porter, D.O.; Marek, T.H.; Howell, T.A.; Stewar, B.A. Identifying and evaluating a suitable index for agricultural drought monitoring in the Texas High Plains. *J. Am. Water Resour. Assoc.* **2015**, *51*, 807–820. [CrossRef]
27. Potop, V.; Mozny, M.; Soukup, J. Drought evolution at various time scales in the lowland regions and their impact on vegetable crops in the Czech Repubic. *Agric. For. Meteor.* **2012**, *156*, 121–133. [CrossRef]
28. Potopova, V.; Stepanek, P.; Mozny, M.; Turkott, L.; Soukup, J. Performance of the standardised precipitation evapotranspiration index at various lags for agricultural drought risk assessment in the Czech Republic. *Agric. For. Meteor.* **2015**, *202*, 26–38. [CrossRef]
29. Ogaya, R.; Barbeta, A.; Basnou, C.; Penuelas, J. Satellite data as indicators of tree biomass growth and forest dieback in a Mediterranean holm oak forest. *Ann. For. Sci.* **2015**, *72*, 135–144. [CrossRef]
30. Klesse, S.; Ettold, S.; Frank, D. Integrating tree-ring and inventory-based measurements of aboveground biomass growth: Research opportunities and carbon cycle consequences from a large snow breakage event in the Swiss Alps. *Eur. J. For. Res.* **2016**, *135*, 297–311. [CrossRef]
31. Liu, S.; Zhang, Y.; Cheng, F.; Hou, X.; Zhao, S. Response of grassland degradation to drought at different time-scales in Qinghai Province: Spatio-temporal characteristics, correlation, and implications. *Rem. Sen.* **2017**, *9*, 1329. [CrossRef]
32. Barnes, M.L.; Moran, M.S.; Scott, R.L.; Kolb, T.E.; Ponce-Campos, G.E.; Moore, D.J.P.; Ross, M.A.; Mitra, B.; Dore, S. Vegetation productivity responds to sub-annual climate conditions across semiarid biomes. *Ecosphere* **2016**, *7*, e0.119. [CrossRef]
33. Knapp, A.K.; Carroll, C.J.W.; Denton, E.M.; La Pierre, K.J.; Collins, S.L.; Smith, M.D. Differential sensitivity to regional-scale drought in six central US grasslands. *Oecologia* **2015**, *177*, 949–957. [CrossRef] [PubMed]
34. Northup, B.K.; Daniel, J.A. Impact of climate and management on species composition of southern tallgrass prairie in Oklahoma. In Proceedings of the 1st National Conference on Grazing Lands, Las Vega, NV, USA, 5–8 December 2000.
35. Norhtup, B.K.; Schneider, J.M.; Daniel, J.A. The effects of management and precipitation on forage composition of a southern tallgrass prairie. In Proceedings of the 15th Conference on Biometeorology and Aerobiology, Kansas City, MO, USA, 27 October–1 November 2002.
36. Knapp, A. PAB02 Biweekly Measurement of Aboveground Net Primary Productivity on an Unburned and Annually Burned Watershed. *Environ. Data Initiat.* **2018**. Available online: http://129.130.186.12/content/pab02-biweekly-measurement-aboveground-net-primary-productivity-unburned-and-annually-burned (accessed on 17 April 2019).
37. Shapiro, S.S.; Wilk, M.B. An analysis of variance test for normality (complete samples). *Biometrika* **1965**, *52*, 591–611. [CrossRef]
38. Slifker, J.F.; Shapiro, S.S. The Johnson system: Selection and parameter estimation. *Technometrics* **1980**, *22*, 239–246. [CrossRef]

39. Tukey, J.W. The Problem of Multiple Comparisons. In *Multiple Comparisons, 1948–1983*; Volume 8 of The Collected Works of John W. Tukey. Unpublished manuscript; Braun, H.I., Ed.; Chapman & Hall: London, UK, 1994; pp. 1–300.
40. Allen, R.G.; Pereira, L.S.; Raes, D.; Smith, M. *Crop Evapotranspiration–Guidelines for Computing Crop Water Requirements*; FAO Irrigation and Drainage Paper 56; Food and Agriculture Organization: Rome, Italy, 1998.
41. Johnson, R.A.; Wichern, D.W. *Applied Multivariate Statistical Analysis*, 2nd ed.; Prentice-Hall Inc.: Englewood Cliffs, NJ, USA, 1988.
42. Lawrence, J. *Introduction to Neural Networks: Design, Theory, and Applications*, 6th ed.; California Scientific Software: Nevada City, CA, USA, 1994.
43. Abrams, M.D.; Knapp, A.K.; Hulbert, L.C. A ten-year record of aboveground biomass in a Kansas tallgrass prairie: Effects of fire and topographic position. *Am. J. Bot.* **1986**, *73*, 1509–1515. [CrossRef]
44. Wiles, L.J.; Dunn, G.; Printz, J.; Patton, B.; Nyren, A. Spring precipitation as a predictor for peak standing crop of mixed-grass prairie. *Rangel. Ecol. Manag.* **2011**, *64*, 215–222. [CrossRef]
45. Andales, A.A.; Derner, J.D.; Ahuja, L.R.; Hart, R.H. Strategic and tactical prediction of forage production in northern mixed-grass prairie. *Rangel. Ecol. Manag.* **2006**, *59*, 576–584. [CrossRef]
46. Nippert, J.B.; Knapp, A.K.; Briggs, J.M. Intra-annual rainfall variability and grassland productivity: Can the past predict the future? *Plant Ecol.* **2006**, *184*, 65–74. [CrossRef]
47. Vincente-Serrano, S.M.; Cuadrat-Prats, J.M.; Romo, A. Early prediction of crop production using drough indices at different time-scales and remote sensing data: Application in the Ebro Valley (north-east Spain). *Int. J. Remote Sens.* **2006**, *27*, 511–518. [CrossRef]
48. Wang, J.; Rich, P.M.; Price, K.P.; Kettle, W.D. Relations between NDVI, grassland production, and crop yield in the central Great Plains. *Geocarto Int.* **2005**, *20*, 5–11. [CrossRef]

© 2019 by the authors. Licensee MDPI, Basel, Switzerland. This article is an open access article distributed under the terms and conditions of the Creative Commons Attribution (CC BY) license (http://creativecommons.org/licenses/by/4.0/).

Article

Climate Effects on Tallgrass Prairie Responses to Continuous and Rotational Grazing

Yuting Zhou [1,*], Prasanna H. Gowda [2], Pradeep Wagle [2], Shengfang Ma [3,4], James P. S. Neel [2], Vijaya G. Kakani [1] and Jean L. Steiner [2]

[1] Department of Plant and Soil Sciences, Oklahoma State University, Stillwater, OK 74078, USA; v.g.kakani@okstate.edu
[2] USDA-ARS Grazinglands Research Laboratory, El Reno, OK 73036, USA; Prasanna.Gowda@ARS.USDA.GOV (P.H.G.); Pradeep.Wagle@ARS.USDA.GOV (P.W.); Jim.Neel@ARS.USDA.GOV (J.P.S.N.); jean.steiner@hughes.net (J.L.S.)
[3] State Key Laboratory of Remote Sensing Science, Institute of Remote Sensing and Digital Earth, Chinese Academy of Sciences, Beijing 100101, China; shengfangma@gmail.com
[4] University of Chinese Academy of Sciences, Beijing 100049, China
* Correspondence: yuting.zhou@okstate.edu; Tel.: +1-405-845-2102

Received: 31 March 2019; Accepted: 28 April 2019; Published: 30 April 2019

Abstract: Cattle grazing is an important economic activity in the tallgrass prairie systems in the Great Plains of the United States. Tallgrass prairie may respond differently to grazing management (e.g., high and low grazing intensity) under variable climate conditions. This study investigated the responses of two replicated (rep a and rep b) tallgrass prairie systems to continuous (C) and rotational (R) grazing under different climate conditions over a decade (2008–2017). The enhanced vegetation index (EVI) and gross primary productivity (GPP) were compared between grazing systems (C vs. R), while EVI was compared among paddocks under rotational grazing to show the impacts of time since grazing. The average EVI in rep a was usually higher than that in rep b which could be explained by different land characteristics (e.g., soil types) associated with different landscape positions. Similar to EVI, GPP was usually higher in rep a than rep b. The average growing season EVI and GPP were higher in rotational grazing than continuous grazing in rep b but not in rep a. The average EVI of paddocks in rotational grazing systems only converged in the growing season-long drought year (2011). In other years, EVI values varied from year to year and no paddock consistently outperformed others. The variations in EVI among rotational grazing paddocks in both reps were relatively small, indicating that rotational grazing generated an even grazing pressure on vegetation at annual scale. Overall, climate and inherent pasture conditions were the major drivers of plant productivity. However, the stocking rate in continuous grazing systems were reduced over years because of deteriorating pasture conditions. Thus, the results indirectly indicate that rotational grazing improved grassland productivity and had higher stocking capacity than continuous grazing systems under variable climate conditions. Adaptive grazing management (adjustment in stocking rates and season of use to adapt to changing climatic conditions) instead of a fixed management system might be better for farmers to cooperate with changing climatic conditions.

Keywords: tallgrass prairie; grazing management; climate; enhanced vegetation index (EVI); gross primary production (GPP)

1. Introduction

Tallgrass prairie systems are important forage sources for beef cattle in the Great Plains of the United States [1] where cattle production is a major revenue for farmers in the region [2]. Cattle grazing also acts as an effective way to maintain the prairie landscape [3,4]. There are two main grazing management systems nowadays: continuous grazing and rotational grazing. The cattle herd remains in the same pasture in continuous grazing management, while the herd is alternately moved to different paddocks in rotational grazing management. In contrast to year-long (or growing season) continuous grazing, rotational grazing has been recommended as an effective tool to maximize livestock production and maintain sustainability of the operations since the mid-20th century [5–9]. However, there has been a long history of debate over continuous versus rotational grazing by both rangeland managers and research scientists across the world that is yet not resolved [10–23].

The overall assumption behind multi-paddock rotational grazing is that it can reduce selective grazing (spatially uneven distribution of livestock grazing) [24] and allow rest to vegetation in the paddocks, during different periods of the year, to help improve species composition and/or productivity [12,21,25]. The disadvantages of the rotational grazing system include trampling damage in high intensity grazing paddocks and waste of feed values in the rested pastures because forage quality decreases after plant maturity [12,26,27]. In comparison, continuous grazing has lower stocking rate and can promote biodiversity for conservation goals [28]. However, selective grazing and underutilization of forage production in the peak growing season can occur in continuous grazing [18,25]. The long-lasting grazing management debate indicates the complexity of direct comparison between continuous and rotational grazing. A wide range of variation exists in vegetation structure, composition, productivity, prior condition, and climate. Climate interacts with multiple objectives (e.g., production, wildlife, soil, and water conservation) and capabilities among ranchers to make responses of grassland to grazing management variable in time and space. Studies that compare continuous and rotational grazing in the Great Plains have not generated consistent conclusions due to varied landscape types, experimental settings, and rancher objectives. Several of them found no statistically significant differences in vegetation responses to continuous and rotational grazing [7,8,15,29–31], while some studies favored rotational grazing [25,32]. Thus, it is still not clear how vegetation will respond to different grazing management strategies in the tallgrass prairie systems of the Great Plains over years under variable climate conditions.

The responses of vegetation to continuous and rotational grazing are usually examined from the perspectives of vegetation canopy structure, species composition, primary productivity, and forage quality using field survey data [14–16,18,23,29,30,32–34]. Field data collection is time-consuming and labor intensive. Little work has been done using periodic and long-term satellite remote sensing data to examine the responses of vegetation to different grazing management systems [35,36]. The major difficulty of utilizing satellite remote sensing in grazing studies lies in the relatively coarse spatial resolution satellite data compared to the size of paddocks. For example, the widely used Moderate Resolution Imaging Spectroradiometer (MODIS) surface reflectance product has a spatial resolution of ~500 m, which might be larger than a single paddock in a rotational grazing system. Landsat has a spatial resolution of 30 m in visible and near infrared (NIR) bands, which makes it more suitable to study the impacts of different grazing management on vegetation phenology. However, Landsat has a longer revisit time than MODIS (eight-day vs. daily) which makes it challenging for studying vegetation productivity. Thus, the combination of MODIS and Landsat data might be a better option to examine the responses of grassland phenology and productivity to varied grazing management under variable climate conditions.

In order to study the impacts of different grazing management under variable climate conditions, this study examined the vegetation phenology and productivity of tallgrass prairies in two pairs of continuous (C) and rotational (R) grazing systems during a span of ten years (2008–2017). All paddocks in the experiment were located in the same area and thus exposed to the same climatic conditions (e.g., air temperature and precipitation). There were large annual and seasonal variations in air temperature and precipitation during the study period, which provide the opportunity to investigate the interactions of climate and grazing management on vegetation phenology and productivity. This paper focuses on the responses of vegetation phenology and productivity to climate and grazing management systems. Other aspects of these systems are discussed in Steiner et al., 2019; Starks et al., 2019; Ma et al., etc [37–39].

2. Materials and Methods

2.1. Study Area and Experiment Design

The study was conducted (Figure 1) at the United States Department of Agriculture–Agricultural Research Service (USDA–ARS), Grazinglands Research Laboratory (GRL) in El Reno, Oklahoma. It consists of two replicates (rep a and rep b) of continuous (Ca and Cb) versus rotational grazing (Ra and Rb) systems. The experiment was initiated in 2009. There were about 20 cow–calf pairs within each of the continuous grazing pastures (Ca and Cb) year-round. In comparison, around 27 cow–calf pairs were allocated in 10 paddocks of the rotational grazing systems (Ra and Rb) following the same routine each year. Cow–calf pairs in the rotational grazing systems grazed paddocks in 7 to 10 day grazing bouts depending on the vegetation conditions. The herd sizes in the continuous and rotational grazing systems were $18(\pm 3.4)$ and $26(\pm 2.5)$ heard year^{-1}, respectively. The average stocking rates for Ca, Cb, and paddocks in Ra and Ra were 0.26, 0.25, 3.36, and 3.07 heard ha^{-1}, respectively. Though the stocking rate was adjusted in continuous grazing paddocks considering pasture conditions (details about the experiment can be found in Steiner et al., 2019 in the same issue). The size of each paddock is presented in Table 1.

In the study region, the long term (1981–2010) annual average air temperature (Ta) is 15.14 °C and average annual precipitation (P) is 91.3 cm. According to data from the Soil Survey Geographic Database (https://websoilsurvey.sc.egov.usda.gov/App/HomePage.htm), multiple soil types exist in the study area (Figure S1). Most of the paddocks were typical tallgrass prairies with big bluestem (*Andropogon gerardii* Vitman) and Indian grass (*Sorghastrum nutans* Nash) as dominant species. However, there were some trees and brushy vegetation which were not edible for cattle (mainly buckbrush, *Ceanothus cuneatus* Nutt.), especially in Ca (Figure S2). Large proportions of Rb-4 and Rb-7 were usually covered by ponds (Figure 1).

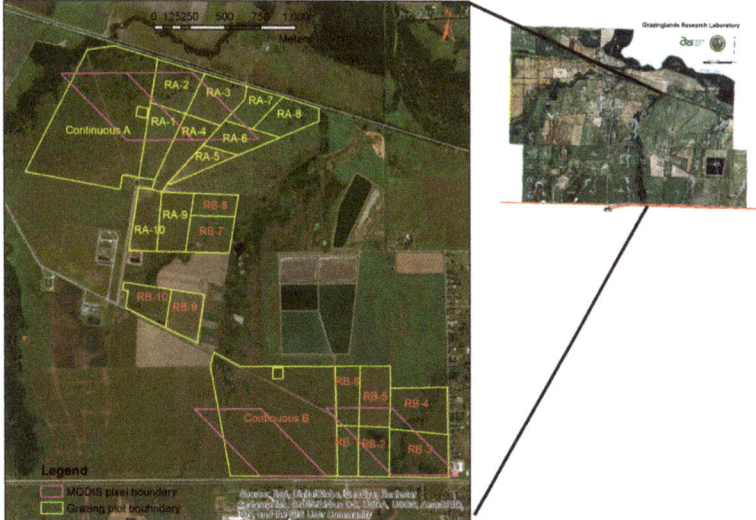

Figure 1. Location of the study area overlapping with MODIS pixels. The right-hand panel shows the location of the study area within the United States Department of Agriculture-Agricultural Research Service (USDA-ARS), Grazinglands Research Laboratory (GRL).

Table 1. The size of paddocks in the grazing experiment: replications a and b of continuous (Ca and Cb) and rotational (Ra and Rb) grazing.

Plot Name	Area (ha)	Plot Name	Area (ha)
Ca	58.6	Cb	62.7
Ra-1	9.3	Rb-1	5.1
Ra-2	9.2	Rb-2	7.5
Ra-3	9.5	Rb-3	12.9
Ra-4	9.2	Rb-4	12.2
Ra-5	4.4	Rb-5	9.0
Ra-6	6.7	Rb-6	6.7
Ra-7	7.6	Rb-7	9.1
Ra-8	5.9	Rb-8	6.6
Ra-9	8.7	Rb-9	7.0
Ra-10	7.6	Rb-10	6.7

2.2. Climate Data

The daily average Ta and total P were collected from the Oklahoma Mesonet El Reno site (35°32′54″ N, 98°2′11″ W) (http://mesonet.org/), which is located two km away from the study area. The daily Ta and P data were aggregated to monthly average Ta and sum P for 2008–2017 to show the seasonality. The annual average Ta and total P for the study period were calculated to indicate the dynamics of climate at the annual scale.

2.3. Landsat Images and the Enhanced Vegetation Index (EVI)

As Landsat 8 data (started from 2013) does not cover the whole study period (2008–2017), we also used the Landsat 7 data, which has been operating without the scan line corrector (SLC) since 2003. Fortunately, our study area is located within the central part of the image tile (Path28/Row35) and is not affected by the strips in Landsat 7 imagery caused by SLC-off. Landsat 7 and Landsat 8 surface reflectance products (Tier 2) covering the study area were downloaded from the USGS EarthExplorer (http://earthexplorer.usgs.gov) to provide an 8-day temporal resolution and 30 m spatial resolution.

False color composite images around early May and late August were used to represent the vegetation conditions in late spring and late summer of each year, depending on the data availability and its quality (Landsat 7 or Landsat 8). A commonly used vegetation index, the enhanced vegetation index (EVI) [40], was calculated using the Landsat images following the equation below (Equation (1)). EVI is linearly correlated with the green leaf area index (LAI) in crop fields [41] and it remains sensitive to canopy variations in high chlorophyll content [40]. Only clear pixels were included in analysis based on the quality assessment band.

$$EVI = 2.5 \times \frac{\rho_{nir} - \rho_{red}}{\rho_{nir} + 6.0 \times \rho_{red} - 7.5 \times \rho_{blue} + 1} \quad (1)$$

where ρ_{blue}, ρ_{red}, and ρ_{nir} represent surface reflectance from blue, red, and near infrared bands (841–876 nm), respectively.

2.4. Gross Primary Production (GPP) from the Vegetation Photosynthesis Model

The vegetation photosynthesis model (VPM) [42] was used to calculate the annual gross primary production (GPP) from 2008 to 2017. The VPM is a light use efficiency model which estimates GPP as a product of actual light use efficiency (ε_g) and absorbed photosynthetically active radiation (APAR) by chlorophyll ($APAR_{chl}$).

$$GPP_{VPM} = \varepsilon_g \times APAR_{chl} \quad (2)$$

$$APAR_{chl} = fPAR_{chl} \times PAR \quad (3)$$

where $fPAR_{chl}$ is set equal to EVI. The ε_g is derived by down-regulating the theoretical maximum light use efficiency (ε_0) with scalars of temperature (T_{scalar}) and water (W_{scalar}) stresses [42–44].

$$\varepsilon_g = \varepsilon_0 \times T_{scalar} \times W_{scalar} \qquad (4)$$

The VPM can be driven by either site level data [45–47] or remote sensing data [48]. The GPP data for the two pairs of MODIS pixels in two replicates (Figure 1) were collected from the global VPM-GPP (8-day and 500 m resolution) product [48] and the annual sums of GPP were calculated for each year.

2.5. Statistical Analysis

The statistical analysis included two parts: inter-treatment comparison (C versus R) and intra-treatment comparison (paddocks among R systems). For the inter-treatment comparison, we treated all the paddocks in the R system as a whole and compared the difference between C and R. For the intra-treatment comparison, we considered all paddocks in the R system as independent units and investigated the variations among them. The average, maximum, minimum, and standard deviation of EVI (σ_{EVI}) values were included in both inter-treatment and intra-treatment comparisons. The growing season was considered as May–September and all of the statistical values were derived for the growing season.

3. Results

3.1. Annual and Seasonal Variations of Ta and P

There were large variations in annual Ta and P during the study years (Figure 2). Compared to the long-term annual P, 2011, 2012, 2014, and 2016 were dry years, while 2013, 2015, and 2017 were wet years. Compared to the long-term annual average Ta, 2011, 2012, and 2016 were hot years, while 2013 and 2014 were cool years.

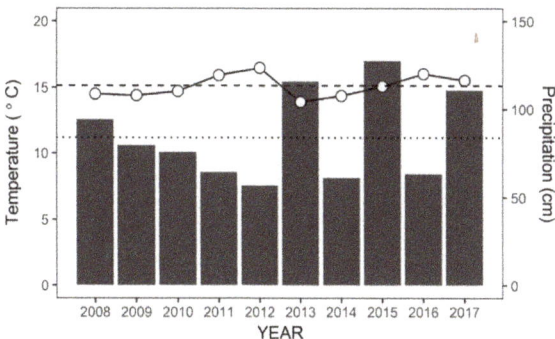

Figure 2. Annual average air temperature (line with open circles) and total precipitation (vertical bars). Long dash line indicates long term annual average temperature while dot line is for long term annual total precipitation.

Figure 3 shows the monthly distributions of Ta and P to illustrate drought development. Summer was very hot in 2011 and 2012, especially in the month of July. In comparison, 2013 and 2015 had relatively cool summers (Figure 3a). Although the total P in 2011 was higher than in 2012, a large proportion of P (26.5%) occurred in November, a part of the non-growing season (Figure 3b), and the growing season was relatively dry and hot. Warmer spring and a good amount of P in March–April of 2012 facilitated vegetation growth in early spring. However, the peak growing months (May–August) were very dry. The dry and hot conditions made 2012 a severe summer drought year. Both 2014 and 2016 were moderate dry years, with the latter hotter than the former.

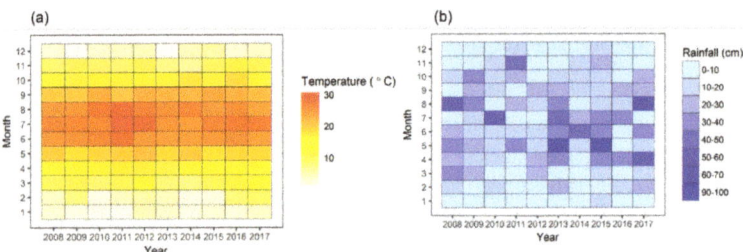

Figure 3. Seasonal distributions of air temperature (**a**) and precipitation (**b**).

3.2. A glimpse of the Continuous and Rotational Grazing Paddocks in Late Spring and Late Summer

Figure 4 shows the false color composite images in late spring and late summer of 2008–2017. Before the experiment in 2008, the study area was heterogeneous. Ca was actually composed of two sub-paddocks. Areas that greened-up early (early May) usually entered senescence stage early (late August). Starting from 2009, most of the paddocks became synchronized in terms of greenness. The paddocks were generally green in early May and remained green until late August. However, the paddocks lost greenness relatively earlier in late August in 2011, 2012, and 2016 due to dry and hot conditions. The size of ponds in Rb-4 and Rb-7 also indicated severe droughts in 2011 and 2012 when they almost dried up.

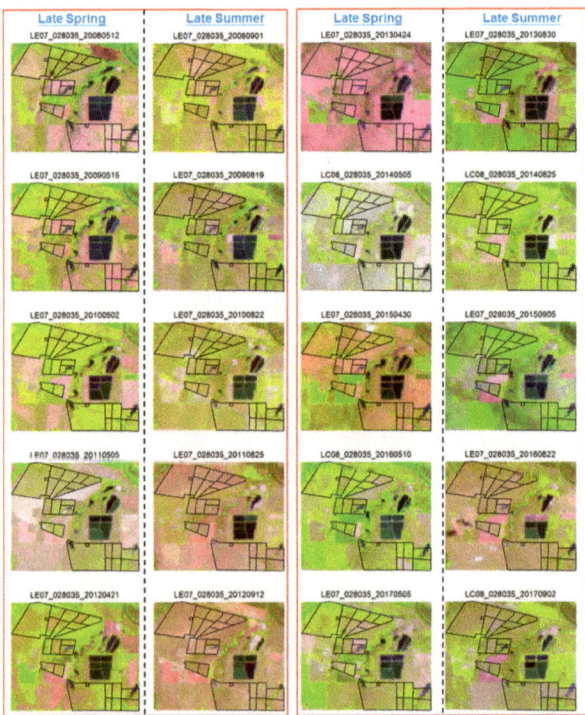

Figure 4. False color composite images (shortwave infrared (SWIR-2), near infrared (NIR), and red bands as RGB) for continuous and rotational grazing paddocks around early May and late August. Green color represents green vegetation and pink or brown color represent dry grasses or soil. The title of each panel includes three parts separated by an underscore: the sensor (LE07 for Landsat 7 ETM+ and LC08 for Landsat 8 OLI/TIRS combined), the path/row number, and acquisition date of the image.

3.3. The Dynamics of EVI Values in the Continuous Versus Rotational Grazing Paddocks

The average growing season EVI is a good indicator of vegetation production during the growing season. The initial condition of average growing season EVI values (Figure 5) varied largely during the prior year (2008) and the first year of the experiment (2009). After that, climate was the major driver in affecting the average growing season EVI. The growing season average EVI values were larger than 0.5 in wet years (2013 and 2015). In severe drought years (2011 and 2012), all treatments had low average EVI values. Treatment differences were larger in moderate drought years (2014 and 2016). Even though 2012 was a drier and hotter year than 2016, the average EVI in 2012 and 2016 were comparable because of early greening up of vegetation and more growth by May in 2012 (Figure 2). The average growing season EVI in Rb was consistently larger than in Cb. However, the average growing season EVI in Ra could be similar to, higher, or lower than Ca in different years.

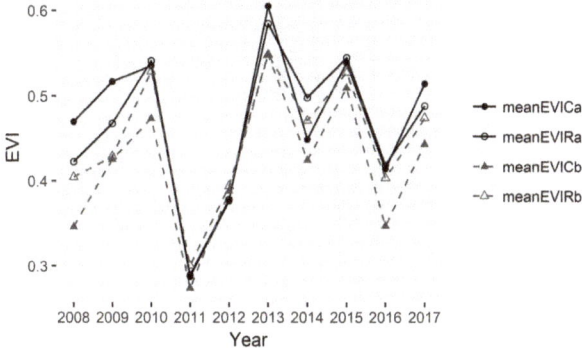

Figure 5. Growing season average EVI from Landsat 7 for different treatments in rep a and rep b. Solid lines are for rep a and dashed lines are for rep b. Solid and hollow shapes represent continuous grazing and rotational grazing.

The maximum growing season EVI indicates the peak vegetation growth during the growing season. Rep a consistently had larger growing season maximum EVI values than rep b during the study period (Figure 6 top), indicating higher photosynthesis capability in rep a than rep b. Maximum EVIs were largest in wet years (2013 and 2015) and lowest in severe drought years (2011 and 2012), showing the significant control of climate conditions on maximum productivity in all treatments. The maximum EVI values in 2012 were higher than those in 2016 because peak EVI occurred before severe drought in 2012. The maximum EVI values only tended to converge in very wet or dry years, otherwise they had large ranges. The maximum EVI values in Rb were larger than in Cb in majority of cases. The differences in maximum EVI between Ra and Ca were smaller than between Rb and Cb.

The minimum growing season EVI can represent the low photosynthesis capability caused by climate or other stresses on vegetation. The growing season minimum EVI (Figure 6 bottom) converged more than maximum EVI, with larger differences in moderate drought years (2014 and 2016). In contrast to what it was observed in maximum EVI, the minimum EVI was much lower in 2012 than in 2016, suggesting a collapse of vegetation condition because of the severe summer drought. The minimum EVI values in both rotational grazing treatments were larger than their counterparts (Rb larger than Cb and Ra larger than Ca) in most years.

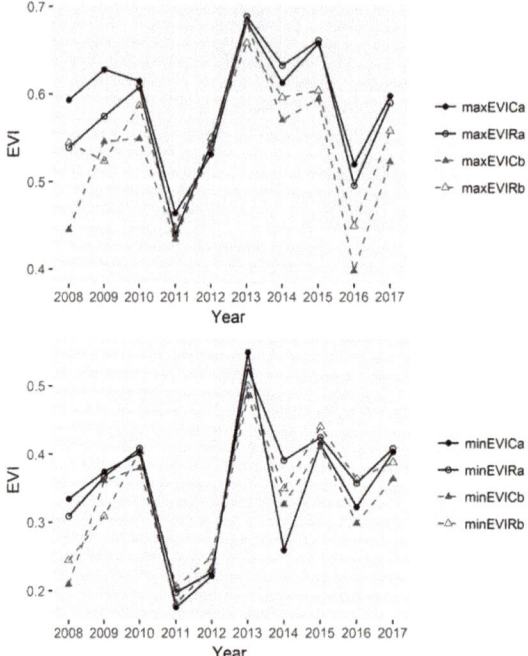

Figure 6. Growing season maximum and minimum EVI from Landsat 7 for different treatments.

The growing season standard deviation of EVI (σ_{EVI}) shows the variation of vegetation production during the growing season. Rep a had larger growing season σ_{EVI} than rep b (Figure 7). Ca had larger σ_{EVI} than Ra in most years while the differences between Cb and Rb were small throughout the years. Ca usually had the largest σ_{EVI} because of its mixture of grass, brush, and trees, which had different seasonality.

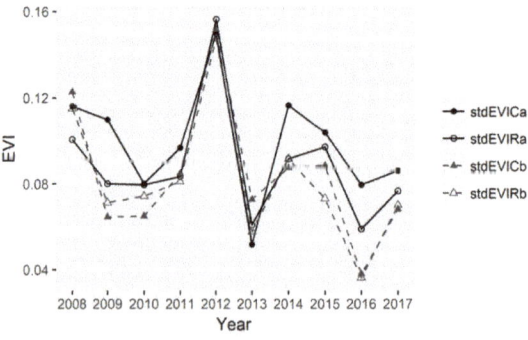

Figure 7. Growing season standard deviation of EVI from Landsat 7 for different treatments.

3.4. Dynamics of GPP in the Continuous Versus Rotational Grazing Paddocks

The dynamics of the annual sum of GPP (Figure 8) were similar to that of the average EVI (Figure 5) as EVI is a major input parameter in the calculation of GPP. Similar to average EVI, all treatments had the lowest GPP in 2011 due to the severe drought during the whole growing season. The GPP in 2012 and 2016 were comparable in spite of higher drought and heat stresses in the former because of early greening up of vegetation and more growth by May in 2012. The GPP in Rb was consistently larger than in Cb, especially while the GPP in Ra and Ca were similar.

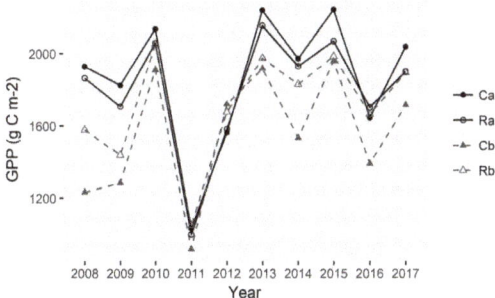

Figure 8. Annual sum of GPP in different treatments.

3.5. Variations in EVI among Paddocks in the Rotational Grazing Systems

Variations in EVI among paddocks in Ra and Rb were used to examine the impacts of time since grazing within an individual paddock. Overall, the variations in growing season average EVI for paddocks in both reps (Ra and Rb) were relatively small (Figure 9). Ra had lower variation in growing season average EVI than Rb, indicating the differences among pastures in rep a were smaller than in rep b. The average EVI converged only in a very dry year (2011). Rb-4 and Rb-7 usually had lower average EVI values than other paddocks because large proportions of them were covered by ponds (Figure 1) in non-drought years. The ranges of maximum EVI within a year were large and no paddocks consistently outperformed than others (Figure 10 left panels). Similarly, as found in inter-treatment comparison (Figure 6), minimum EVI among paddocks converged more than maximum EVI (Figure 10 right panels). The σ_{EVI} in Ra and Rb were the largest in 2012 (Figure 11) caused by the severe summer drought. In wet years (2013 and 2015) Ra had lower σ_{EVI} than Rb.

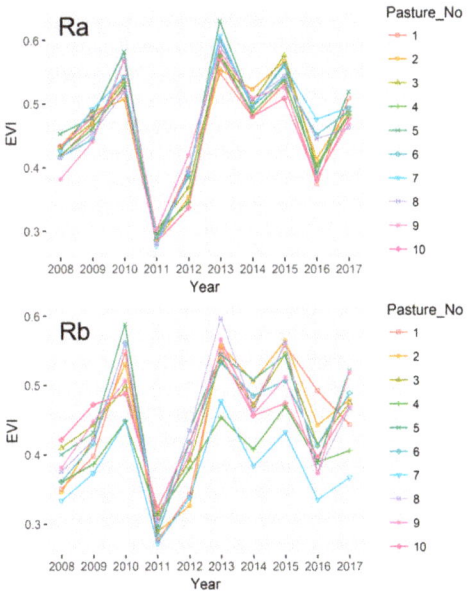

Figure 9. Growing season average EVI from Landsat 7 for paddocks in rotational grazing.

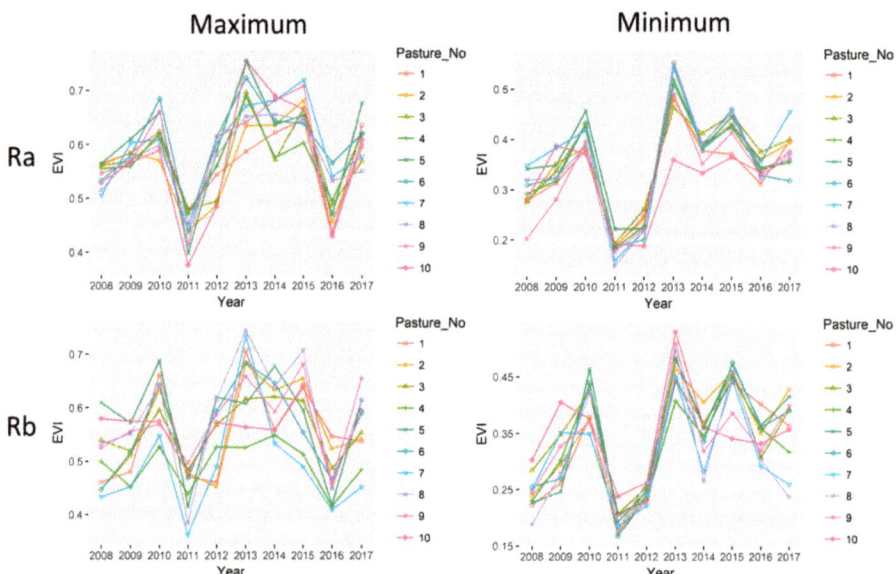

Figure 10. Growing season maximum and minimum EVI from Landsat 7 for paddocks in rotational grazing.

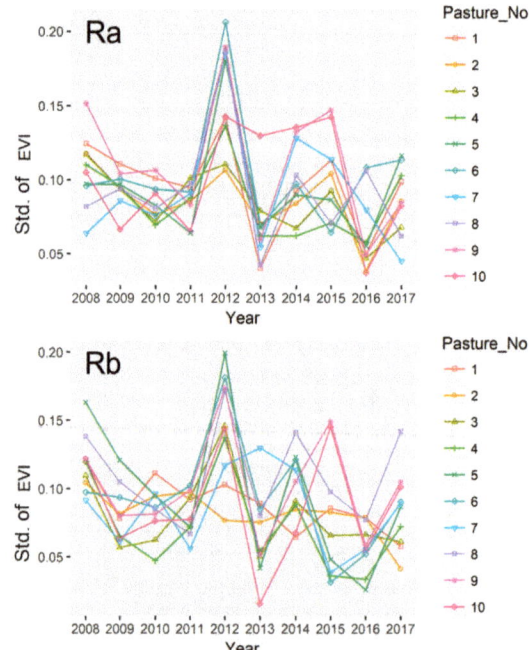

Figure 11. Growing season standard deviation EVI from Landsat 7 for paddocks in rotational grazing.

4. Discussion

This study provided a direct comparison of vegetation productivity in continuous and rotational grazing management systems in tallgrass prairies of the Southern Great Plains. The large climate variation during the study period allowed us to examine the differences in response to climate under continuous and rotational grazing management. Many studies indicated that multi-paddock rotational grazing can improve grassland productivity by allowing post-grazing recovery [25,32,49]. Our study found that the rotational grazing had higher vegetation production in rep b but not in rep a. Both average growing season EVI and GPP were higher in rep a than rep b probably caused by their inherent differences such as soil type. Additionally, our results indicated that productivity of tallgrass prairie was greatly controlled by climate, more specifically by the seasonality of air temperature and precipitation. This finding is also supported by other studies that compared the continuous and rotational grazing in the same experiment from perspectives of aboveground biomass (Starks et al. 2019) and landscape pattern (Ma et al. 2019), which were also presented in this issue.

The initial experiment design was to put around 25 cow–calf pairs in each of the treatments. However, the stocking rate in the continuous grazing system was adjusted according to the paddock conditions (considering woody encroachment, climate, species composition and desirable species, and grass growth). With reduced stocking rate in continuous grazing systems, the vegetation production was still similar with rotational grazing systems that had higher stocking rates. Thus, the concept of adaptive grazing management [19,21,23,50–53], adjusting grazing management depending on the vegetation and climate conditions, might be better to help ranchers to understand the necessity of adjusting their grazing routine to achieve sustainable and profitable usage of pastures.

Since little differences in vegetation growth were detected between Ca and Ra, it indirectly demonstrated that rotational grazing has higher grazing capacity, as Ca had a reduced stocking rate. Variations in EVI among paddocks in rotational grazing systems were small (Figure 11), indicating their similarity in vegetation growth. The relative small variations among paddocks in different years demonstrated that rotational grazing generated evenly distributed grazing pressure on tallgrass prairies in different climate conditions. Even with a higher stocking rate, the rotational grazing management generated similar vegetation responses with continuous grazing and an evenly grazed vegetation pattern. The results indicate the rotational grazing may support higher livestock production while maintaining sustainable usage of the tallgrass prairie pastures in the region.

The interaction of climate variability and grazing management makes the study complicated. There were very dry (2011 and 2012) and wet years (2013 and 2015) during the study period. Even the drought years had different characteristics. For example, 2011 was a growing season-long drought year and 2012 was a summer drought and hot year. The seasonal distribution of Ta and P generated different impacts on vegetation. Their dynamics combined with different grazing management controlled the grassland phenology and productivity. Thus, it is hard to conclude a general comparison between C and R for the entire study period. The average EVI only converged in very dry years (2011 and 2012) among treatments (Figure 5) and large variations exist in other years. Although 2012 was a drier and hotter year than 2016, they had comparable EVI. This difference was caused by the seasonal distribution of Ta and P. There was more precipitation in spring 2012 (March–April) than in 2016, which facilitated the vegetation growth in spring and early summer in 2012. It was hot and dry in summer 2012. Favorable vegetation growth in spring and a hot and dry summer induced a flash drought and quickly suppressed vegetation growth [45,54,55]. The largest σ_{EVI} in 2012 (Figure 7) was caused by good vegetation growth in spring and poor vegetation growth in summer. The larger average EVI in rep a than in rep b could be attributed to their original characteristics, such as soil types (Figure S1).

Besides comparing the difference between C and R, we also examined the differences among rotational grazing paddocks in both reps (Figures 9–11). Except in severe drought years (2011 and 2012), large variations existed among individual paddocks in rotational grazing systems. Similar with the inter-treatment comparison, their EVI variations were mostly controlled by climate conditions. The variations in Ra were consistently smaller than in Rb, which supported the idea that the difference between rep a and rep b was caused by innate differences, such as soil type. This is also demonstrated by Ma et al. (2019) in the same issue from the shrub encroachment perspective.

5. Conclusions

Through examining the responses of vegetation to continuous and rotational grazing, this study demonstrated that vegetation growth in tallgrass prairies were mainly controlled by the seasonality of precipitation and temperature. The average EVI values only converged in a very dry year in inter- and intra-treatments comparisons. The vegetation growth was also highly related to the original pasture conditions (e.g., soil types), as rep a consistently had higher EVI values than rep b. The intra-treatment analysis showed that paddocks in rotational grazing systems had relatively small variations in different years, indicating that the rotational grazing created an even grazing pressure on grassland. The growing season average EVI and GPP were higher in Rb than in Cb. Although there were no significant differences in EVI and vegetation production between Ra and Ca, rotational grazing might increase vegetation productivity as the stocking rate in C had been reduced because of the deteriorating pasture conditions. The study suggests that an adaptive grazing management approach might better help farmers to adapt to ever-changing climatic conditions (e.g., annual and seasonal variations in air temperature and precipitation).

Supplementary Materials: The following are available online at http://www.mdpi.com/2073-4395/9/5/219/s1, Figure S1: Soil types in study area. Data is from the Soil Survey Geographic Database, Figure S2: Brush and tree in the continuous grazing pasture in rep a.

Author Contributions: Conceptualization, Y.Z. and P.H.G.; methodology, Y.Z., P.H.G., and P.W.; formal analysis Y.Z., S.M., and V.G.K.; investigation: Y.Z., S.M., and J.L.S.; data curation: J.P.S.N.; writing—original draft preparation: Y.Z., S.M., and V.G.K.; writing—review and editing: P.H.G., P.W., J.P.S.N., V.G.K., and J.L.S.; visualization: Y.Z., P.H.G., P.W., and S.M.; funding acquisition, V.G.K., P.H.G., and J.L.S.

Funding: This research was funded by USDA-NIFA's Agriculture and Food Research Initiative (AFRI), grant number 2013-69002.

Conflicts of Interest: The authors declare no conflict of interest.

References

1. Samson, F.B.; Knopf, F.L.; Ostlie, W.R. Great Plains ecosystems: Past, present, and future. *Wildl. Soc. Bull.* **2004**, *32*, 6–15. [CrossRef]
2. Cunfer, G. *On the Great Plains: Agriculture and Environment*; Texas A&M University Press: College Station, TX, USA, 2005.
3. Westoby, M.; Walker, B.; Noy-Meir, I. Opportunistic management for rangelands not at equilibrium. *J. Range Manag.* **1989**, *42*, 266–274. [CrossRef]
4. Fuhlendorf, S.D.; Engle, D.M. Restoring Heterogeneity on Rangelands: Ecosystem Management Based on Evolutionary Grazing Patterns: We propose a paradigm that enhances heterogeneity instead of homogeneity to promote biological diversity and wildlife habitat on rangelands grazed by livestock. *BioScience* **2001**, *51*, 625–632.
5. Hubbard, W.A. Rotational grazing studies in western Canada. *J. Range Manag.* **1951**, *4*, 25–29. [CrossRef]
6. Hyder, D.N.; Sawyer, W. Rotation-deferred grazing as compared to season-long grazing on sagebrush-bunchgrass ranges in Oregon. *J. Range Manag.* **1951**, *4*, 30–34. [CrossRef]
7. McIlvain, E.; Savage, D. Eight-year comparisons of continuous and rotational grazing on the Southern Plains Experimental Range. *J. Range Manag.* **1951**, *4*, 42–47. [CrossRef]
8. Rogler, G.A. A twenty-five year comparison of continuous and rotation grazing in the Northern Plains. *J. Range Manag.* **1951**, *4*, 35–41. [CrossRef]
9. Sampson, A.W. A Symposium on Rotation Grazing in North America. *J. Range Manag.* **1951**, *4*, 19–24. [CrossRef]
10. Sampson, A.W. *Range Improvement by Deferred and Rotation Grazing*; Series: Bulletin of the U.S. Department of Agriculture No. 34; U.S. Department of Agriculture: Washington, DC, USA, 1913.

11. Leo, B.M. A Variation of Deferred Rotation Grazing for Use under Southwest Range Conditions. *J. Range Manag.* **1954**, *7*, 152–154.
12. Heady, H.F. Continuous vs. specialized grazing systems: A review and application to the California annual type. *J. Range Manag.* **1961**, *14*, 182–193. [CrossRef]
13. Bryant, F.C.; Dahl, B.E.; Pettit, R.D.; Britton, C.M. Does short-duration grazing work in arid and semiarid regions? *J. Soil Water Conserv.* **1989**, *44*, 290–296.
14. White, M.R.; Pieper, R.D.; Donart, G.B.; Trifaro, L.W. Vegetational response to short-duration and continuous grazing in southcentral New Mexico. *J. Range Manag.* **1991**, *44*, 399–403. [CrossRef]
15. Gillen, R.L.; Gillen, R.L.; McCollum III, F.T.; Tate, K.W.; Hodges, M.E. Tallgrass prairie response to grazing system and stocking rate. *J. Range Manag.* **1998**, *51*, 139–146. [CrossRef]
16. Jacobo, E.J.; Rodríguez, A.M.; Bartoloni, N.; Deregibus, V.A. Rotational Grazing Effects on Rangeland Vegetation at a Farm Scale. *Rangel. Ecol. Manag.* **2006**, *59*, 249–257. [CrossRef]
17. Briske, D.D.; Derner, J.D.; Brown, J.R.; Fuhlendorf, S.D.; Teague, W.R.; Havstad, K.M.; Gillen, R.L.; Ash, A.J.; Willms, W.D. Benefits of rotational grazing on rangelands: An evaluation of the experimental evidence. *Rangel. Ecol. Manag.* **2008**, *61*, 3–17. [CrossRef]
18. Briske, D.D.; Derner, J.D.; Brown, J.R.; Fuhlendorf, S.D.; Teague, W.R.; Havstad, K.M.; Gillen, R.L.; Ash, A.J.; Willms, W.D. Rotational Grazing on Rangelands: Reconciliation of Perception and Experimental Evidence. *Rangel. Ecol. Manag.* **2008**, *61*, 3–17. [CrossRef]
19. Budd, B.; Thorpe, J. Benefits of Managed Grazing: A Manager's Perspective. *Rangelands* **2009**, *31*, 11–14. [CrossRef]
20. Briske, D.D.; Sayre, N.F.; Huntsinger, L.; Fernández-Giménez, M.; Budd, B.; Derner, J.D. Origin, Persistence, and Resolution of the Rotational Grazing Debate: Integrating Human Dimensions into Rangeland Research. *Rangel. Ecol. Manag.* **2011**, *64*, 325–334. [CrossRef]
21. Teague, R.; Provenza, F.; Kreuter, U.; Steffens, T.; Barnes, M. Multi-paddock grazing on rangelands: Why the perceptual dichotomy between research results and rancher experience? *J. Environ. Manag.* **2013**, *128*, 699–717. [CrossRef] [PubMed]
22. Wang, T.; Teague, W.R.; Park, S.C. Evaluation of Continuous and Multipaddock Grazing on Vegetation and Livestock Performance—A Modeling Approach. *Rangel. Ecol. Manag.* **2016**, *69*, 457–464. [CrossRef]
23. Becker, W.; Kreuter, U.; Atkinson, S.; Teague, R. Whole-Ranch Unit Analysis of Multipaddock Grazing on Rangeland Sustainability in North Central Texas. *Rangel. Ecol. Manag.* **2017**, *70*, 448–455. [CrossRef]
24. Norton, B.E.; Barnes, M.; Teague, R. Grazing Management Can Improve Livestock Distribution. *Rangelands* **2013**, *35*, 45–51. [CrossRef]
25. Teague, W.; Dowhower, S.L.; Baker, S.A.; Haile, N.; DeLaune, P.B.; Conover, D.M. Grazing management impacts on vegetation, soil biota and soil chemical, physical and hydrological properties in tall grass prairie. *Agric. Ecosyst. Environ.* **2011**, *141*, 310–322. [CrossRef]
26. Heitschmidt, R.; Dowhower, S.; Walker, J. Some effects of a rotational grazing treatment on quantity and quality of available forage and amount of ground litter. *J. Range Manag.* **1987**, *40*, 318–321. [CrossRef]
27. Hart, R.H.; Samuel, M.J.; Test, P.S.; Smith, M.A. Cattle, vegetation, and economic responses to grazing systems and grazing pressure. *J. Range Manag.* **1988**, *41*, 282–286. [CrossRef]
28. Hickman, K.R.; Hartnett, D.C.; Cochran, R.C.; Owensby, C.E. Grazing management effects on plant species diversity in tallgrass prairie. *Rangel. Ecol. Manag.* **2004**, *57*, 58–66. [CrossRef]
29. Wood, M.K.; Blackburn, W.H. Vegetation and soil responses to cattle grazing systems in the Texas Rolling Plains. *J. Range Manag.* **1984**, *37*, 303–308. [CrossRef]
30. Derner, J.D.; Hart, R.H. Grazing-induced modifications to peak standing crop in northern mixed-grass prairie. *Rangel. Ecol. Manag.* **2007**, *60*, 270–276.
31. Derner, J.D.; Hart, R.H. Livestock and vegetation responses to rotational grazing in short-grass steppe. *West. North Am. Nat.* **2007**, *67*, 359–367. [CrossRef]
32. Cassels, D.M.; Gillen, R.L.; Ted McCollum, F.; Tate, K.W.; Hodges, M.E. Effects of grazing management on standing crop dynamics in tallgrass prairie. *J. Range Manag.* **1995**, *48*, 81–84. [CrossRef]
33. Martin, S.C.; Ward, D.E. Perennial grasses respond inconsistently to alternate year seasonal rest. *Rangel. Ecol. Manag. J. Range Manag. Arch.* **1976**, *29*, 346. [CrossRef]
34. Reardon, P.O.; Merrill, L.B. Vegetative response under various grazing management systems in the Edwards Plateau of Texas. *J. Range Manag.* **1976**, *29*, 195–198. [CrossRef]

35. Kawamura, K.; Akiyama, T.; Yokota, H.O.; Tsutsumi, M.; Yasuda, T.; Watanabe, O.; Wang, S. Quantifying grazing intensities using geographic information systems and satellite remote sensing in the Xilingol steppe region, Inner Mongolia, China. *Agric. Ecosyst. Environ.* **2005**, *107*, 83–93. [CrossRef]
36. Numata, I.; Roberts, D.A.; Chadwick, O.A.; Schimel, J.; Sampaio, F.R.; Leonidas, F.C.; Soares, J.V. Characterization of pasture biophysical properties and the impact of grazing intensity using remotely sensed data. *Remote Sens. Environ.* **2007**, *109*, 314–327. [CrossRef]
37. Ma, S.; Zhou, Y.; Gowda, P.; Chen, L.; Steiner, J.; Starks, P.; Neel, J. Evaluating the impacts of continuous and rotational grazing on tallgrass prairie landscape using high spatial resolution imagery. *Agronomy* **2019**, in press.
38. Starks, P.J.; Steiner, J.L.; Neel, J.P.S.; Turner, K.E.; Northup, B.K.; Brown, M.A.; Gowda, P.H. Assessment of the SPEI as a potential management tool for grasslands. *Agronomy* **2019**, in press.
39. Steiner, J.L.; Starks, P.J.; Neel, J.P.S.; Northup, B.K.; Gowda, P.H.; Brown, M.A.; Coleman, S. Managing Tallgrass Prairies for Productivity and Ecological Function: A Long Term Grazing Experiment in the Southern Great Plains, USA. *Agronomy* **2019**, in press.
40. Huete, A.; Didan, K.; Miura, T.; Rodriguez, E.P.; Gao, X.; Ferreira, L.G. Overview of the radiometric and biophysical performance of the MODIS vegetation indices. *Remote Sens. Environ.* **2002**, *83*, 195–213. [CrossRef]
41. Boegh, E.; Søgaard, H.; Broge, N.; Hasager, C.B.; Jensen, N.O.; Schelde, K.; Thomsen, A. Airborne multispectral data for quantifying leaf area index, nitrogen concentration, and photosynthetic efficiency in agriculture. *Remote Sens. Environ.* **2002**, *81*, 179–193. [CrossRef]
42. Xiao, X.; Hollinger, D.; Aber, J.; Goltz, M.; Davidson, E.A.; Zhang, Q.; Moore III, B. Satellite-based modeling of gross primary production in an evergreen needleleaf forest. *Remote Sens. Environ.* **2004**, *89*, 519–534. [CrossRef]
43. Xiao, X.; Zhang, Q.; Braswell, B.; Urbanski, S.; Boles, S.; Wofsy, S.; Moore III, B.; Ojima, D. Modeling gross primary production of temperate deciduous broadleaf forest using satellite images and climate data. *Remote Sens. Environ.* **2004**, *91*, 256–270. [CrossRef]
44. Wagle, P.; Xiao, X.; Torn, M.S.; Cook, D.R.; Matamala, R.; Fischer, M.L.; Jin, C.; Dong, J.; Biradar, C. Sensitivity of vegetation indices and gross primary production of tallgrass prairie to severe drought. *Remote Sens. Environ.* **2014**, *152*, 1–14. [CrossRef]
45. Zhou, Y.; Xiao, X.; Wagle, P.; Bajgain, R.; Mahan, H.; Basara, J.B.; Dong, J.; Qin, Y.; Zhang, G.; Luo, Y. Examining the short-term impacts of diverse management practices on plant phenology and carbon fluxes of Old World bluestems pasture. *Agric. For. Meteorol.* **2017**, *237–238*, 60–70. [CrossRef]
46. Bajgain, R.; Xiao, X.; Basara, J.; Wagle, P.; Zhou, Y.; Mahan, H.; Gowda, P.; McCarthy, H.R.; Northup, B.; Neel, L.; et al. Carbon dioxide and water vapor fluxes in winter wheat and tallgrass prairie in central Oklahoma. *Sci. Total Environ.* **2018**, *644*, 1511–1524. [CrossRef]
47. Doughty, R.; Xiao, X.; Wu, X.; Zhang, Y.; Bajgain, R.; Zhou, Y.; Qin, Y.; Zou, Z.; McCarthy, H.R.; Friedman, J.; et al. Responses of gross primary production of grasslands and croplands under drought, pluvial, and irrigation conditions during 2010–2016, Oklahoma, USA. *Agric. Water Manag.* **2018**, *204*, 47–59. [CrossRef]
48. Zhang, Y.; Xiao, X.; Wu, X.; Zhou, S.; Zhang, G.; Qin, Y.; Dong, J. A global moderate resolution dataset of gross primary production of vegetation for 2000–2016. *Sci. Data* **2017**, *4*, 170165. [CrossRef]
49. Teague, W.; Dowhower, S.; Waggoner, J. Drought and grazing patch dynamics under different grazing management. *J. Arid Environ.* **2004**, *58*, 97–117. [CrossRef]
50. Díaz-Solís, H.; Grant, W.E.; Kothmann, M.M.; Teague, W.R.; Díaz-García, J.A. Adaptive management of stocking rates to reduce effects of drought on cow–calf production systems in semi-arid rangelands. *Agric. Syst.* **2009**, *100*, 43–50. [CrossRef]
51. Teague, W.R.; Provenza, F.; Norton, B.; Steffens, T.; Barnes, M.; Kothmann, M.; Roath, R. Benefits of multi-paddock grazing management on rangelands: Limitations of experimental grazing research and knowledge gaps. In *Grasslands: Ecology, Management and Restoration*; Schröder, H., Ed.; Nova Science Publishers: Hauppauge, NY, USA, 2009; pp. 41–80.
52. Torell, L.A.; Murugan, S.; Ramirez, O.A. Economics of Flexible Versus Conservative Stocking Strategies to Manage Climate Variability Risk. *Rangel. Ecol. Manag.* **2010**, *63*, 415–425. [CrossRef]
53. Roche, L.M.; Cutts, B.B.; Derner, J.D.; Lubell, M.N.; Tate, K.W. On-Ranch Grazing Strategies: Context for the Rotational Grazing Dilemma. *Rangel. Ecol. Manag.* **2015**, *68*, 248–256. [CrossRef]

54. Otkin, J.A.; Anderson, M.C.; Hain, C.; Svoboda, M.; Johnson, D.; Mueller, R.; Tadesse, T.; Wardlow, B.; Brown, J. Assessing the evolution of soil moisture and vegetation conditions during the 2012 United States flash drought. *Agric. For. Meteorol.* **2016**, *218*, 230–242. [CrossRef]
55. Zhou, Y.; Xiao, X.; Zhang, G.; Wagle, P.; Bajgain, R.; Dong, J.; Basara, J.; Jin, C.; Anderson, M.C.; Hain, C.; et al. Quantifying agricultural drought in tallgrass prairie region in the U.S. Southern Great Plains through analysis of a water-related vegetation index from MODIS images. *Agric. For. Meteorol.* **2017**, *246* (Suppl. C), 111–122. [CrossRef]

© 2019 by the authors. Licensee MDPI, Basel, Switzerland. This article is an open access article distributed under the terms and conditions of the Creative Commons Attribution (CC BY) license (http://creativecommons.org/licenses/by/4.0/).

Article

Conservation of Soil Organic Carbon and Nitrogen Fractions in a Tallgrass Prairie in Oklahoma

Alan J. Franzluebbers [1],*, Patrick J. Starks [2] and Jean L. Steiner [2,3,†]

1. USDA-Agricultural Research Service, 3218 Williams Hall, NCSU Campus Box 7620, Raleigh, NC 27695, USA
2. USDA-Agricultural Research Service, Grazinglands Research Laboratory, 7207 W. Cheyenne Street, El Reno, OK 73036, USA; patrick.starks@ars.usda.gov (P.J.S.); jlsteiner@k-state.edu (J.L.S.)
3. Agronomy Department, Kansas State University, Manhattan, KS 66506, USA
* Correspondence: alan.franzluebbers@ars.usda.gov; Tel.: +1-919-515-1973
† The author is retired.

Received: 14 March 2019; Accepted: 18 April 2019; Published: 20 April 2019

Abstract: Native grasslands in the Great Plains of North America have mostly disappeared in the past century due to agricultural expansion. A grazing study was established on Paleustolls and Argiustolls supporting a remnant, but historically grazed tallgrass prairie in central Oklahoma. Stocking method of beef cattle was differentiated into continuous and rotational treatments (10 sub-paddocks) in 2009 and these treatments continued until present. Soil was sampled in 2009 and 2012 at depths of 0–6, 6–12, 12–20, and 20–30 cm and in 2017 at depths of 0–15 and 15–30 cm. Total, particulate, microbial biomass, and mineralizable C and N fractions were highly stratified with depth, having 2–10 times greater concentration at a depth of 0–6 cm as that at 20–30 cm. Strong associations existed among most of these soil organic C and N fractions, given the large range that resulted from sampling at multiple depths. No discernable differences in soil organic C and N fractions occurred due to stocking method at any sampling time or depth. Evidence for biological nitrification inhibition suggested a mechanism for conservation of available N with less opportunity for loss. In addition, strong association of available N with biologically active C indicated slow, but sustained release of N that was strongly coupled to C cycling. We conclude that stocking method had a neutral effect on conservation of already high antecedent conditions of soil organic C and N fractions during the first 8 years of differentially imposed management.

Keywords: nitrogen mineralization; soil organic carbon; soil-test biological activity; stocking method; total soil nitrogen

---

1. Introduction

Grasslands were historically widespread ecosystems throughout the Great Plains of North America [1]. Both mass conversion of grasslands to cropland and neglect of remaining grasslands reduced their favor on landscapes in the 20th century [2,3]. Since the turn of the new millennium, greater recognition of the importance of grasslands in conserving soil, promoting biodiversity, stabilizing farming communities, and providing a wealth of natural ecosystem services has led to renewed interest in how grasslands function [4,5]. An important aspect of regaining full functionality of grasslands has focused on how livestock are stocked and allowed to graze available forages [6–8].

Grazing lands typically have greater soil organic C and N contents than other agricultural land uses, despite often relegated to poorer positions of the landscape [9]. Grasslands are ecosystems with the vast majority of C stored belowground in organic matter [10,11]. However, grazing livestock are an important regulator of how C and N in grasslands are partitioned in the ecosystem [12]. Grazing alters N and P cycling by transforming plant nutrients into microbial-enhanced animal feces and by increasing potential N loss through volatilization and leaching [13,14]. Overgrazing of native rangelands can cause significant loss of soil organic C and greater in situ CO_2 emission [15]. Soil organic C and associated biologically active components of microbial biomass and potentially mineralizable C were greater under pasture than under cultivated cropland in the surface 0–20 cm depth in Texas [16]. Across a diversity of studies in the southeastern USA, soil organic C sequestration under pastures was 0.84 ± 0.11 Mg C ha^{-1} year^{-1} [17]. Among mature grasslands, the presence of livestock may or may not always lead to changes in soil organic C [5,18].

Grazing of perennial forages stimulates regrowth with subsequent impacts on root turnover and storage of C and N in organic matter [18]. Grazing may also remove sufficient older forage to allow greater rates of photosynthesis and subsequent deposition of C into soil [19]. Stocking rate can be a factor in how much C and N is stored in soil as a result of how rapidly forage can recover from defoliation [20].

Rotational stocking has been recently promoted over continuous stocking to allow forage stands to rest and accumulate larger quantities of biomass more often during the year prior to initiation of grazing by livestock [21]. The objective of rotational stocking is to achieve efficient and more effective defoliation of forages to optimize pasture productivity and persistence [22]. Continuous stocking allows livestock unlimited and uninterrupted access to a pasture for as long as the manager desires. If grazed year-round, continuously stocked pastures often have hay fed to livestock during dormant periods on the same pasture that livestock graze when forage is growing. Under extreme conditions, continuous stocking can lead to dominance of the most grazing-resistant forage and/or predominance of undesirable plant species that livestock do not consume. Significant debate still exists in the scientific community as to if, how, and to what extent rotational stocking methods might improve upon functioning of grazing lands compared with continuous stocking [23,24]. From a research survey comparing farms with and without rotational stocking in Texas, soil organic matter and aggregate stability were greater in multi-paddock systems than in heavy continuous grazing [21]. In the surface 5 cm of soil, soil organic C was significantly greater with high-density rotational stocking than with continuous stocking on a previously degraded rangeland in South Africa [25]. However, among a dozen paired sites in Australia, no differences in soil organic C could be detected between rotational and continuous stocking [26].

Our goal was to contribute to this scientific discourse by documenting ecosystem-relevant effects of stocking method on soil C and N fractions in a temperate, mostly native grassland ecosystem relic in central Oklahoma of the U.S. Great Plains region. Our hypothesis was that rotational stocking would lead to improvements in soil organic C and N fractions relative to continuous stocking in response to greater residual forage mass, surface residue cover, and deeper and more vigorous rooting.

2. Materials and Methods

2.1. Experimental Conditions

The experiment was initiated in 2009 on existing pastures at the Grazinglands Research Laboratory in El Reno, Oklahoma (35°32′46″ N, 98°0′37″ W). Four experimental units were assigned one of two treatments, i.e., continuous and rotational stocking. A total of ~346 ha was included in the study on Norge silt loam (fine-silty, mixed, active, thermic Udic Paleustolls), Pond Creek silt loam (fine-silt, mixed, superactive, thermic Pachic Argiustolls), Kirkland-Pawhuska complex (fine, mixed, superactive, thermic Udertic Paleustolls), and Bethany silt loam (fine, mixed, superactive, thermic Pachic Paleustolls). Soil was sampled in April–May 2009 to a depth of 30 cm and contained 22 ± 4% clay and 31 ± 5% sand. Soil was sampled again in February–March 2012 and June–July 2017. Different sampling times were a consequence of weather and labor availability. Long-term mean annual temperature is 15.5 °C and mean annual precipitation is 801 mm.

The 346 ha area was divided into two experimental units of ~52 ha each that were assigned to continuous stocking and two experimental units of ~91 ha each that were assigned to rotational stocking. Replicates with rotational stocking were further divided into 10 nearly equally sized sub-paddocks.

Stocking rate was ~0.25 head ha^{-1} with cow-calf pairs [27]. Continuous stocking had livestock present from March to October, while rotational stocking had several (1–3) week-long grazing events per year on any particular sub-paddock.

2.2. Soil Sampling and Analyses

In 2009, 40 soil sampling sites were located within each replicate using a stratified sampling design, where the strata were composed of soil mapping units in the area (downloaded from the NRCS Geospatial Data Gateway: https://datagateway.nrcs.usda.gov/GDGOrder.aspx) [27]. Number of sampling sites for a given strata within a field replicate was based on an area weighted average of the soil mapping units. Sampling points within a given strata were randomly located and their positions recorded by a hand-held GPS device. Soil sampling in subsequent years was guided by these GPS coordinates. Soil cores were extracted from sampling locations using a hydraulically operated probe equipped with a 30 cm long barrel and inside diameter of 4 cm. Cores were divided into depth increments (0–6, 6–12, 12–20, and 20–30 cm in 2009 and 2012 and 0–15 and 15–30 cm in 2017), air-dried, and placed in a sealable plastic bag for later processing.

Dried soil was sieved to pass a screen with 4.75 mm openings prior to all analyses. For total organic C and total soil N, a representative 20–30 g subsample was ball-milled to a fine powder and ~1 g of sample was analyzed with dry combustion using a Leco TruSpec on samples collected in 2009 and 2012 and using a Leco TruMac on samples collected in 2017. Using the ball-milled subsample, residual inorganic N was determined from a filtered extract of a 10 g subsample shaken with 20 mL of 2 M KCl for 30 min by salicylate-nitroprusside (NH_4-N) and hydrazine reduction (NO_3-N) autoanalyzer techniques [28].

All other analyses were determined on coarsely-sieved soil using a standardized laboratory approach according to the following [29]. In all three years of evaluation, two subsamples of soil were incubated in tandem for determination of potential C and N mineralization, soil microbial biomass C, and particulate organic C and N. Two subsamples were weighed into volume-delimited 60 mL glass bottles—two 20 g subsamples in 2009, two 40 g subsamples in 2012, and two 50 g subsamples in 2017. Dried soil was wetted to 50% water-filled pore space after determining the volume of lightly packed soil to the nearest 5 mL. Both subsamples were placed into the same 1 L canning jar along with a screw-cap vial containing 10 mL of 1 M NaOH to trap CO_2 and a vial of water to maintain humidity. Jars were incubated at 25 °C for up to 24 days. Alkali traps were replaced at 3 and 10 days of incubation. Evolved CO_2-C was determined by titration with 1 M HCl in the presence of excess $BaCl_2$ and vigorous stirring to a phenolphthalein endpoint. At 10 days, one of the subsamples was removed from the incubation jar, fumigated with $CHCl_3$ under vacuum for 1 day, vapors removed, placed into a separate canning jar along with vials of alkali and water, and incubated at 25 °C for 10 more days. Potential C mineralization was calculated from the cumulative evolution of CO_2 during 24 days of incubation. Soil-test biological activity was from the flush of CO_2 that occurred in the first 3 days of incubation. Basal soil respiration was calculated from the assumed near-linear rate of C mineralization from 10 to 24 days of incubation (i.e., as potential steady-state respiration rate). Potential N mineralization was determined from the difference in inorganic N concentration between 0 and 24 days of incubation. Inorganic N (NH_4-N + NO_2-N + NO_3-N) at the end of 24 days of incubation was determined from the filtered extract of a 10 g subsample of dried (55 °C for ≥2 days) and sieved (<2 mm) soil that was shaken with 20 mL of 2 M KCl for 30 min using salicylate-nitroprusside and hydrazine reduction autoanalyzer techniques. Particulate organic matter was isolated from the dried subsample (55 °C for ≥2 days) that was incubated for microbial biomass C determination. The entire subsample was transferred to a 125 mL Nalgene screw-top bottle, shaken with 0.01 M $Na_4P_2O_7$ overnight for ~16 h, and the suspension passed over a screen with 0.053 mm openings to collect particulate organic matter. Material on the screen was washed with a stream of water until effluent became clear and transferred to a glass drying vessel. Samples were dried in an oven at 55 °C for 1 day past visual dryness. Dried samples were weighed to determine sand concentration, ball-milled to a fine powder, and analyzed for C and N concentrations with dry combustion.

2.3. Statistical Analyses

Soil C and N properties were analyzed using the general linear model procedure of SAS v. 9.4 for each year separately. Grazing method was the main effect and soil depth increment was considered a repeated measure with a separate error term. Means were separated using least significant difference (LSD) with alpha set at 0.05. In 2012, soil was collected from each of 10 separate paddocks within each of the two field replicates of both treatments. Mean values across the 10 subunits were computed, such that a total of four experimental units with four sampling depths (4 × 4 = 16 observations) was statistically analyzed the same way in 2009 and 2012. Only two depth increments (0–15 and 15–30 cm) were sampled in 2017 for a total of eight observations. In 2012, mean square error from within-paddock variation (i.e., 10 locations within a paddock across 2 treatments and 4 depths) was computed and compared with mean square error from between-paddock variation (i.e., 2 replications across 2 treatments and 4 depths).

3. Results and Discussion

3.1. Total Organic C and N

Total soil N was not affected by grazing management during any year of sampling and at any soil depth, as shown in Table 1. However, total soil N was highly stratified with soil depth, and remained so over time. Concentration of total soil N at the surface depth of 0–6 cm was nearly double that at the 6–12-cm depth and was 2.5–3 times greater than concentration at the 20–30 cm depth. Total soil N concentration of the 0–15 cm depth in 2017 was 1.8–2.0 times greater than at the 15–30 cm depth. These data clearly show that protection of surface soil from erosion is important in preserving N in soil. Within-paddock variation of total soil N was 3.4 times greater than between-paddock variation, suggesting that representative sampling of each paddock from multiple cores was a necessary step in characterizing soil condition.

Table 1. Total soil N (g N kg^{-1} soil) as affected by stocking method, soil depth, and year of sampling.

Soil Depth (cm)	2009		2012		2017 *	
	Continuous	Rotational	Continuous	Rotational	Continuous	Rotational
0–6	3.05	2.84	2.57	2.49	1.98	1.82
6–12	1.55	1.52	1.37	1.38		
12–20	1.23	1.22	1.15	1.16	0.99	1.02
20–30	1.01	0.97	1.00	1.00		
LSD ($p < 0.05$)	0.40		0.08		0.39	

LSD: Least significant difference ($p < 0.05$) is among all means within a year of sampling. * Sampling in 2017 was 0–15 and 15–30 cm depths.

Total organic C followed a similar pattern of declining concentration with a depth like that of total soil N (data not shown). Soil C:N ratio was greater in the surface 0–6 cm depth than other depths, and this was likely due to deposition of partially decomposed surface residue inputs. In 2009, soil C:N ratio was 15.6 ± 0.4 g g^{-1} at a depth of 0–6 cm and 13.3 ± 0.5 g g^{-1} at all lower depths. In 2012, within-paddock variation of total organic C was 3.3 times greater than between-paddock variation.

3.2. Particulate Organic C and N

Grazing management did not significantly affect particulate organic C at any soil depth, as shown in Table 2. Particulate organic C was more stratified with depth than total organic C or total soil N. As a portion of total organic C, particulate organic C was 341 ± 48 g kg^{-1} at a depth of 0–6 cm in

2009 and 2012 and was 109 ± 20 g kg^{-1} at depths of 6–30 cm. Such large stratification of particulate organic C with soil depth has been observed previously in managed pastures of Georgia [30]. Similarly, in cropland managed with no-tillage following termination of long-term pasture, the portion of total organic C as particulate organic C was between 300 and 400 g kg^{-1} at a depth of 0–6 cm, while it was 200 to 300 g kg^{-1} under conventional tillage [31]. Keeping soil undisturbed and allowing it to accumulate decomposing residues leads to an enrichment of particulate organic matter near the surface. From observations in Georgia, return of residues to the soil surface via dung deposition and trampling with grazing led to greater particulate organic C and portion of total organic C as particulate organic C compared with haying [30]. In the current study, stocking method did not appear to alter the balance between particulate and total organic C.

Table 2. Particulate organic C (g C kg^{-1} soil) as affected by stocking method, soil depth, and year of sampling.

Soil Depth (cm)	2009		2012		2017 *	
	Continuous	Rotational	Continuous	Rotational	Continuous	Rotational
0–6	18.9	16.4	10.3	9.9	4.5	4.3
6–12	2.6	2.9	2.0	2.0		
12–20	1.6	1.7	1.4	1.5	0.9	1.0
20–30	1.2	1.1	1.1	1.1		
LSD (p < 0.05)	4.9		1.2		1.4	

LSD: Least significant difference (p < 0.05) is among all means within a year of sampling. * Sampling in 2017 was 0–15 and 15–30 cm depths.

3.3. Soil Microbial Biomass and Activity

Soil microbial biomass C was not affected by grazing management regime in any particular year of sampling, but a shift with time was trending towards greater microbial biomass C near the soil surface with rotational stocking compared with continuous stocking, as shown in Table 3. Although not directly comparable, soil microbial biomass C levels were similar in magnitude as a study with newly developed bermudagrass pasture in Georgia [32]. In 2012, within-paddock variation of soil microbial biomass C was 5.0 times greater than between-paddock variation.

Table 3. Soil microbial biomass C (mg C kg^{-1} soil) as affected by stocking method, soil depth, and year of sampling.

Soil Depth (cm)	2009		2012		2017 *	
	Continuous	Rotational	Continuous	Rotational	Continuous	Rotational
0–6	1306	915	1171	1274	719	800
6–12	647	670	560	588		
12–20	333	410	472	510	260	297
20–30	383	448	401	389		
LSD (p < 0.05)	370		129		193	

LSD: Least significant difference (p < 0.05) is among all means within a year of sampling. * Sampling in 2017 was 0–15 and 15–30 cm depths.

Basal soil respiration was highly stratified with depth, like other soil C and N properties, but was generally not affected by stocking method, as shown in Figure 1. However, at a depth of 12–20 cm in 2009, basal soil respiration was greater under rotational than continuous stocking. The lack of consistency in this effect over time suggests that it may have been an artefact of sampling/analysis technique or random variation within the field. All years of data clearly showed the strong depth stratification, even when changing from narrow soil depth increments to broader depth increments. In 2012, within-paddock variation of basal soil respiration was 3.6 times greater than between-paddock variation.

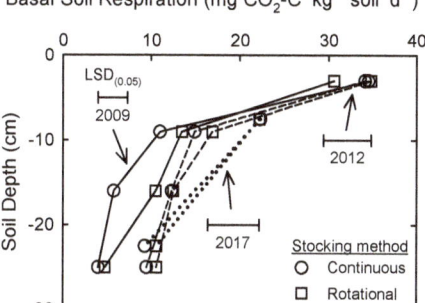

Figure 1. Basal soil respiration as affected by stocking method, soil depth, and year of sampling. Least significant difference ($p < 0.05$) is among all means within a year of sampling.

Soil-test biological activity was not affected by stocking method and was gradually stratified with depth like many other soil C and N properties, as shown in Table 4. Soil-test biological activity was highly associated with soil microbial biomass C, as well as with other indicators of soil microbial activity like basal soil respiration and net N mineralization, as shown in Figure 2. Strong association among these biological indicators has been documented previously [33]. Soil-test biological activity was also highly associated with total soil N ($r^2 = 0.86$, $p < 0.001$). Soil-test biological activity has recently been suggested as a simple, rapid indicator of biologically derived soil N that can supply greenhouse-grown [34] and field-grown forages [35] with N. Proportional to total soil N, net N mineralization during 24 days of aerobic incubation at standard temperature and water conditions released 40 g N kg^{-1} total soil N. This would be a considerable amount of N that could supply forage plants with N on an annual basis. Assuming an average bulk density of the surface 12 cm of 1.1 Mg m^{-3} and using the average total soil N of the surface 12 cm of 2.10 g N kg^{-1}, then this soil would have contained 2767 kg N ha^{-1}. If 40 g N kg^{-1} soil (4%) were considered mineralizable, then 111 kg N ha^{-1} could be expected to be mineralized from inherent soil conditions in a growing season. Considering soil from 12–30 cm with bulk density of 1.2 Mg m^{-3} and 1.09 g N kg^{-1}, then an additional 2360 kg N ha^{-1} would be present with expected mineralization of 94 kg N ha^{-1}. A total of 205 kg N ha^{-1} mineralized from organic matter could support 13.7 Mg ha^{-1} of forage that had an average N concentration of 15 g kg^{-1}. Assuming this forage was consumed and only a small fraction removed in animal carcass, the majority of this N could be effectively recycled year after year. However, significant losses of N could occur through NH$_3$ volatilization from urine deposits, as well as denitrification and leaching if sufficient nitrate were present.

Table 4. Soil-test biological activity (mg CO$_2$-C kg^{-1} soil 3 day^{-1}) as affected by stocking method, soil depth, and year of sampling.

Soil Depth (cm)	2009		2012		2017 *	
	Continuous	Rotational	Continuous	Rotational	Continuous	Rotational
0–6	575	565	535	581	497	475
6–12	280	297	278	293		
12–20	207	225	235	241	203	210
20–30	189	174	207	209		
LSD ($p < 0.05$)	72		60		161	

LSD: Least significant difference ($p < 0.05$) is among all means within a year of sampling. * Sampling in 2017 was 0–15 and 15–30 cm depths.

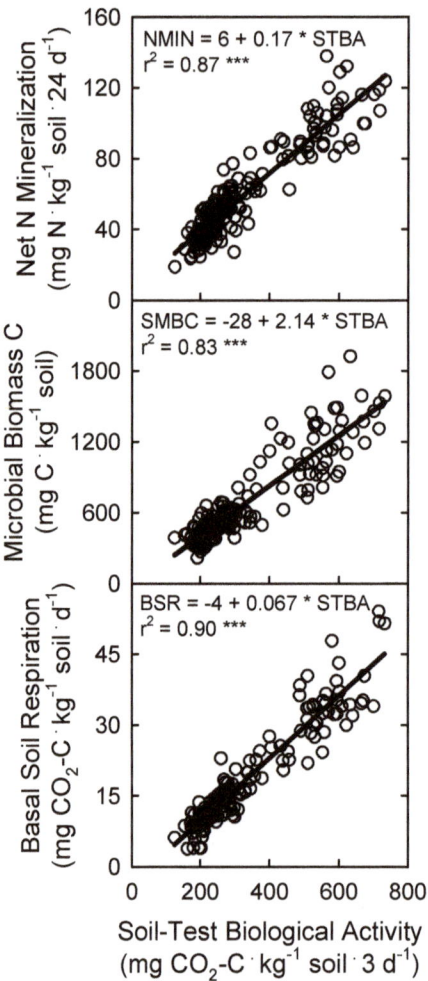

Figure 2. Association of soil-test biological activity to other indicators of soil biology, including basal soil respiration, soil microbial biomass C, and net N mineralization during aerobic incubation. *** indicates significance of association at $p < 0.001$.

3.4. Residual Inorganic N and Net N Mineralization

Residual inorganic N in soil was predominately in the form of NH_4-N and barely detectable in the form of NO_3-N, as shown in Table 5. Low residual soil nitrate in this pasture ecosystem would suggest that either rapid plant uptake limited accumulation of nitrate in soil or that nitrification was slow or limited in this soil. Interestingly, evidence for limited nitrification potential was observed during the laboratory incubation, in which apparent nitrification estimates during the 24 day aerobic incubation were only ~50% of mineralized N and declined with soil depth, as shown in Table 6. Nitrification activity is typically high in modern agricultural production systems, but may be much more limited in low-N-input production systems using traditional crop rotations and diversity of crops [36]. Biological nitrification inhibition is a potential mechanism that might have prevailed in this grassland ecosystem, which was dominated by warm-season perennial grasses with relatively low N input. In Brazil under *Brachiaria humidicola* pasture, very low nitrate accumulation occurred following urine application, suggesting significant biological nitrification inhibition and subsequently reduced N_2O emission [37]. Another possibility is that frequently water-limited conditions at this location in Oklahoma may chronically impede nitrifying activity. Whatever the mechanism for this low nitrifying activity, it may be an effective strategy to avoid N loss in an otherwise N limited natural ecosystem that relies on organic matter accumulation and internal N cycling for maintaining productivity.

Table 5. Residual inorganic N (separated into NH_4-N and NO_3-N components) as affected by stocking method, soil depth, and year of sampling.

Soil Depth (cm)	2009		2012		2017 *	
	Continuous	Rotational	Continuous	Rotational	Continuous	Rotational
Residual soil ammonium (mg NH_4-N kg^{-1} soil)						
0–6	7	14	24	23	15	9
6–12	8	9	14	14		
12–20	14	7	10	10	9	6
20–30	11	6	8	8		
LSD ($p < 0.05$)	8		2		3	
Residual soil nitrate (mg NO_3-N kg^{-1} soil)						
0–6	1	1	1	1	1	1
6–12	1	1	2	2		
12–20	1	1	1	1	<1	<1
20–30	1	1	1	1		
LSD ($p < 0.05$)	1		<1		1	

LSD: Least significant difference ($p < 0.05$) is among all means within a year of sampling. * Sampling in 2017 was 0–15 and 15–30 cm depths.

Table 6. Apparent nitrification of mineralized N (mg NO_3-N accumulation mg^{-1} mineralized N) as affected by stocking method, soil depth, and year of sampling.

Soil Depth (cm)	2009		2012		2017 *	
	Continuous	Rotational	Continuous	Rotational	Continuous	Rotational
0–6	0.44	0.58	0.57	0.51	0.12	0.15
6–12	0.25	0.19	0.08	0.05		
12–20	0.12	0.12	0.04	0.04	0.03	0.03
20–30	0.11	0.09	0.04	0.04		
LSD ($p < 0.05$)	0.34		0.14		0.11	

LSD: Least significant difference ($p < 0.05$) is among all means within a year of sampling. * Sampling in 2017 was 0–15 and 15–30 cm depths.

Further evidence for inhibition of nitrification was from strong net N mineralization that corresponded proportionally to variations in C mineralization, as shown in Figure 2. During the 24 day incubation period, significant accumulation of NH_4-N occurred at all depths, averaging 49 ± 14 mg NH_4-N kg^{-1} soil at a depth of 0–6 cm in both 2009 and 2012 and 43 ± 10 mg NH_4-N kg^{-1} soil at depths of 6–30 cm. Significant nitrification occurred at the 0–6 cm depth with additional accumulation of 57 ± 23 mg NO_3-N kg^{-1} soil, but limited nitrification could be detected at depths of 6–30 cm with only 5 ± 6 mg NO_3-N kg^{-1} soil accumulated during the 24 day incubation. Net N mineralization during 24 days of incubation was strongly stratified with soil depth, like many other soil C and N fractions, as shown in Table 7. The soil-handling process of oven-drying prior to rewetting may have altered nitrification potential somewhat, but it was not considered a major factor since apparent nitrification in cropland soils using the same laboratory protocols was 0.92 ± 0.13, 0.71 ± 0.27, and 0.53 ± 0.29 mg NO_3-N mg^{-1} total N mineralized at depths of 0–10, 10–20, and 20–30 cm, respectively [33].

Table 7. Net N mineralization (mg N kg^{-1} soil 24 day^{-1}) as affected by stocking method, soil depth, and year of sampling.

Soil Depth (cm)	2009		2012		2017 *	
	Continuous	Rotational	Continuous	Rotational	Continuous	Rotational
0–6	110	120	97	99	89	82
6–12	76	65	55	55		
12–20	46	54	46	47	33	38
20–30	32	37	33	34		
LSD ($p < 0.05$)	35		7		3	

LSD: Least significant difference ($p < 0.05$) is among all means within a year of sampling. * Sampling in 2017 was 0–15 and 15–30 cm depths.

4. Conclusions

Stocking method on tallgrass prairie pasture during an 8 year period had little discernable effect on soil C and N fractions within the surface 30 cm of the soil profile. This may have been due to the lack of recent soil disturbance and high antecedent concentration of soil properties prior to initiation of this study. All soil organic C and N properties were highly stratified with depth. One curious finding was the occurrence of what appeared to be significant biological nitrification inhibition, which may have provided substantial conservation of N for internal cycling of N from soil to forages, and ultimately for livestock protein intake and animal performance, and subsequent deposition back onto pasture with urine and feces. We conclude that in this intermediate period of grassland evaluation, rotational stocking of livestock did not lead to changes in total, particulate, microbial biomass, or mineralizable fractions of C and N as compared with continuous stocking. Of key importance in this study was the preservation of soil properties with both grazing management approaches. We documented strong association among biologically active fractions of soil organic C and N.

Author Contributions: Conceptualization, P.J.S. and J.L.S.; methodology, A.J.F.; formal analysis, A.J.F.; investigation, P.J.S. and J.L.S.; resources, A.J.F., P.J.S., and J.L.S.; writing—original draft preparation, A.J.F.; writing—review and editing, P.J.S. and J.L.S.; supervision, J.L.S.; project administration, P.J.S. and J.L.S.

Funding: This research was a contribution from the Long Term Agroecosystem Research (LTAR) network. USDA-Agricultural Research Service and USDA-National Institute of Food and Agriculture (AFRI Grant 2012-02355 and 2013-69002-23146) provided financial support.

Acknowledgments: We thank Ellen Leonard, Ashtyn Mizelle, and Erin Silva for sound technical support with laboratory analyses. We are grateful for the dedicated field support of G.P. King, D. Walker, and D. Gassen.

Conflicts of Interest: The authors declare no conflict of interest.

References

1. Stefferud, A. (Ed.) *Grass: The 1948 Yearbook of Agriculture*; U.S. Govt. Print. Office: Washington, DC, USA, 1948; 892p.
2. Savory, A.; Butterfield, J. *Holistic Management*, 2nd ed.; Island Press: Washington, DC, USA, 1999; 616p.
3. Wedin, W.F.; Fales, S.L. (Eds.) *Grassland: Quietness and Strength for a New American Agriculture*; American Society for Agronomy, Crop Science Society of America, Soil Science Society of America: Madison, WI, USA, 2009; 256p.
4. Franzluebbers, A.J. Ecosystem services from forages. In *Cool Forages: Advanced Management of Temperate Forages*; Bittman, S., Hunt, D., Eds.; Pacific Field Corn Association: Agassiz, BC, Canada, 2013; pp. 2–6.
5. Sollenberger, L.E.; Kohmann, M.M.; Dubeux, J.C.B.; Silveira, M.L. Grassland management affects delivery of regulating and supporting ecosystem services. *Crop Sci.* **2019**, *59*, 441–459. [CrossRef]
6. Briske, D.D.; Derner, J.D.; Brown, J.R.; Fuhlendorf, S.D.; Teague, W.R.; Havstad, K.M.; Gillen, R.L.; Ash, A.J.; Willms, W.D. Rotational grazing on rangelands: Reconciliation of perception and experimental evidence. *Rangel. Ecol. Manag.* **2008**, *61*, 3–17. [CrossRef]

7. Franzluebbers, A.J.; Paine, L.K.; Winsten, J.R.; Krome, M.; Sanderson, M.A.; Ogles, K.; Thompson, D. Well-managed grazing systems: A forgotten hero of conservation. *J. Soil Water Conserv.* **2012**, *67*, 100A–104A. [CrossRef]
8. Nelson, C.J. (Ed.) *Conservation Outcomes from Pastureland and Hayland Practices: Assessment, Recommendations, and Knowledge Gaps*; Allen Press, Inc.: Lawrence, KS, USA, 2012; 362p.
9. Franzluebbers, A.J. Grass roots of soil carbon sequestration. *Carbon Manag.* **2012**, *3*, 9–11. [CrossRef]
10. Parton, W.J.; Stewart, J.W.B.; Cole, C.V. Dynamics of C, N, P and S in grassland soils: A model. *Biogeochemistry* **1988**, *5*, 109–131. [CrossRef]
11. Burke, I.C.; Yonker, C.M.; Parton, W.J.; Cole, C.V.; Flach, K.; Schimel, D.S. Texture, climate and cultivation effects on soil organic matter content in U.S. grassland soils. *Soil Sci. Soc. Am. J.* **1989**, *53*, 800–805. [CrossRef]
12. Steinfeld, H.; Wassenaar, T. The role of livestock production in carbon and nitrogen cycles. *Ann. Rev. Environ. Resour.* **2007**, *32*, 271–294. [CrossRef]
13. Taboada, M.A.; Rubio, G.; Chaneton, E.J. Grazing impacts on soil physical, chemical and ecological properties in forage production systems. In *Soil Management: Building a Stable Base for Agriculture*; Hatfield, J.L., Sauer, T.J., Eds.; American Society for Agronomy, Soil Science Society of America: Madison, WI, USA, 2011; pp. 301–320.
14. Soussana, J.-F.; Lemaire, G. Coupling carbon and nitrogen cycles for environmentally sustainable intensification of grasslands and crop-livestock systems. *Agric. Ecosyst. Environ.* **2014**, *190*, 9–17. [CrossRef]
15. Abdalla, K.; Mutema, M.; Chivenge, P.; Everson, C.; Chaplot, V. Grassland degradation significantly enhances soil CO_2 emission. *Catena* **2018**, *167*, 284–292. [CrossRef]
16. Franzluebbers, A.J.; Hons, F.M.; Zuberer, D.A. In situ and potential CO_2 evolution from a Fluventic Ustochrept in southcentral Texas as affected by tillage and cropping intensity. *Soil Tillage Res.* **1998**, *47*, 303–308. [CrossRef]
17. Franzluebbers, A.J. Achieving soil organic carbon sequestration with conservation agricultural systems in the southeastern United States. *Soil Sci. Soc. Am. J.* **2010**, *74*, 347–357. [CrossRef]
18. Hewins, D.B.; Lyseng, M.P.; Schoderbek, D.F.; Alexander, M.; Willms, W.D.; Carlyle, C.N.; Chang, S.X.; Bork, E.W. Grazing and climate effects on soil organic carbon concentration and particle-size association in northern grasslands. *Sci. Rep.* **2018**, *8*, 1336. [CrossRef] [PubMed]
19. Gomez-Casanovas, N.; DeLucia, N.J.; Bernacchi, C.J.; Boughton, E.H.; Sparks, J.P.; Chamberlain, S.D.; DeLucia, E.H. Grazing alters net ecosystem C fluxes and the global warming potential of a subtropical pasture. *Ecol. Appl.* **2018**, *28*, 557–572. [CrossRef] [PubMed]
20. Wright, A.L.; Hons, F.M.; Rouquette, F.M. Long-term management impacts on soil carbon and nitrogen dynamics of grazed bermudagrass pastures. *Soil Biol. Biochem.* **2004**, *36*, 1809–1816. [CrossRef]
21. Teague, W.R.; Dowhower, S.L.; Baker, S.A.; Haile, N.; DeLaune, P.B.; Conover, D.M. Grazing management impacts on vegetation, soil biota and soil chemical, physical and hydrological properties in tall grass prairie. *Agric. Ecosyst. Environ.* **2011**, *141*, 310–322. [CrossRef]
22. Sollenberger, L.E.; Agouridis, C.T.; Vanzant, E.S.; Franzluebbers, A.J.; Owens, L.B. Prescribed grazing on pasturelands. In *Conservation Outcomes from Pastureland and Hayland Practices: Assessment, Recommendations, and Knowledge Gaps*; Nelson, C.J., Ed.; Allen Press, Inc.: Lawrence, KS, USA, 2012; pp. 111–204.
23. Briske, D.D.; Sayer, N.F.; Huntsinger, L.; Fernandez-Gimenez, M.; Budd, B.; Derner, J.D. Origin, persistence, and resolution of the rotations grazing debate: Integrating human dimensions into rangeland research. *Rangel. Ecol. Manag.* **2011**, *64*, 325–334. [CrossRef]
24. Teague, R.; Provenza, F.; Kreuter, U.; Steffens, T.; Barnes, M. Multi-paddock grazing on rangelands: Why the perceptual dichotomy between research results and rancher experience? *J. Environ. Manag.* **2013**, *128*, 699–717. [CrossRef] [PubMed]
25. Chaplot, V.; Dlamini, P.; Chivenge, P. Potential of grassland rehabilitation through high density-short duration grazing to sequester atmospheric carbon. *Geoderma* **2016**, *271*, 10–17. [CrossRef]
26. Sanderman, J.; Reseigh, J.; Wurst, M.; Young, M.-A.; Austin, J. Impacts of rotational grazing on soil carbon in native grass-based pastures in southern Australia. *PLoS ONE* **2015**, *10*, e0136157. [CrossRef] [PubMed]
27. Steiner, J.L.; Starks, P.J.; Neel, J.P.S.; Northup, B.; Turner, K.E.; Gowda, P.; Coleman, S.; Brown, M. *Managing Tallgrass Prairies for Productivity and Ecological Function: A Long Term Grazing Experiment in the Southern Great Plains, USA*; USDA-Agricultural Research Service: Raleigh, NC, USA, 2019.

28. Bundy, L.G.; Meisinger, J.J. Nitrogen availability indices. In *Methods of Soil Analysis, Part 2*; Weaver, R.W., Angle, J.S., Bottomley, P.J., Eds.; Soil Science Society of America: Madison, WI, USA, 1994; pp. 951–984.
29. Franzluebbers, A.J.; Stuedemann, J.A. Early response of soil organic fractions to tillage and integrated crop-livestock production. *Soil Sci. Soc. Am. J.* **2008**, *72*, 613–625. [CrossRef]
30. Franzluebbers, A.J.; Stuedemann, J.A. Particulate and non-particulate fractions of soil organic carbon under pastures in the Southern Piedmont USA. *Environ. Pollut.* **2002**, *116*, S53–S62. [CrossRef]
31. Franzluebbers, A.J.; Stuedemann, J.A. Temporal dynamics of total and particulate organic carbon and nitrogen in cover crop grazed cropping systems. *Soil Sci. Soc. Am. J.* **2014**, *78*, 1404–1413. [CrossRef]
32. Franzluebbers, A.J.; Stuedemann, J.A. Bermudagrass management in the Southern Piedmont USA. III. Particulate and biologically active soil carbon. *Soil Sci. Soc. Am. J.* **2003**, *67*, 132–138. [CrossRef]
33. Franzluebbers, A.J.; Pershing, M.R.; Crozier, C.; Osmond, D.; Schroeder-Moreno, M. Soil-test biological activity with the flush of CO_2: I. C and N characteristics of soils in corn production. *Soil Sci. Soc. Am. J.* **2018**, *82*, 685–695. [CrossRef]
34. Franzluebbers, A.J.; Pershing, M.R. Soil-test biological activity with the flush of CO_2: II. Greenhouse growth bioassay from soils in corn production. *Soil Sci. Soc. Am. J.* **2018**, *82*, 696–707. [CrossRef]
35. Franzluebbers, A.J.; Pehim-Limbu, S.; Poore, M.H. Soil-test biological activity with the flush of CO_2: IV. Fall-stockpiled tall fescue yield response to applied nitrogen. *Agron. J.* **2018**, *110*, 2033–2049. [CrossRef]
36. Subbarao, G.V.; Yoshihashi, T.; Worthington, M.; Nakahara, K.; Ando, Y.; Sahrawat, K.L.; Rao, I.M.; Lata, J.-C.; Kishii, M.; Braun, H.-J. Suppression of soil nitrification by plants. *Plant Sci.* **2015**, *233*, 155–164. [CrossRef] [PubMed]
37. Byrnes, R.C.; Nunez, J.; Arenas, L.; Rao, I.; Trujillo, C.; Alvarez, C.; Arango, J.; Rasche, F.; Chirinda, N. Biological nitrification inhibition by Brachiaria grasses mitigates soil nitrous oxide emissions from bovine urine patches. *Soil Biol. Biochem.* **2017**, *107*, 156–163. [CrossRef]

© 2019 by the authors. Licensee MDPI, Basel, Switzerland. This article is an open access article distributed under the terms and conditions of the Creative Commons Attribution (CC BY) license (http://creativecommons.org/licenses/by/4.0/).

Article

Synergistic and Antagonistic Effects of Poultry Manure and Phosphate Rock on Soil P Availability, Ryegrass Production, and P Uptake

Patricia Poblete-Grant [1,2,3], Philippe Biron [3], Thierry Bariac [3], Paula Cartes [1,4], María de La Luz Mora [1,*] and Cornelia Rumpel [3,*]

1. Center of Plant, Soil Interaction and Natural Resources Biotechnology, Scientific and Biotechnological Bioresource Nucleus (BIOREN-UFRO), Universidad de La Frontera, 4780000 Temuco, Chile; patty.grant87@gmail.com (P.P.-G.); paula.cartes@ufrontera.cl (P.C.)
2. Programa de Doctorado en Ciencias de Recursos Naturales, Universidad de La Frontera, Avenida Francisco Salazar, 01145, P.O. Box 54-D, 4780000 Temuco, Chile
3. CNRS, SU, Institute of Ecology and Environmental Sciences of Paris (IEES, UMR SU-UPEC-CNRS-INRA-IRD), Campus AgroParis Tech, 78850 Thiverval-Grignon, France; philippe.biron@inra.fr (P.B.); bariac.thierry@numericable.fr (T.B.)
4. Departamento de Ciencias Químicas y Recursos Naturales, Facultad de Ingeniería y Ciencias, Universidad de La Frontera, Avenida Francisco Salazar 01145, 4780000 Temuco, Chile
* Correspondence: mariluz.mora@ufrontera.cl (M.d.L.L.M.); cornelia.rumpel@inra.fr (C.R.); Tel.: +56-45-2734191 (M.d.L.L.M.); +33-1-30815479 (C.R.)

Received: 15 February 2019; Accepted: 8 April 2019; Published: 15 April 2019

Abstract: To maintain grassland productivity and limit resource depletion, scarce mineral P (phosphorus) fertilizers must be replaced by alternative P sources. The effect of these amendments on plant growth may depend on physicochemical soil parameters, in particular pH. The objective of this study was to investigate the effect of soil pH on biomass production, P use efficiency, and soil P forms after P amendment application (100 mg kg^{-1} P) using poultry manure compost (PM), rock phosphate (RP), and their combination (PMRP). We performed a growth chamber experiment with ryegrass plants (*Lolium perenne*) grown on two soil types with contrasting pH under controlled conditions for 7 weeks. Chemical P fractions, biomass production, and P concentrations were measured to calculate plant uptake and P use efficiency. We found a strong synergistic effect on the available soil P, while antagonistic effects were observed for ryegrass production and P uptake. We conclude that although the combination of PM and RP has positive effects in terms of soil P availability, the combined effects of the mixture must be taken into account and further evaluated for different soil types and grassland plants to maximize synergistic effects and to minimize antagonistic ones.

Keywords: Poultry manure; phosphate rock; ryegrass; plant biomass; phosphorus uptake; phosphorus availability

1. Introduction

Fertilization of grasslands with mineral phosphorus (P) fertilizer is a common practice in many regions of the world to maintain productivity, especially on soils with high P retention [1–3]. In order to reduce the use of scarce phosphate rock (RP), alternative P fertilizers need to be found [4,5]. In this context, poultry manure, an abundant organic waste material from the growing broiler industry [6], is known for its high P content [7]. Its transformation through composting into organic amendments, and their subsequent application in grassland systems may be a promising strategy [8,9] to reduce the use of mineral fertilizers. Several studies showed that plant nutrient uptake and the biomass of several plants could be significantly increased using poultry manure compost (PM) [10,11]. The application of PM led to changes in soil P forms and phosphatase activity [12]. However, despite its positive effects on plant nutrient availability and biomass production, the application of PM may lead to the simultaneous introduction of contaminants [13] and could also lead to a loss of P to waterways following long term application.

Therefore, nowadays, the use of PM in combination with RP has been considered as good practice to limit the use of both materials without compromising plant requirements [14]. However, the fertilizer value of both substrates may depend on the soil reaction. For example, RP efficiency may be limited in soils with high pH due to its low dissolution rate [15], while in soils with acid pH, RP may lead to further acidification [16]. The efficiency of PMRP mixture for increasing wheat and chili yield and P uptake has already been demonstrated for acid and alkaline soils [17,18]. However, no study has been carried out to investigate quantitatively the synergetic or antagonistic effects of the combined application of both materials.

In this study, we carried out a growth chamber experiment to investigate the effect of the combined use of PM and RP as compared to their application as a single amendment in soils with similar properties, but contrasting in pH. The objective of the study was to determine the effect of PM application, alone or combined with RP, on ryegrass biomass production, P use efficiency, and soil P availability in an acid and an alkaline soil. We hypothesized that the soils' and plants' response to the combined use of PM and RP in terms of biomass production and P use efficiency may depend on the soil reaction and that the mixture will have additional effects as compared to the use of PM and RP as a single amendment. Moreover, we hypothesized that the combined use of PM and RP will ameliorate P availability and biomass production as compared to their use as a single amendment.

2. Materials and Methods

2.1. Materials

We used two silty soils (50–60% silt): A Neoluvisol with a pH of 6.1 (moderately acid soil) and a carbonated Luvisol, with a pH of 8.5 (alkaline soil) (Table 1) according to the French Référentiel Pédologique 2008 [19]. Both soils showed similar texture, organic matter content, and soil forming processes (lixiviation). They were differentiated by pH and also their initial Olsen P concentration. They are part of the French observatory SOERE PRO (https://www6.inra.fr/valor-pro/SOERE-PRO-les-sites). The Neoluvisol is located in Eastern France at Colmar, and the Luvisol is located in northwestern France in Brittany at Le Rheu. We sampled the first 0 to 30 cm. of the control plots without fertilization at the two sites. After sampling, the soils were transported to the laboratory, air dried, and sieved at 2 mm. The plant species used was ryegrass (*Lolium perenne*), a typical pasture plant used for grazing systems.

Table 1. Soil physical and chemical characterization.

Soil Type	pH	C_{org} g kg^{-1}	C/N	P olsen mg kg^{-1}	K_2O Cmol + kg^{-1}	Clay %	Silt	Sand
Moderately acid	6.1	11.9	10	60	0.32	14.6	68.3	16.1
Alkaline	8.5	12.1 *	10	11	0.26	20.7	59.8	6.8

* $CaCO_3$ = 128 g kg^{-1}.

PM compost was provided by KOMECO B.V in pellet form with a dry matter content of 880 g kg^{-1} with an organic matter content of 600 g kg^{-1}. Phosphorus content was 13.2 g kg^{-1} d.w. The material contained 42% organic P and 58% of inorganic P (Table 2). RP was bought from 'Les comptoirs de Jardin' and was derived from bones with 30% P and 50% calcium. It was provided in powder form with 90% of the particles smaller than 0.16 mm.

Table 2. Inorganic (Pi) and organic P (Po) in fractions sequentially extracted from poultry manure compost PM.

H$_2$O		NaHCO$_3$		NaOH		HCl		Residual
				P mg kg^{-1}				
Pi	Po	Pi	Po	Pi	Po	Pi	Po	Pt
187 ± 8	122 ± 43	202 ± 3	149 ± 30	47 ± 4	72 ± 13	181 ± 9	97 ± 8	145 ± 7

2.2. Growth Chamber Experiment

The experiment was carried out in pots in the RUBIC V biogeochemical reactor—Servathin, Carrières-sous-Poissy France—for 7 weeks. To account for the contrasting bulk densities, we amended 490 g of the moderately acid soil and 550 g of the alkaline soil per treatment, with four replicates with poultry manure compost (PM), phosphate rock (RP), or their mixture consisting of 70% PM and 30% RP (PMRP). In total, 100 mg P per kg^{-1} soil d.w. were added to each treatment. The amendments were supplied in the form of a dry powder. We added 13.30 g of PM and 0.80 g RP to the pots with a single amendment and 9.31 g of PM and 0.23 g of RP to the pots with amendment mixtures. To account for N and K supplied by PM (262 mg N and 221 mg K per kg d.w. soil), we added the corresponding amounts of K and N in the form of KCl and NH$_4$NO$_3$ to all other treatments, including the control. The PM application was equivalent to 9.8 Mg ha^{-1} when applied in the mixture with RP and 14 Mg ha^{-1} when applied as a single amendment. The RP application was equivalent to 0.25 Mg ha^{-1} when applied in a mixture with PM and 0.8 Mg ha^{-1} when applied as a single amendment.

After addition of the amendments, the soils were thoroughly mixed in plastic bags, added to each pot, and brought to field capacity with tap water. After one day, a total of 97 ryegrass seeds were added to each pot. Seeds were sown on the surface and covered superficially with soil material. Plants were grown at 24 °C (day temperature) and at 17 °C (night temperature) with a day length (light intensity of 650 µmol m^{-2} s^{-1}) of 8 h for the first 13 days, and 11 h until the end of the experiment. Soil moisture was maintained at 40% of the available field capacity by watering regularly. Air humidity was 75% to 65, % respectively, for day and night conditions.

After 7 weeks, shoots and roots were separated from soil and their fresh biomass was weighed. Thereafter, biomass was dried at 65 °C for 48 hours. Oven-dried plant material was ground to pass through 20-mesh (0.84 mm) sieves. Microbial biomass was determined using 5 g of fresh samples. The remaining soil masses were oven-dried at 40 °C and sieved at 2 mm. An aliquot was ground for further analyses. All data is expressed on a dry weight basis.

2.3. Soil Analysis

Phosphorus forms based on P solubility were measured using a modified Hedley fractionation scheme [20] with successive chemical P extraction from soluble to residual fractions. Briefly, 1 g of dry and sieved soil was extracted sequentially by shaking for 16 h with 30 mL of (1) distilled water, (2) 0.5 M NaHCO$_3$ at pH 8.5, (3) 0.1 M NaOH, and (4) 1 M HCl. Each suspension was centrifuged at 10,000 rpm for 10 min and the supernatants were recovered and analyzed for total P (Pt) and inorganic P (Pi). Organic P (Po) was determined by difference. The residues were dried at 60 °C and used for subsequent extractions. Residual P remaining after the last step was extracted with 1 M sulphuric acid (H$_2$SO$_4$) during 24 h, after calcination of the residue for 1 h at 550 °C. Inorganic P was determined in the solutions by the ammonium molybdate-ascorbic acid method [21]. Total P was determined by taking aliquots from supernatants for digestion using potassium persulfate (K$_2$S$_2$O$_8$) and 2.5 M sulphuric acid (H$_2$SO$_4$). Fraction one and two represent readily available P. Moderately available P is found in fraction 3, less available P in fraction 4, whereas residual P represents unavailable P. Pt, Po and Pi of bulk soil were calculated as sum of Hedley fractions.

Total organic C and N concentrations of soil samples were determined using an elemental analyzer (Variopyrocube, Elementar, Langensebold, Hesse, Germany). The acid soil was carbonate free and total soil C therefore corresponded to organic C. For the alkaline soil (carbonated soil), HCl-fumigation was performed before elemental analyses to remove inorganic C [22]. Microbial biomass P was determined by the chloroform fumigation-extraction method [23]. Briefly, 5 g of fresh soil were extracted with 0.5 M NaHCO$_3$ before and after fumigation with CHCl$_3$. Total P was determined in the solutions by the ammonium molybdate-ascorbic acid method [21]. Microbial biomass P was calculated as the difference between fumigated and non-fumigated soil and multiplied by a factor of 0.40 [23].

2.4. Biomass Analysis

Total N and C concentrations were measured using an elemental analyzer (Variopyrocube, Elementar, Langensebold, Hesse, Germany). Total shoot P contents were analyzed by calcination followed by acid recovery using inductive coupled plasma mass spectrometry (iCAPTM Q ICP-MS, Thermo ScientificTM, Waltham, MA, USA). The P uptake (mg) was calculated as a product of the shoots' or roots' nutrient concentrations (mg g^{-1}) and shoot or root biomass (g). Nutrient use efficiency (PUE) was also calculated according to Baligar et al. [24] as follows:

$$PUE = \frac{P \text{ uptake in treatment } (mg) - P \text{ uptake in control}(mg)}{\text{total P applied}} \times 100 \quad (1)$$

2.5. Synergistic and Antagonistic Effect of Mixture

Based on the quantities of PM and RP applied as single amendments or in mixture (see 2.2), we calculated the additional effect of the PMRP mixture on the soil available P, biomass production, and P uptake resulting from the combined application of PM and RP as compared to their use as a single amendment. This calculation is justified by the observation made by many others that PM would lead to enhanced mineralization of RP due to the release of organic acids (exp. 18). However, this additional effect was never quantified. We used Equation (2) to calculate the additional effect, i.e., change of PMRP as compared to the sum of single amendments:

$$\% \text{ change} = \frac{\text{observed result} - \text{expected result}}{\text{expected result}} \times 100 \quad (2)$$

The observed result was corrected for the control in order to obtain the effect of the amendments. The expected result for PMRP was obtained as a sum of the PM and RM after multiplication of the observed results in the two treatments with a coefficient accounting for the different proportions of the two amendments used in the mixture, i.e., 70% and 30%.

2.6. Statistical Analysis

The experiment was arranged in a completely randomized design with four replicates. Data were checked for normality (Shapiro-Wilk test) and homogeneity of variance (Levene test) were determined before analyses. Statistical differences of means (95% significance level) were analyzed using two-way analyses of variance (two-way ANOVA). Post hoc tests with the function Tukey-test were made for the explanatory variables independently when the ANOVA detected significant differences. The relationships between soil available P and plant parameters were tested by Pearson correlation analyses. Statistical testing was done using the statistical program R Foundation for Statistical Computing Version 1.1.456 (R Development Core Team 2009–2018); effects were deemed significant at $p \leq 0.05$. Principal component analysis (PCA) was performed using the package, Factoextra; we consider one for the soil P fractions and a second one for the plant P uptake and biomass.

3. Results

3.1. Total Soil C, N, and P Concentrations

After the end of the experiment, PM treatment led to significantly increased total soil C concentrations in the moderately acid and alkaline soil by 57 and 29%, respectively (Table 3). The RP treatment showed no differences as compared to control, while its combination with PM increased the total C concentrations in both soils, with significant effects only in the moderately acid soil. Total N concentration was increased by 29% using PM in the alkaline soil. In the moderately acid soil, PM treatment increased total soil N concentration by 57%, whereas a lower increase was noted in the PMRP treatment (30%). Total P increased in both soils with PM amendment.

Table 3. Soil parameters in control and soil amended with poultry manure compost (PM), phosphate rock (RP), and their combination (PMRP) after 7 weeks of ryegrass growth.

Soil Type		C	N	C/N	Pt	C:Po	C:Pi	N:Pi	N:Po
		g kg^{-1}			mg kg^{-1}				
Moderate acid	Control	10.5 Ca *	1.2 Da	8.9 Aa	100.2 Ca	344.9 Ba	192.2 Bb	21.6 Cb	38.6 Ba
	RP	11.0 Ca	1.4 Ca	7.7 Ba	97.3 Ca	314.4 Cb	194.4 Bb	25.3 Bb	40.9 Bb
	PM	16.5 Aa	2.0 Aa	8.5 Aa	132.1 Aa	356.4 Ba	255.9 Aa	30.2 Aa	42.1 Ba
	PMRP	13.7 Ba	1.7 Ba	8.3 Aa	118.0 Ba	414.9 Aa	210.5 Aa	25.3 Bb	49.8 Aa
Alkaline	Control	11.5 Ba	1.3 Ba	9.2 Aa	103.2 Ba	365.3 Ba	241.9 Aa	26.3 Ba	39.8 Ba
	RP	12.0 Ba	1.7 Ab	7.2 Aa	103.5 Ba	399.7 Aa	239.2 Aa	33.3 Aa	55.6 Aa
	PM	14.8 Aa	1.7 Ab	8.8 Ba	127.2 Aa	383.2 Ba	244.1 Aa	28.1 Ba	44.3 Ba
	PMRP	13.4 ABa	1.6 Aa	8.7 Aa	120.8 Aa	275.0 Cb	290.1 Aa	33.5 Aa	31.8 Bb

* Upper case letters denote significant differences ($p \leq 0.05$) between treatments for one soil. Lower case letters denote significanct differences ($p \leq 0.05$) between soils for one treatment.

The C:N decreased in both soils amended with RP as compared with PM, PMRP, and the control (Table 3). In the moderately acid soil, PMRP treatments showed the highest N:Po and C:Po, while for the alkaline soil, the use of the mixture led to an increase of N:Pi, while C:Po and N:Po decreased. C:Pi was increased by both PM and PMRP treatments in the moderately acid soil only. N:Pi increased in all treatments in the moderately acid soil, while in the alkaline soil, only PMRP and RP increased this ratio significantly as compared with the control (Table 3).

3.2. Soil Phosphorus Forms

Concentrations of readily available fractions (extractable with water and NaHCO$_3$) ranged between 5.20 to 14.94 mg P kg^{-1} soil for inorganic and 6.07 to 27.58 mg P kg^{-1} soil for organic P (Figure 1). The moderately available fraction (NaOH extractable) ranged between 6.07 and 27.58 mg P kg^{-1} soil. Soils treated with PM and PMRP increased significantly inorganic and organic P concentrations in the readily available fraction as compared with the control and RP treatment. In the alkaline soil, RP amendment increased readily available inorganic and organic P on average 1.2-fold as compared with the control. No differences were found for the moderately acid soil (Figure 1). In the other fractions, the P concentrations ranged between 22.52 to 39.73 and 1.20 to 14.30 mg P kg^{-1} soil as the inorganic and organic form, respectively.

All amendments increased organic P in the less available P fraction in both soils, whereas residual P was increased with regards to the control only by PM amendments.

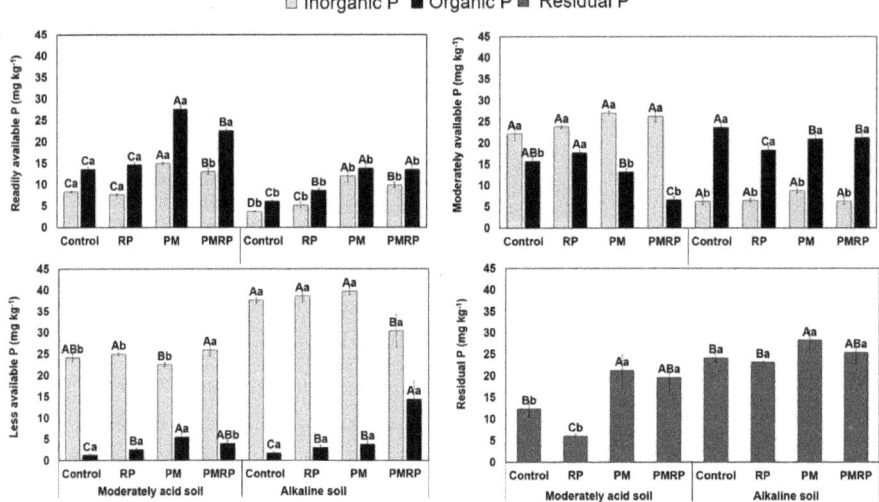

Figure 1. Inorganic and organic P concentration of fractions extracted from control and soil amended with poultry manure compost (PM), phosphate rock (RP), and their combination (PMRP) after 7 weeks of ryegrass growth. Upper case letters denote significant differences ($p \leq 0.05$) between treatments for one soil. Lower case letters denote significant differences ($p \leq 0.05$) between soils for one treatment.

3.3. Microbial Biomass P

Microbial biomass P ranged between 7.55 to 35.10 mg P kg^{-1} soil, and it was significantly increased with all P amendments as compared to the control (Figure 2). With PM amendment, microbial biomass P concentrations increased greatly as compared to RP amendment by 81% to 93% in moderately acid and alkaline soil, respectively (Figure 2). In the moderately acid soil, PM and its combination with RP induced the greatest increases of microbial biomass P (81–100%) as compared with RP while for the alkaline soil, only PM as a single amendment increased MBP with regards to RP.

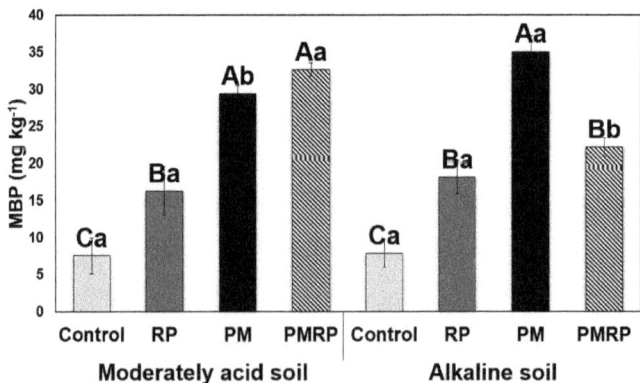

Figure 2. Microbial biomass phosphorus (MBP) in in control and soil amended with poultry manure compost (PM), phosphate rock (RP), and their combination (PMRP) after 7 weeks of ryegrass growth. Upper case letters denote significant differences ($p \leq 0.05$) between treatments for one soil. Lower case letters denote significant differences ($p \leq 0.05$) between soils for one treatment.

3.4. Shoot and Root Biomass Production

Plants grown in the moderately acid soil showed greater root biomass in all treatments compared to those grown in the alkaline soil, whereas shoot biomass was similar for both soils, except for the RP treatment.

The dry weight of plants cultivated in moderately acid soil treated with different P amendments and their combination was significantly higher ($p \leq 0.05$) compared to the control (Figure 3). The PM and RP increased shoot biomass similarly (~2.5 fold) as compared with the control. Root biomass increased by 1.2 to 3.3-fold in both soils treated with either PM alone or combined with RP.

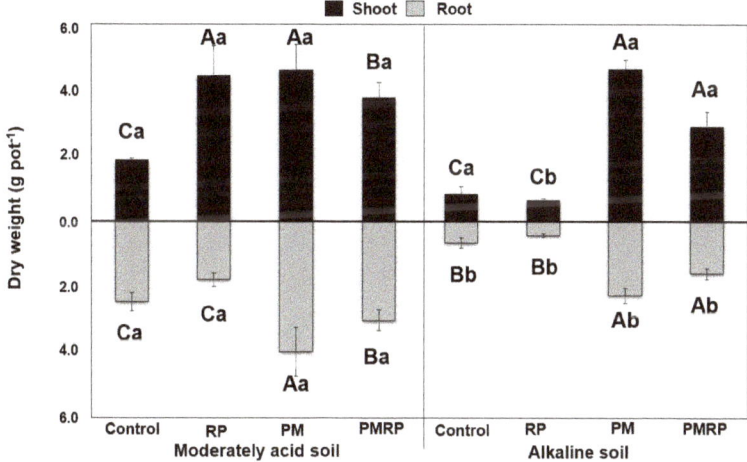

Figure 3. Dry weight of shoots and roots in control and soil amended with poultry manure compost (PM), phosphate rock (RP), and their combination (PMRP) after 7 weeks of ryegrass growth. Upper case letters denote significant differences ($p \leq 0.05$) between treatments for one soil. Lower case letters denote significant differences ($p \leq 0.05$) between soils for one treatment.

In the alkaline soil, no difference in dry matter production was observed for the RP treatment with respect to the control. In contrast, PM alone and its combination with RP significantly increased shoot biomass (5.6 and 3.5-fold) and root biomass (3.3 and 2.3-fold) as compared to the control (Figure 3).

3.5. Shoot and Root P Concentrations and Uptake

Shoot and root P concentrations and root P uptake increased significantly with all amendments as compared to the control (Table 4). Shoot P uptake showed no change compared to the control for both soils. The highest increase was noted for PM and PMRP treatments, which showed on average 37% to 48% higher root and shoot P concentrations than plants of the control treatment (Table 4). This is in line with a higher P uptake (Table 4). Shoot P concentration and uptake was higher in all treatments except RP in the moderately acid soil as compared to the alkaline one. Despite differences in uptake, root P concentrations were similar for both soils.

Table 4. P concentration and uptake of roots and shoots and P use efficiency after 7 weeks of ryegrass growth in control soil and soil amended with poultry manure compost (PM), phosphate rock (RP), and their combination (PMRP).

Soil Type	Treatment	P conc		P uptake		P Use Efficiency
		Shoot	Root	Shoot	Root	
		g kg^{-1}		mg		% of input
Moderate acid	Control	7.2 ^{Da}	1.4 ^{Da}	3.6 ^{Ba}	3.6 ^{Ca}	-
	RP	12.5 ^{Ca}	2.2 ^{Ca}	3.5 ^{Ba}	4.6 ^{Ca}	6 ^{Ca}
	PM	24.2 ^{Aa}	3.8 ^{Aa}	4.9 ^{Aa}	14.8 ^{Aa}	28 ^{Aa}
	PMRP	16.6 ^{Ba}	2.8 ^{Ba}	5.0 ^{Aa}	8.7 ^{Ba}	14 ^{Ba}
Alkaline	Control	1.4 ^{Db}	1.7 ^{Ba}	2.6 ^{Bb}	1.1 ^{Cb}	-
	RP	4.9 ^{Cb}	2.1 ^{Ba}	3.0 ^{Ba}	2.8 ^{Ba}	5 ^{Ca}
	PM	17.7 ^{Ab}	3.3 ^{Aa}	3.8 ^{Ab}	9.7 ^{Ab}	25 ^{Aa}
	PMRP	11.3 ^{Bb}	3.0 ^{Aa}	3.8 ^{Ab}	6.9 ^{Aa}	16 ^{Ba}

Upper case letters denote significant differences ($p \leq 0.05$) between treatments for one soil. Lower case letters denote significant differences ($p \leq 0.05$) between soils for one treatment.

Plant P use efficiency of the added P sources and their combinations ranged from 5% to 6 % for RP to a maximum of ~28% for PM alone (Table 4). Plant use efficiency of the mixture, PMRP, was in between these values. Few differences were noted between soils.

3.6. Relationship between Soil and Plant Parameters

In the alkaline soil, the readily available inorganic and organic P fractions, moderately available Po, residual P, and microbial biomass P were strongly correlated with shoot and root biomass and nutrient uptake (Table 5).

Table 5. Relationship between soil and plant parameters after 7 weeks of ryegrass growth in control and soil amended with poultry manure compost (PM), phosphate rock (RP), and their combination (PMRP). *denotes significant correlation coefficients ($p \leq 0.05$), $n = 16$.

Soil	Soil Parameters	Shoot Biomass	Root Biomass	P Uptake	
				Shoot	Root
Moderately acid soil	readily-Pi	0.41	0.80 *	0.87 *	0.92 *
	readily-Po	0.51 *	0.78 *	0.95 *	0.96 *
	moderately-Pi	0.61 *	0.59 *	0.83 *	0.83 *
	moderately-Po	−0.07	−0.43	−0.37	−0.39
	Less avail.-Pi	−0.10	−0.44	−0.40	−0.45
	less-avail. Po	0.71 *	0.61 *	0.93 *	0.89 *
	Residual-P	0.19	0.70 *	0.63 *	0.74 *
	Microbial P	0.62 *	0.61 *	0.83 *	0.78 *
	Total soil N	0.65 *	0.69 *	0.94 *	0.91 *
	Total soil C	0.52 *	0.71 *	0.89 *	0.88 *
Alkaline soil	readily-Pi	0.94 *	0.93 *	0.96 *	0.91 *
	readily-Po	0.84 *	0.82 *	0.89 *	0.84 *
	moderately-Pi	0.75 *	0.67 *	0.75 *	0.67 *
	moderately-Po	0.09	0.16	−0.18	−0.13
	Less avail.-Pi	−0.03	−0.14	−0.03	−0.09
	less-avail. Po	0.30	0.38	0.33	0.39
	Residual-P	0.81 *	0.80 *	0.70 *	0.66 *
	Microbial P	0.88 *	0.84 *	0.95 *	0.88 *
	Total soil N	0.45	0.31	0.60 *	0.51
	Total soil C	0.81 *	0.80 *	0.82 *	0.77 *

In the acid soil, inorganic and organic P in the readily available fraction and moderately available-Pi showed a positive correlation with root biomass and root P uptake (Table 5). Residual P was correlated with root biomass and less available organic P was somewhat correlated with shoot biomass and root P uptake. In this soil, the strongest correlations were noted for P uptake in the root and shoot with readily available P, moderately available Pi, less available Po, microbial P, total C, and total N.

To investigate the effect of the different types of amendments on soil and plant parameters, we performed PCA analyses. The first two PCA components explained 82% of the total variance of soil P fractions (Figure 4A). The individual representation of treatments on the factor map showed spatial separation of treatments in both soils, with a tendency to the formation of two groups, one for the PM and PMRP treatments and a second one for the control and RP (Figure 4A). Both groups were separated along the 2nd axis related to the contribution of less available organic P and residual P. Additionally, acid and alkaline soils could be separated according to soil P forms along the first axis. Alkaline soils were thus associated with less and moderately available Po and less available Pi, while acid soils were associated with readily available Pi and Po and moderately available Pi (Figure 4A).

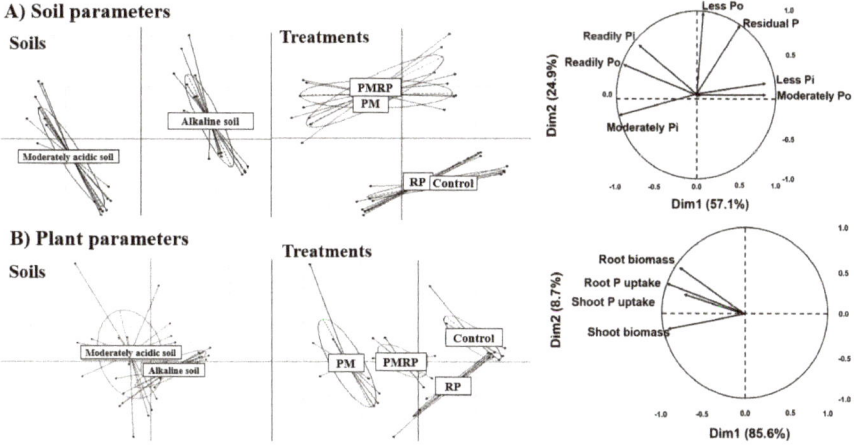

Figure 4. Principal component analysis of soil P forms (**A**) and plant (**B**) variables studied as the response to different soil types and treatments.

For the PCA performed with plant parameters, the first two components explained 94% of the total variability. The individual representation of treatments on the factor map (Figure 4B) showed spatial separation of all the treatments of both soils. Treatments with poultry manure were associated with P uptake and aboveground and belowground biomass in the positive direction. Plant parameters did not differentiate acid and alkaline soils.

3.7. Synergistic and Antagonistic Effects between PM and RP on Soil and Plant Parameters

To elucidate the effects of the amendment mixture, PMRP, on plant and soil parameters, we compared the expected values as the sum of the single use of PM and RP to the observed value for the mixture. Differences were interpreted in terms of synergistic and antagonistic effects. While the mixture had a similar synergistic effect on P availability in both soils, their effect on plant parameters was strongly dependent on the soil type (Table 6). For the alkaline soil, the mixture had no or a slightly antagonistic effect on plant biomass and P uptake, while in the moderately acid soil, the mixture had an antagonistic effect on P uptake, and a synergistic effect on root and shoot biomass.

Table 6. Synergistic and antagonistic effect on soil and plant parameters as the response of the combined application of poultry manure compost and phosphate rock applied at a P rate of 100 mg kg^{-1} soil after 7 weeks of ryegrass growth. +SD.

			Observed	Expected	Mixture Effect (%)
Moderately acid soil	Soil	P availability (mg kg^{-1})	137.1 ± 21.1	132.5 ± 4.9	9.9 ± 1.8
	Shoot	Biomass (g)	1.9 ± 0.5	1.6 ± 0.4	25.5 ± 3.9
		P uptake (mg)	9.4 ± 1.3	12.3 ± 1.2	−23.6 ± 1.8
	Root	Biomass (g)	1.16 ± 0.3	0.65 ± 0.2	91.3 ± 4.9
		P uptake (mg)	1.5 ± 1.1	3.97 ± 0.9	−59.6 ± 3.7
Alkaline soil	Soil	P availability (mg kg^{-1})	133.5 ± 15.0	114.4 ± 9.5	17.5 ± 3.0
	Shoot	Biomass (g)	2.1 ± 0.5	2.38 ± 0.2	−13.1 ± 1.9
		P uptake (mg)	9.97 ± 0.3	11.6 ± 0.5	−13.6 ± 2.6
	Root	Biomass (g)	0.78 ± 0.1	0.81 ± 0.1	−3.3 ± 1.2
		P uptake (mg)	5.5 ± 2.1	5.82 ± 1.2	1.8 ± 0.2

4. Discussion

4.1. Impact of Organic and Inorganic P Amendments on C, N, and P Stoichiometry and Microbial Biomass P

Treatments with poultry manure compost (PM and PMRP) showed increased soil C, N, and P concentrations, most probably related to organic matter input through the amendment as well as higher biomass production (Figure 3). All amendments changed to some extent soil organic matter stoichiometric ratios. These ratios determine the interlinkage between biochemical cycles of C, N, and P, providing information about nutrient availability following SOM decomposition and stabilization processes. The effect of the amendments on stoichiometric ratios was dependent on the type of amendment and also the soil type. Differences in stoichiometric ratios in the two soil types suggest that the amendment effect could be soil pH dependent.

Amendments increased P incorporation into the microbial biomass. Soil type had an effect on the microbial response to the application of the mixture (PMRP). It was interesting to note that while in the moderately acid soil, microbial biomass P was similar in PM and PMRP, in the alkaline soil, microbial biomass P was lower in PMRP compared to PM (Figure 2). Similar negative effects of inorganic fertilizers on soil microorganisms have been reported before [25]. The results of this study show an antagonistic effect of both materials when applied in combination to alkaline soil. In view of the importance of soil microorganisms for the maintenance of soil fertility and the sustainability of grassland ecosystems and their role for P immobilization, especially in soils with low C:P ratios, such as the ones of the present study [26], we suggest that the soil reaction may be an important criterion to consider when organic P fertilizers are applied in combination with inorganic ones.

4.2. Impact of Organic and Inorganic P Amendments on Nutrient Uptake and Biomass Production and Soil P Forms

Our results, indicated that shoot biomass increased with all P sources, except for RP in alkaline soil (Figure 3). A contrasting effect of RP on plant growth depending on the soil reaction has been observed before and the higher efficiency of RP under acid soil conditions is well documented (e.g., [18]). As many studies have shown [12], biomass production was strongly correlated to readily available inorganic P ($r = 0.94$ for shoot biomass in the alkaline soil and for root biomass in both soils) (Table 6). Readily available soil P may be already present in amendments [27] or may have been mineralized from organic forms after amendment addition [28]. Our data indicate that while large amounts of readily available P were added with PM (Table 2), in both soils, P uptake was correlated to microbial biomass P, less available organic P, and/or residual P, C, and N (Table 5). This suggests that the mineralization of organic matter is an important process for biomass production after amendment addition. Soil reaction may influence the importance of the latter process, as much stronger correlations between those parameters were noted in the acid soil as compared to the alkaline one. The importance of the soil reaction for soil P forms is further illustrated by the PCA analyses of soil parameters (Figure 4A).

However, this is different for plant parameters, which were differentiated by treatments, but not soil types (Figure 4B).

Our data show that using 14 Mg ha^{-1} PM as a single amendment supplied sufficient nutrients to meet plant requirements well above the critical N, P, and K concentration in shoots for producing 90% of the maximum ryegrass yield, which are 18, 3.4, and 28 mg g^{-1} [10]. Since PM is a very rich animal manure compost, it may help to build up soil productivity better than other amendments due to additional effects [29]. This is especially true for the moderately acidic soil, which was not deficient in plant available P as indicated by the plant P concentrations of the control treatment. However, we found that in both soils, PM addition increased the shoot and root P uptake as compared to the RP when used as single amendments (Table 4). Moreover, PM led to higher plant P use efficiency than RP (Table 4). This could be an indication that there are nutrient limitations other than N, P, and K occurring, which were counterbalanced with PM. Moreover, it is also possible that PM led to the stimulation of microbial activity by promoting P uptake [21].

4.3. Synergistic and Antagonistic Effects of the Combined Application of PM and RP

Our results showed that combining RP with PM may be highly efficient in increasing the soil available P concentrations, thereby enhancing biomass production. While, in general, RP has low concentrations of available P [30], various studies [31–34] have demonstrated that P availability from RP may be increased by co-application of organic amendments. For example, Sohail et al. [35] showed that combining RP with compost increases soil P availability, and Abbasi and co-workers [18] demonstrated that the release of P to the soil increased by 80% compared with when RP was combined with PM.

In this study, our data revealed strong synergistic effects in terms of P availability independent of the soils' pH (Table 6). In contrast, biomass production showed a strong synergistic response in the moderately acid soil and an antagonistic response in the alkaline soil. Soil type dependent antagonistic effects were also observed for P uptake. Root P uptake was lower than expected in the moderately acid soil, while it was similar to the expected value in the alkaline soil (Table 6). This might be explained by a higher immobilization of P in the microbial biomass in moderately acid soil as compared to the alkaline one (Figure 3). Microbial P may become available and thus microbial immobilization as well as the antagonistic effect might be transitory. Another explanation of these contrasting effects may be related to the initial P status of the soil. As the moderately acid soil was not P deficient, the increased availability of P following PM and RP did not foster uptake. We suggest that synergistic and antagonistic effects in different soil types should be evaluated and taken into consideration, when elaborating new fertilizer strategies through the combination of organic and inorganic fertilizers.

5. Conclusions

Our study showed that the P distributed between inorganic and organic P fractions was greatly affected by the type of amendment and soil type. Poultry manure compost increased highly soil P availability, consequently improving above and belowground plant biomass production on both soil types, while phosphate rock amendment had limited effects on soil P fractions and positive effects on aboveground plant biomass production only in moderately acidic soil. In general, the influence of amendment type on soil parameters was limited and mainly related to the organic matter input. In contrast, the soil amendment type had a strong effect on plant parameters.

We found synergistic effects of the combined use of PM and RP for soil available P in both soils. For plant parameters, synergistic and antagonistic effects were soil type dependent. We therefore suggest that fertilizer strategies through the combination of organic and inorganic fertilizers must be tested in different soil types by quantifying their synergistic and antagonistic effects. Moreover, the use of poultry manure compost, alone or combined with phosphate rock, could be a strategy to replace inorganic fertilizers and should be tested in long-term field experiments.

Author Contributions: Data curation, P.P.-G.; Formal analysis, P.P.-G.; Funding acquisition, C.R., P.C. and M.d.L.L.M.; Investigation, C.R.; Methodology, P.B. and T.B.; Resources, T.B. and C.R.; Software, P.B.; Supervision, M.d.L.L.M. and C.R.; Visualization, P.P.-G.; Writing—original draft, P.P.-G.; Writing, review, and editing, P.P.-G., P.B., P.C., M.d.L.L.M. and C.R.

Funding: This research was funded by CNRS under the framework of the EC2CO project LOMBRICOM, FONDECYT projects 1181050 and 1161326, and Conicyt scholarship n° 21150715.

Acknowledgments: The authors gratefully acknowledge the IEES laboratory for technical support. We also thank the SOERE-PRO for providing the soils used in this study and the KOMECO company for providing the poultry manure compost.

Conflicts of Interest: The authors declare no conflict of interest.

References

1. Redel, Y.; Cartes, P.; Demanet, R.; Poblete-Grant, P.; Bol, R.; Mora, M.L.; Velásquez, G. Assessment of phosphorus status influenced by Al and Fe compounds in volcanic grassland soils. *J. Soil Sci. Nutr.* **2016**, *16*, 490–506. [CrossRef]
2. Velásquez, G.; Calabi-Floody, M.; Poblete-Grant, P.; Rumpel, C.; Demanet, R.; Condron, L.; Mora, M. Fertilizer effects on phosphorus fractions and organic matter in Andisols. *J. Soil Sci. Nutr.* **2016**, *16*, 294–304. [CrossRef]
3. Rumpel, C.; Crème, A.; Ngo, P.; Velásquez, G.; Mora, M.; Chabbi, A. The impact of grassland management on biogeochemical cycles involving carbon, nitrogen and phosphorus. *J. Soil Sci. Nutr.* **2015**, *15*, 353–371. [CrossRef]
4. Cordell, D.; Drangert, J.-O.; White, S. The story of phosphorus: Global food security and food for thought. *Glob. Environ. Chang.* **2009**, *19*, 292–305. [CrossRef]
5. Reijnders, L. Phosphorus resources, their depletion and conservation, a review. *Resour. Conserv. Recycl.* **2014**, *93*, 32–49. [CrossRef]
6. Food and Agriculture Organization of the United Nations. *FAO Food Outlook. Biannual Report on Global Food Markets*; Food and Agriculture Organization of the United Nations: Roma, Italy, 2018.
7. Pagliari, P.H.; Laboski, C.A.M. Investigation of the Inorganic and Organic Phosphorus Forms in Animal Manure. *J. Environ. Qual.* **2012**, *41*, 901. [CrossRef] [PubMed]
8. Calabi-Floody, M.; Medina, J.; Rumpel, C.; Condron, L.M.; Hernandez, M.; Dumont, M.; Mora, M.D.L.L. Smart Fertilizers as a Strategy for Sustainable Agriculture. *Adv. Agron.* **2018**, *147*, 119–157. [CrossRef]
9. Redding, M.; Lewis, R.; Kearton, T.; Smith, O. Manure and sorbent fertilisers increase on-going nutrient availability relative to conventional fertilisers. *Sci. Total Environ.* **2016**, *569*, 927–936. [CrossRef]
10. Evers, G.W. Ryegrass-Bermudagrass Production and Nutrient Uptake when Combining Nitrogen Fertilizer with Broiler Litter. *Agron. J.* **2002**, *94*, 905–910. [CrossRef]
11. Pederson, G.A.; Brink, G.E.; Fairbrother, T.E. Nutrient Uptake in Plant Parts of Sixteen Forages Fertilized with Poultry Litter: Nitrogen, Phosphorus, Potassium, Copper, and Zinc. *Agron. J.* **2002**, *94*, 895–904. [CrossRef]
12. Waldrip, H.M.; He, Z.; Erich, M.S. Effects of poultry manure amendment on phosphorus uptake by ryegrass, soil phosphorus fractions and phosphatase activity. *Boil. Fertil. Soils* **2011**, *47*, 407–418. [CrossRef]
13. Foust, K.; Phillips, M.; Hull, K.; Ychorova, D. Changes in Arsenic, Copper, Iron, Manganese and Zinc Levels Resulting from the Application of Poultry Litter to Agricultural Soils. *Toxics* **2018**, *6*, 28. [CrossRef]
14. Song, K.; Xue, Y.; Zheng, X.; Lv, W.; Qiao, H.; Qin, Q.; Yang, J. Effects of the continuous use of organic manure and chemical fertilizer on soil inorganic phosphorus fractions in calcareous soil. *Sci. Rep.* **2017**, *7*, 327. [CrossRef]
15. Zapata, F.; Roy, R.N. *Utilización de Las Rocas Fosfóricas Para Una Agricultura Sostenible*; Food and Agriculture Organization of the United Nations: Roma, Italy, 2007; p. 15.
16. Rajan, S.S.S.; Fox, R.L.; Upsdell, M.; Saunders, W.M.H. Influence of pH, time and rate of application on phosphate rock dissolution and availability to pastures. *Nutr. Cycl. Agroecosyst.* **1991**, *28*, 85–93. [CrossRef]
17. Abbasi, M.K.; Mansha, S.; Rahim, N.; Ali, A. Agronomic Effectiveness and Phosphorus Utilization Efficiency of Rock Phosphate Applied to Winter Wheat. *Agron. J.* **2013**, *105*, 1606. [CrossRef]
18. Abbasi, M.K.; Musa, N.; Manzoor, M. Mineralization of soluble P fertilizers and insoluble rock phosphate in response to phosphate-solubilizing bacteria and poultry manure and their effect on the growth and P utilization efficiency of chilli (*Capsicum annuum* L.). *Biogeosciences* **2015**, *12*, 4607–4619. [CrossRef]
19. Baize, D.; Girard, M.C. *Référentiel Pédologique*; Editions Quae: Versailles, France, 2008. (In French)

20. Hedley, M.J.; Stewart, J.W.B.; Chauhan, B.S. Changes in Inorganic and Organic Soil Phosphorus Fractions Induced by Cultivation Practices and by Laboratory Incubations. *Soil Sci. Soc. Am. J.* **1982**, *46*, 970–976. [CrossRef]
21. Murphy, B.; Riley, J.P. A modified single solution method for the determination of phosphate in natural waters. *Anal. Chim. Acta.* **1962**, *27*, 31–36. [CrossRef]
22. Harris, D.; Horwáth, W.R.; van Kessel, C. Acid fumigation of soils to remove carbonates prior to total organic carbon or carbon-13 isotopic analysis. *Soil Sci. Soc. Am. J.* **2001**, *65*, 1853–1856. [CrossRef]
23. Brookes, P.C.; Powlson, D.S.; Jenkinson, D.S. Measurement of microbial biomass phosphorus in soil. *Soil Biol. Biochem.* **1982**, *14*, 319–329. [CrossRef]
24. Baligar, V.C.; Fageria, N.K.; He, Z.L. Nutrient use efficiency in plants. *Commun. Soil Sci. Plant Anal.* **2001**, *32*, 921–950. [CrossRef]
25. Lupwayi, N.; Lea, T.; Beaudoin, J.; Clayton, G. Soil microbial biomass, functional diversity and crop yields following application of cattle manure, hog manure and inorganic fertilizers. *Can. J. Soil Sci.* **2005**, *85*, 193–201. [CrossRef]
26. Zhang, L.; Ding, X.; Peng, Y.; George, T.S.; Feng, G. Closing the Loop on Phosphorus Loss from Intensive Agricultural Soil: A Microbial Immobilization Solution? *Front. Microbiol.* **2018**, *9*, 1–4. [CrossRef]
27. Giles, C.D.; Cade-Menun, B.J.; Liu, C.W.; Hill, J.E. The short-term transport and transformation of phosphorus species in a saturated soil following poultry manure amendment and leaching. *Geoderma* **2015**, *257–258*, 134–141. [CrossRef]
28. Singh, A.K.; Sarkar, A.K.; Kumar, A.; Singh, B.P. Effect of Long-term Use of Mineral Fertilizers, Lime and Farmyard Manure on the Crop Yield, Available Plant Nutrient and Heavy Metal Status in an Acidic Loam soil. *J. Indian Soc. Soil Sci.* **2009**, *57*, 362–365.
29. Agbede, T.M.; Ojeniyi, S.O. Tillage and poultry manure effects on soil fertility and sorghum yield in southwestern Nigeria. *Soil Tillage Res.* **2009**, *104*, 74–81. [CrossRef]
30. Kaleeswari, R.K.; Subramanian, S. Chemical reactivity of phosphate rocks—A review. *Agric. Rev.* **2001**, *22*, 121–126.
31. Ghosh, P.K.; Tripathi, A.K.; Bandyopadhyay, K.K.; Manna, M.C. Assessment of nutrient competition and nutrient requirement in soybean/sorghum intercropping system. *Eur. J. Agron.* **2009**, *31*, 43–50. [CrossRef]
32. Arcand, M.M.; Schneider, K. Plant- and microbial-based mechanisms to improve the agronomic effectiveness of phosphate rock: A review. *Ann. Braz. Acad. Sci.* **2006**, *78*, 791–807. [CrossRef]
33. Antil, R.S.; Singh, M. Effects of organic manures and fertilizers on organic matter and nutrients status of the soil. *Arch. Agron. Soil Sci.* **2007**, *53*, 519–528. [CrossRef]
34. Akande, M.O.; Adediran, J.A.; Oluwatoyinbo, F.I. Effects of rock phosphate amended with poultry manure on soil available P and yield of maize and cowpea. *Afr. J. Biotechnol.* **2005**, *4*, 444–448.
35. Qureshi, S.A.; Rajput, A.; Memon, M.; Solangi, M.A. Nutrient composition of rock phosphate enriched compost from various organic wastes. *E3 J. Sci. Res.* **2014**, *2*, 47–51.

© 2019 by the authors. Licensee MDPI, Basel, Switzerland. This article is an open access article distributed under the terms and conditions of the Creative Commons Attribution (CC BY) license (http://creativecommons.org/licenses/by/4.0/).

Article

Modeling Carbon and Water Fluxes of Managed Grasslands: Comparing Flux Variability and Net Carbon Budgets between Grazed and Mowed Systems

Nicolas Puche [1], Nimai Senapati [2], Christophe R. Flechard [3], Katia Klumpp [4], Miko U.F. Kirschbaum [5] and Abad Chabbi [1,6,*]

1. INRA, Versailles-Grignon, UMR ECOSYS, Bâtiment EGER, 78850 Thiverval-Grignon, France; nicolas.puche@inra.fr
2. Rothamsted Research, Department of Plant Sciences, West Common, Harpenden, Herts AL5 2JQ, UK; nimai.senapati@rothamsted.ac.uk
3. INRA UMR 1069 SAS, 65 rue de Saint-Brieuc, 35042 Rennes, France; christophe.flechard@inra.fr
4. INRA, VetAgro Sup, UMR 874 Ecosystème Prairial, 63100 Clermont Ferrand, France; katja.klumpp@inra.fr
5. Manaaki Whenua—Landcare Research, Private Bag 11052, Palmerston North 4442, New Zealand; KirschbaumM@landcareresearch.co.nz
6. INRA, Centre de recherche Nouvelle-Aquitaine-Poitiers, URP3F, 86600 Lusignan, France
* Correspondence: abad.chabbi@inra.fr; Tel.: +33-(0)1-3081-5289 or +33-(0)6-8280-0285

Received: 2 March 2019; Accepted: 7 April 2019; Published: 10 April 2019

Abstract: The CenW ecosystem model simulates carbon, water, and nitrogen cycles following ecophysiological processes and management practices on a daily basis. We tested and evaluated the model using five years eddy covariance measurements from two adjacent but differently managed grasslands in France. The data were used to independently parameterize CenW for the two grassland sites. Very good agreements, i.e., high model efficiencies and correlations, between observed and modeled fluxes were achieved. We showed that the CenW model captured day-to-day, seasonal, and interannual variability observed in measured CO_2 and water fluxes. We also showed that following typical management practices (i.e., mowing and grazing), carbon gain was severely curtailed through a sharp and severe reduction in photosynthesizing biomass. We also identified large model/data discrepancies for carbon fluxes during grazing events caused by the noncapture by the eddy covariance system of large respiratory losses of C from dairy cows when they were present in the paddocks. The missing component of grazing animal respiration in the net carbon budget of the grazed grassland can be quantitatively important and can turn sites from being C sinks to being neutral or C sources. It means that extra care is needed in the processing of eddy covariance data from grazed pastures to correctly calculate their annual CO_2 balances and carbon budgets.

Keywords: grassland; eddy covariance; carbon cycling; grazing; mowing; CenW model

1. Introduction

Managed grasslands and rangelands represent ~70% of global agricultural area [1], which is 25% of the Earth's ice-free land surface [2]. The soils of these agroecosystems contain ~20% of the world's soil organic carbon (SOC) stocks, which implies that they play a significant role in the global carbon and water cycles [3–7]. In Europe, grasslands cover 22% of the land area [8], where management practices and climate strongly influence their C sequestration rates. Average annual estimates of carbon balances of temperate grasslands for EU countries ranged from being a C source of 45 kg C ha^{-1} year^{-1} to a C sink of 400 kg C ha^{-1} year^{-1} [9]. Hence, these managed agroecosystems may contribute to the mitigation of climate change [7,10–12]. However, these ecosystems are particularly complex and difficult to investigate because of the wide range of management and environmental conditions that they are exposed to, leading to a large variability in their CO_2 source/sink capacity [5,13–17].

Most of the vegetation growing on pastoral lands is used to either feed animals directly (grazing), or it is harvested and used to feed animals at other times or locations (mowing). Grasslands managed through mowing are fundamentally different to grassland managed through grazing with respect to their above ground biomass removal patterns, export and cycling of carbon, and applications of fertilizer, as more nitrogen is returned to the field during grazing through animals excreta compared to mowing where almost everything is exported from the system [15,18]. In addition, there is large uncertainties about the effects of mowing and grazing on different ecological processes related to their C cycle [16,19,20].

The frequency and intensity of foliage removal and its fate (grazed on site or mowed and exported) have effects on the carbon budgets but also on the nutrient cycling and development of the grassland [7,21,22]. Grazing intensity showed to have significant effect on the soil carbon sequestration potential of grassland ecosystems. Positive C sequestration was reported for light-to-moderate grazing intensities [11,23], while overgrazing or trampling were found to have a negative effect on SOC stocks [24]. Although less studied than grazing systems [25], mowing is usually related to important losses of soil organic carbon unless manure is returned to the paddock because of the export of biomass from the grassland that reduces the amount of C inputs to the agroecosystem [26]. However, previous studies found that soil carbon stocks of mowed grassland could also increase depending on the cutting/harvesting intensity [8,18].

Direct and accurate measurements of small changes in soil organic carbon stocks over short time periods in response to different management practices are difficult to achieve because of the large spatial variability of SOC and of the large C content of the soil relative to the rate of change [27–29]. Despite the uncertainties associated with flux measurements, eddy covariance (EC) is a powerful tool for measuring ecosystem/atmosphere carbon fluxes [8,30–32]. With EC, it is possible to detect changes in net ecosystem exchange (NEE) of carbon at a half-hourly time resolution, which enables estimates to be made of whether land management practices result in systems being net sinks or sources of CO_2 [12,26]. NEE is the balance of gross primary production (GPP) and ecosystem respiration (ER) and it represents the net exchange of CO_2 between the atmosphere and terrestrial ecosystems. For managed ecosystems, the carbon balance has to comprise NEE and C losses (harvested biomass, enteric fermentation, export of animal products, and organic and inorganic C losses through leaching and erosion) as well as nonphotosynthetic carbon gains (organic fertilization), resulting in the net biome productivity (NBP = NEE + carbon export − carbon import (positive value indicates that the ecosystem is a carbon source). NEE is a key variable to determine the carbon balance of an ecosystem and therefore, understanding it responses to environmental change, management, and site characteristics is essential [33–35]. Over seasonal and interannual time scales, NEE in managed grasslands can vary with the frequency, timing and duration of management practices. Mowing and grazing removes photosynthesizing biomass and can thereby temporarily but substantially reduce GPP [18,36,37].

To develop a better understanding of ecosystem processes, or predict the response of ecosystem to climate change and management practices, various process-based (mechanistic) vegetation models have been developed [38–40]. They vary in complexity and can operate at different temporal and spatial scales. They are being used widely to simulate ecosystem carbon and nitrogen dynamics. The reliability of these models highly depend on the quantity and quality of the data used for their calibration [41–43]. Model parameters calibration aims to constrain the uncertainty in model parameter space and optimize the model output of ecosystem–atmosphere CO_2 exchange [44]. Measured CO_2 fluxes were used to constrain model simulation through parameter calibration. Once parameterized, these models allow to simulate separately the constituents of NEE (soil, microbial, plant, and animal respiration and gross primary production) that cannot be measured directly by EC, or to interpolate and extrapolate CO_2 fluxes in time and space. However, insufficient knowledge about underlying processes (e.g., all the processes leading to observed carbon and water fluxes and flows that exists in the real world but are not or are only poorly understood and modeled) as well as parameters and initial conditions uncertainties can lead to bias and uncertainty in model simulations [45]. Also, because interannual responses are usually less well captured by models than daily and seasonal dynamics [46–48], the availability and quality of long-term datasets are crucial to improve model performances [49]. Process-based simulation models are therefore required to gain insight of processes and interactions between managed grasslands C dynamics, climate change, and management practices, in combination with experimental observations, especially for long-term analyses [50,51].

CenW (carbon, energy, nutrients, and water) is a process-based model, running at a daily time step. It was originally developed to simulate the carbon balance of forests over time [52–54]. The soil organic matter module of the model was derived from the CENTURY model [55], which was originally developed for grasslands (more details are given in Section 2.2). Recently, CenW was successfully parameterized and used to simulate carbon and water fluxes of an intensively grazed dairy pasture in New Zealand [56], and to test effects of different climate and management practices on soil carbon stocks and milk production [57].

In this study, we used the CenW model to simulate the seasonal and interannual variability of carbon dioxide and water fluxes of two differently managed grassland fields located in France. The two selected paddocks are part of the Agroecosystem Biogeochemical Cycles and Biodiversity (ACBB) long-term national research infrastructure. They are located only 200 meters apart and are equipped with eddy covariance (EC) flux towers. These sites were either regularly mowed or grazed, and they received different fertilizer doses applied following different application patterns. Within the framework of this study, we focused on the differences between carbon and water fluxes between grasslands under mowing and grazing managements. Flux data from the paired paddocks were used to parameterize and validate the CenW model.

The specific objectives of the present study were to

1. test the ability of the CenW model to simulate water and CO_2 flux dynamics of two temperate grassland ecosystems under mowing and grazing management, respectively;
2. evaluate the model's ability to capture the seasonal and interannual dynamics of CO_2 and water fluxes in response to climate variability (five years) in interaction with two contrasting management practices (mowing and grazing); and
3. determine the effects of mowing and grazing on eddy covariance fluxes and on the CO_2 budget of managed grasslands.

2. Materials and Methods

2.1. Experimental Details

The experimental site is located at the Lusignan INRA (National Institute for Agricultural Research) experimental farm, France (46°25′12.91″ N; 0°07′29.35″ E), which covers ~22 ha (Figure 1). INRA and CNRS (National Centre for Scientific Research) jointly designed the long-term observational study to gain a better understanding of the environmental impacts resulting from different grassland management practices and grassland/cropping rotations. The study site was established on temporary grasslands that were sown in spring 2005 (March–April). Before 2005, part of the observational site was grassland, and the other part was alternated between grass and crop rotations for 17 years. The total surface area of the experimental site was ploughed to establish a base line for the system before sowing grass in 2005. The upper soil horizons are characterized by a loamy texture, classified as Cambisol, whereas lower soil horizons have a clayey texture rich in kaolinite and iron oxides, classified as a Paleo-Ferralsol [58,59]. The sown grasslands consisted of a mix of three grass species (*Lolium perenne, Festuca arundinacea*, and *Dactylis glomerata* L.). For the original experiment, the 22 ha of the study area were divided into four blocks of five 0.4 ha plots and four larger plots of 3 ha to test seven different treatments (shown in different colors in Figure 1) [60]. The treatments relevant for the present work are two of the 3 ha plots with pasture being either mowed or grazed. The towers footprints are crucial in experimental set up of eddy covariance and a detailed footprint study was performed [26]. The wind rose, which is identical for the two paddocks, is reported in Figure 1. Footprint analysis indicated that ~70% of the median percentage of the footprint was in the field, which is a similar fraction than that found in other similar studies [26].

For the present study, two temporary sown grasslands paddocks, each of a size of ~3 ha and of rectangular shape, were equipped with two eddy covariance measurement systems and a meteorological station (Figure 1). One of the paddocks was regularly mowed (P2), with harvested hay exported off-site to feed animals during periods of insufficient vegetation growth (mainly drought periods and during winter). Dairy cows regularly grazed the other paddock (P4), with all animal excreta directly returned to the paddock, except for the fraction that was deposited off site during milking and during the daily transit times from the milking shed to the field. Both paddocks received regular applications of nitrogen fertilizer, with higher rates applied to the mowed than the grazed paddock (Appendix A).

For the two contrasting grassland systems studied here, the dates of mowing and grazing, the length of each grazing event, the animal stocking densities, and timing and amounts of N fertilizer applications varied in the different years. Details are given in Appendix A. Over the 5-year study period (2006–2010) the mowed paddock received 1290 kg N ha^{-1} split into 17 fertilizer applications and was mowed 17 times with 3 cuts per year, except for the wetter than normal summer half-year 2007 (5 cuts). Over the same period, the grazed paddock received 590 kg N ha^{-1} (not including N returned in dung and urine during grazing) over 14 applications and was grazed 37 times with grazing events spread, on average, over 5 consecutive days.

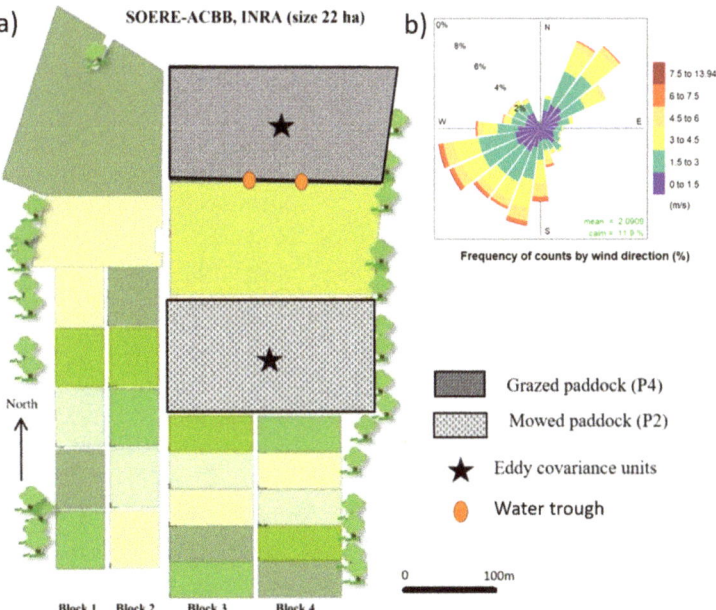

Figure 1. (**a**) Field site layout of the Agroecosystem Biogeochemical Cycles and Biodiversity (ACBB) experimental farm (22 ha). Different colors are used to distinguish different treatments. The treatments relevant for the present work are shown by stars indicating the location of the eddy covariance masts on the two studied paddocks (P2: mowed paddock; P4: grazed paddock). (**b**) The wind rose shows the frequency and intensity of winds blowing from different directions. The length of each "spoke" around the circle is related to the frequency of time that the wind blows from the specified direction.

2.1.1. Meteorological Conditions at the Study Site

Meteorological conditions for the two paddocks were acquired from a weather station coupled to a data logger (CR-10X, Campbell Scientific Inc., Logan, UT) placed 1.9 meters aboveground on the mown paddock [17,26,61]. Briefly, the weather station provided 30 min averaged values of precipitation (SBS500, Campbell Scientific Inc., Logan, UT), air temperature and relative humidity (HMP 45 AC, Vaisala), radiation components (CNR1, Kipp & Zonen), and wind speed (A100L2, Vector Instruments) and direction (W200P, Vector Instruments). Volumetric soil water content data were collected by time domain reflectometry (TDR) probes at 10, 20, 30, 60, 80, and 100 cm depths (CS616, Campbell Scientific Inc., Logan, UT), and soil temperatures were measured at 5, 10, 20, 30, 60, 80, and 100 cm down the soil profile (PT100, Mesurex), but only for the mowed paddock. Soil heat flux was measured at 5-cm-depth (HFP01, Hukseflux), and data were corrected for changes in heat storage in the soil layer above the flux plate [62]. Over the study period (2006–2010), average air temperature and average annual precipitation were 11.2 °C and 774 mm yr^{-1}, respectively. Half-hourly meteorological data, i.e., air temperature, global radiation, humidity, and precipitation, were summed/averaged to daily values to be used as driving variables for the CenW model runs of both paddocks.

The predominant wind direction was from the southwest and a secondary peak from the northeast (Figure 1b).

Throughout the five years of the study, daily maximum air temperature exceeded 25 °C for 10.5% of the time, with a maximum of 35.5 °C. Daily minimum air temperature was negative for 13.3% of the time, with a lowest value of −11.0 °C (data not shown). Summer months (June–September) were hot and dry with average monthly air temperature ranging between 15.6 and 19.4 °C and precipitations between 48 and 71 mm mth^{-1} (Figure 2a). The wettest and coldest month are November (98 mm mth^{-1}) and December (3.6 °C), respectively (Figure 2a). Among the five years of the study, 2006 was the warmest (11.9 °C) and wettest (888 mm yr^{-1}), while 2010 was the coldest (10.5 °C) and driest (697 mm yr^{-1}) year.

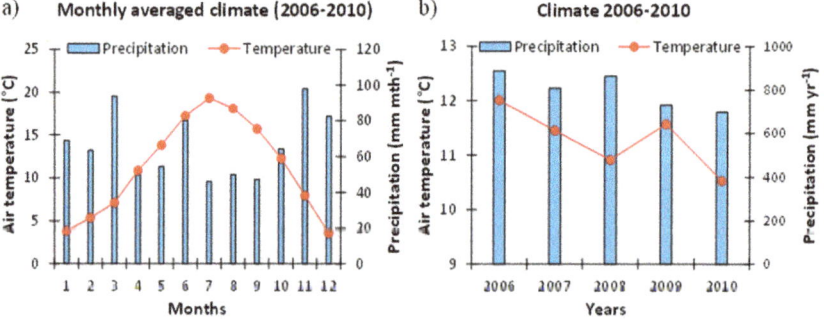

Figure 2. Temporal variation of mean daily air temperature and precipitation over the experimental period (2006–2010) monthly (**a**) and annual (**b**) time scales in Lusignan, France.

2.1.2. Eddy Covariance (EC) Measurements and Processing

We used paired eddy covariance systems for the 2006–2010 period because it provided EC data measurements for both mowed (P2) and grazed (P4) paddocks. The two EC systems recorded raw data at 20 Hz, and EddyPro®software (LI-COR Inc.) was used for postprocessing and the calculation at 30-minute intervals for fluxes of CO_2, momentum, and sensible and latent heat. Each EC unit included a fast response sonic anemometer (Solent R3-50; Gill Instrument, Lymington, UK) and an open-path CO_2-H_2O infrared gas analyzer (LI-7500; LI-Cor Inc., Lincoln, NE, USA) placed at 1.55 m above the ground. In this study, the micrometeorological sign convention is followed, with negative net CO_2 fluxes (NEE) representing the transport from the atmosphere towards the surface (assimilation of CO_2

through photosynthesis) and positive ones indicate that the system is a source of carbon (release of CO_2 through respiration).

Based on previous work using these EC datasets [17,26], flux measurements, and quality checks were done according to the CarboEurope-IP guidelines [63]. The flux footprint distribution and random uncertainty were analyzed. High-frequency loss corrections [18,64] were not considered in flux processing process [26]. The Webb–Pearman–Leuning (WPL) correction [65] was applied except for the self-heating of the IRGA because of the sensor orientation [66]. All years of flux measurements for the two EC towers were quality checked and filtered with a custom R program. The quality check led to the rejection of half-hourly flux observations based on nine criteria:

1. NEE values lower than −35 or higher than 25 $\mu mol\ m^{-2}\ s^{-1}$
2. NEE values higher than 3.5 $\mu mol\ m^{-2}\ s^{-1}$ when PAR was above 400 $\mu mol\ m^{-2}\ s^{-1}$
3. NEE values lower than −2 $\mu mol\ m^{-2}\ s^{-1}$ when PAR was below 25 $\mu mol\ m^{-2}\ s^{-1}$
4. $Rn > 300\ W\ m^{-2}$ and $LE < 0\ W\ m^{-2}$
5. If precipitation > 0 mm
6. If $u^* < 0.1\ m\ s^{-1}$
7. λE values higher than 750 or lower than −100 $W\ m^{-2}$
8. H values higher than 750 or lower than −100 $W\ m^{-2}$
9. Atmospheric CO_2 concentration higher than 650 or lower than 320 ppm, respectively.

Common time series of eddy covariance measurements unavoidably include missing data due to power failures, instrumental malfunctions, or unfavorable micrometeorological conditions that cause the rejection of observations through the filtering process of data. However, complete time series of EC data at the half-hourly timescale are required to be summed to daily, monthly, or annual values [67,68]. Over the five years of EC measurements used in this study, there were gaps for 39.7% and 40.9% of NEE observations in the dataset for the mowed and grazed paddocks, respectively.

Gaps in 30 minutes NEE were filled and NEE was partitioned between GPP and total ecosystem respiration rate (ER) using the online gap-filling and flux partitioning procedure described by Reichstein et al. (2005) [69], hereafter referred to the Reichstein algorithm. This gap-filling method uses an improved, running-window look-up table that utilizes both the covariation of NEE with meteorological conditions and temporal autocorrelation of NEE [70]. In the Reichstein algorithm, ER was modeled using the Lloyd and Taylor equation [71] fitted to air temperature. Following this approach, nighttime ER was first regressed against nighttime air temperature, and this relationship was then used to estimate ER for both nighttime and daytime. GPP was determined by subtracting the parameterized ER from NEE.

2.1.3. Vegetation and Soil Organic Carbon Measurements

Harvested hay production (mowed paddock) was measured after each mowing event (Table A1). The total amount of harvested C was calculated by multiplying hay dry matter weight by the C concentration in biomass. Harvested biomass samples were collected from 6 replicates of 7.5 m² and oven dried at 60 °C. C concentration in the hay was measured in five replicates by dry combustion using a LECO C analyzer (TruSpecR CN Analyser; LECO Corporation, St Joseph, MI, USA).

Aboveground biomass present on the grazed paddock was measured just before and after each grazing event on six replicates within the field. Samples were oven dried and their C concentration measured with the same method than for harvested hay production.

Root biomass of the different treatments was measured once a year in three soil horizons (0–30, 30–60, and 60–90 cm). Each measurement is the average of twelve samples from a 6.5 cm⌀ mechanical auger.

Total SOC and soil profiles physical characteristics were measured before the start of the experiment in early 2005 and then SOC content was measured every three years. Soil physical properties were used to set up the water dynamic procedure of the CenW model and total soil organic carbon content measured in 2005 was use to initialize the model.

2.2. Modeling Details

2.2.1. CenW 4.2 Overview

CenW is an open-source process-based model, combining the major carbon, energy, nutrient, and water fluxes in an ecosystem [52]. For the present work, we used CenW version 4.2, which is available for download, together with its source code and a list of relevant equations available in CENW documentation, version 4.1.1 (*A growth and C balance simulation model*, © 2017). A number of additional routines were added to run the model for managed pastures [56]. A list of relevant parameters is given in Appendix B. The CenW model runs on a daily time step and encompasses major ecosystem processes (canopy photosynthesis, allocation and growth, litterfall, decomposition, autotrophic and heterotrophic respiration), and their relationships to climatic drivers to simulate the behavior of the ecosystem over time, which are further modified through management practices (i.e., mowing, grazing, N fertilizer applications, plowing, and sowing).

The main CO_2 fluxes are photosynthetic carbon gain by plants which is integrated over the whole canopy and the whole daytime period [72] and CO_2 losses through autotrophic respiration by plants and heterotrophic respiration by soil organisms and grazing animals, when they are present on the modeled paddock. CenW simulates soil heterotrophic respiration individually for growth and maintenance. These fluxes are modified by temperature and nutrient and water balances. Plant growth is determined by the dynamic allocation of fixed carbon to the different plant organs, which depends on the plant root/shoot ratio, vegetation type and development stages, and water and nutrient stresses.

The model contains a fully integrated nitrogen cycle as well as a coupled multilayer bucket water model. Water is gained by rainfall and lost through evapotranspiration. Any amount of water exceeding the soil's water-holding capacity is lost by deep drainage beyond the root zone, with important controls by soil depth and water-holding capacity. CenW simulates total evapotranspiration by modeling separately canopy and soil evaporation rates, and plant transpiration. These individual fluxes are calculated using the Penman–Monteith equation, with canopy resistance for calculating transpiration explicitly linked to photosynthetic carbon gain. This module of CenW is particularly important, as it is likely that soil water availability constituted an important constraint on plant productivity over the summer months at the experimental site, which is prone to summer droughts.

The soil organic matter component of CenW is based on the CENTURY model [55], which was originally developed for grasslands. The model includes three soil organic matter pools (active, slow, and resistant) with different potential decomposition rates. Leaves and roots senescence and litter production are controlled by plant type and phenology, and by water, temperature, and specific senescence parameters which depend on plant species. Dead foliage can either fall onto the soil surface and become part of the decomposing litter pool, remain standing for some time where it either decomposes during wet periods, or eventually falls onto the soil surface after some time. It was important to model these processes as the estimates of foliage biomass included a component of dead standing biomass that was not separated out in the data. These processes were modeled by assuming that all senescence, or drought-induced leaf death, initially transferred foliage from a live to a dead foliage pool [56]. The soil is divided into multiple layers and the same calculations driving the behavior of organic matter are applied to all of them, with each layer having its own complement of all organic matter pools. Layers only differ by the amounts and qualities of litter entering each layer. In addition, a small fraction of each pool is transferred to the corresponding pool in the layer below [53,73]. This allows changes in organic matter and C:N ratios in the surface litter layer and with depth in the soil to be simulated.

To effectively model net carbon fluxes in a managed pasture system, it was essential to know the timing of grazing, feed supplementation, and harvesting carried out on each paddock [56]. For the "grazed" paddock of the Lusignan study farm described above, the model assumed (similar to the study of a dairy farm in New Zealand [56]) that cows consumed, at each grazing event, a given amount of above ground biomass [74]. If grazing was spread over several consecutive days, grazing percentages on individual days were adjusted to add to a total of that fixed percentage at the end of the grazing events. Of that feed, 50% was assumed to be lost by respiration [75], 5% as methane [76], and 18% removed as milk solids [15,75,77], with the remaining 27% returned to the paddock as dung and urine. It is also assumed that animal weights remained constant and not added to carbon gains or losses from the paddocks. For the mowed paddock it is assumed in the model that during each harvest event, a given amount (depending on total above ground biomass and cutting height) of aboveground photosynthesizing biomass is cut, and of that amount 95% is exported from the farm with the 5% remaining being left on the pasture as residues.

2.2.2. Model Parameterization and Statistical Analysis

Harvested hay production was measured after each mowing event for the mowed paddock and the amounts of biomass on the grazed paddock were measured just before and after each grazing events. These observations were used to constrain the grazing and harvesting procedures in CenW simulations and measured root biomass and soil water contents at different depth in the soil were used to constrain the soil water extraction of the CenW ecosystem model.

Total SOC content was measured in early 2005 for different soil layers and for the two managed grasslands (mowed and grazed), and these values were used to initialize the CenW model independently for the two paddocks through the spin up of the model simulations until equilibrium conditions between measured and modeled initial SOC stocks were reached.

Model simulations were optimized by selecting a set of parameter values that minimized the residual sums of squares across different EC measurements and ancillary observations. Measurements used for CenW parameterization were daily- and weekly-averaged estimates of evapotranspiration (ET) and net ecosystem exchange (NEE). We separated our five years of eddy covariance data into weekly sets, with one week of daily values used for parameter optimization and the other week for model validation. There are therefore two flux datasets for each paddock, one for model calibration, and one for model verification. CenW uses an automatic parameter optimization routine that worked by changing parameter values within specified boundaries to minimize the residual sums of squares. That was applied to both daily and weekly-averaged data within the data set selected for parameter optimization.

Initial parameter values to run CenW for managed grasslands were retrieved from a previous study where the model was run for a grazed dairy farm of New Zealand [56]. Specific management practices from farm records were implemented in CenW and we used a spin up of the model to initialize soil carbon and nitrogen pools. Then, model simulations were optimized for these paddocks based on a selection of eddy covariance observations (NEE and ET) and ancillary data (amounts of vegetation mowed and grazed and soil water content) by the automatic parameter optimization procedure imbedded in the model that aim to maximize the agreement between model and observations.

The overall goodness of fit was described by the Nash–Sutcliffe model efficiency (EF) [78]:

$$EF = 1 - \frac{\sum (y_o - y_m)^2}{\sum (y_o - \overline{y})^2}, \tag{1}$$

where y_o represents the individual observations, y_m is the corresponding modeled values, and $\overline{y}$ the mean of all observations. EF quantifies both the tightness of the relationship between measured and modeled data and assesses whether there is any consistent bias in the model. Model efficiency values range from minus infinity to 1. High model efficiency can only be achieved when there is a tight relationship with little unexplained random variation and little systematic bias. Negative values of model efficiencies indicate poor model fit and that the mean value of the observation is a better predictor than the model. EF = 0 implies that the mean of the observations is as good a predictor than the model, while positive values indicate that the model is a better predictor than the observed mean. The closer EF is to one, the stronger the agreement between observed and modeled data.

The final sets of parameters values used for the simulations of the mown and grazed paddocks are given in Appendix B.

3. Results

3.1. CenW Performances to Simulate Carbon Dioxide and Water Fluxes of Mown and Grazed Grasslands

Over the five years of the study period, a wide range of climatic conditions (Figure 2) were encountered as well as different management practices like different mowing, grazing, and fertilizer application frequencies (Appendix A) and different stocking rates for the various grazing events. Achieving good model/data agreement is challenging because the model needs to incorporate a wide variety of processes to simulate accurately such complex systems and to capture the variability of fluxes and vegetation dynamics affected by biotic and abiotic factors. The CenW model used only one fixed set of parameters for multiple years and after the calibration of the CenW model for the two grassland sites, daily modeled and observed carbon and water fluxes could be compared.

3.1.1. Carbon Dioxide Fluxes

Comparisons between modeled and measured daily CO_2 fluxes for the two differently managed grasslands are shown in Figure 3, and model efficiencies for model calibration and validation are given in Table 1. In this section, only the best quality data from background periods (outside mowing and grazing events) were used for the comparisons. Observations that would have been affected by mowing and grazing events were omitted from the analyses [56], as well as days when fluxes from the eddy covariance systems had to be gap filled for more than 1/3 of half hourly periods. This selection of only best quality observation was necessary to avoid

1. the calibration of the model with data that strongly depended on another simpler model (i.e., the Reichstein gap-filling and partitioning tool) and
2. to limit the bias that would have resulted from the non or incomplete capture by EC of the large respiratory losses during measurement periods when grazing animals were present around the EC tower or when freshly cut or drying grass was present on the ground during mowing events [56,79].

Table 1. Model efficiencies for six key observations of the two-modeled systems. Only daily NEE and ET were used for the model parameterization. 'Total' refers to the complete data set that included data in both the parameterization and validation data sets.

| | Mowed Paddock | | | | Grazed Paddock | | | |
| | Daily | | | Weekly | Daily | | | Weekly |
	Calibration	Validation	Total	Total	Calibration	Validation	Total	Total
GPP	-	0.85	0.85	0.87	-	0.80	0.80	0.79
ER	-	0.77	0.77	0.78	-	0.72	0.72	0.67
NEE	0.75	0.73	0.74	0.72	0.64	0.66	0.65	0.64
ET	0.82	0.81	0.82	0.87	0.81	0.80	0.80	0.85
Averaged SWC	-	0.85	0.85	0.87	NA	NA	NA	NA
Harvested biomass	-	0.80	-	NA	NA	NA	NA	NA

For the NEE dataset, which was separated into parameterization and validation subsets, R^2, and model efficiencies are slightly higher for the parameterization subset than for the validation one for the two grassland sites. Moreover, in general, model/data agreements (Figure 3 and Table 1) are better for GPP and ER than for NEE as it is easier to model large photosynthesis and respiration fluxes than the relatively small difference between the two (NEE). Model/data agreements are also better for mowing than for grazing, certainly because of the higher complexity of grazed systems compared to mowing. On the one hand, for the mowed paddock, management practices (mowing events and fertilizer applications) are accomplished within a day and evenly applied to the field. While, on the other hand, for the grazed paddock, grazing events last several consecutive days, the stocking density vary for the different events, there are dung and urine patches, there is an uneven reparation of cattle on the paddock, there could be some preferential grazing of plant species, and pasture could be damaged by trampling.

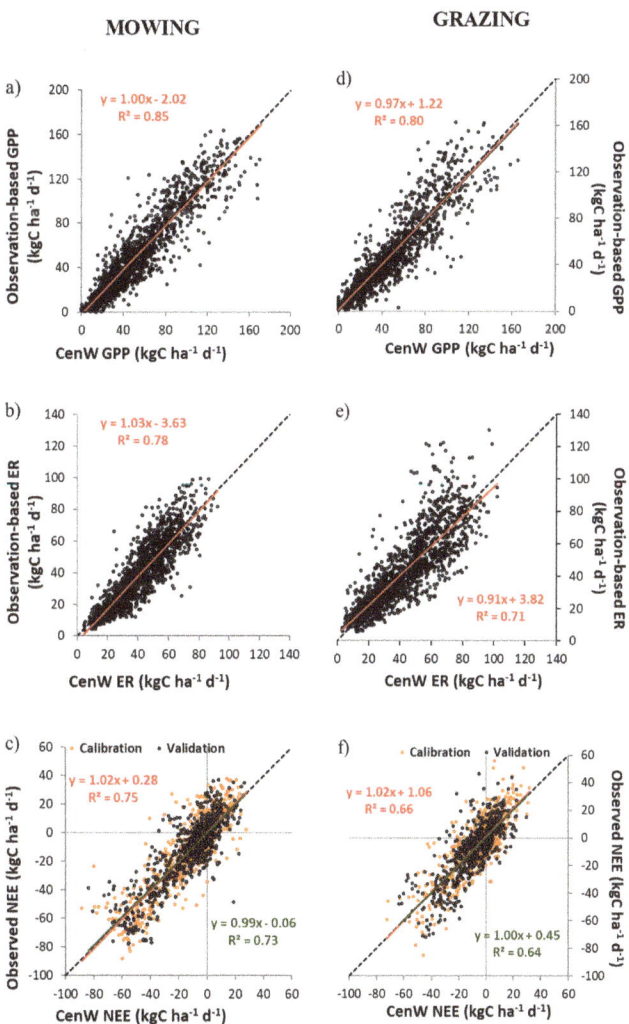

Figure 3. Scatter plots of observed vs. modeled daily carbon fluxes for the mown (panels **a–c**) and grazed (panels **d–f**) paddocks, for background measurement periods outside management (grazing, mowing) events. For NEE, negative numbers refer to carbon gain by the system. Corresponding model efficiencies for each comparison are given in Table 1.

After parameterizing the model with good quality data over the full length of the study period (2006–2010), we obtained good agreement between modeled and observed carbon fluxes. Figure 3a,b shows the comparison of daily modeled GPP against their observation-based counterparts for the mowed and grazed paddocks, respectively. Good agreement was shown for both managed grasslands by the slopes close to 1, small intercepts of the linear regressions, and high correlation coefficients (R^2 = 0.80–0.90). For the mowed paddock, model efficiencies for GPP were 0.86 and 0.88 for daily and weekly averaged fluxes, respectively (Table 1). Slightly lower, but still good EF was also found for GPP of the grazed paddock with daily and weekly model efficiencies of 0.80 and 0.79, respectively (Table 1).

The model also showed good performance in simulating daily and weekly averaged ecosystem respiration rates for both sites. For instance, the daily EF and R^2 were 0.79 and 0.85 for the mowed and 0.73 and 0.71 for the grazed paddocks, respectively (Figure 3b,e). The grazed paddock had higher ER values than the mowed paddock and there was more scatter in the model/data comparison (Figure 3b,e) according to the lower values of R^2 for daily and weekly comparisons.

The comparison of modeled NEE with EC measurements showed that the CenW model performed well in capturing the variability in NEE in background conditions. Across the two studied grassland sites, the CenW model explained between 65 and 74% of the variation in daily NEE for the mowed and grazed paddocks, respectively (Figure 3c,f and Table 1). For the mowed paddock, the coefficients of determination for daily and weekly averaged net carbon fluxes were 0.74 and 0.77, respectively (Table 1). The agreement between observed and modeled NEE for the grazed paddock was lower than for the mown paddock with daily and weekly R^2 of 0.65 and 0.71, respectively. For the two managed grasslands, the overall seasonal and annual variations in NEE are reasonably well modeled and consistent agreement between modeled and measured NEE was achieved with daily model efficiencies of 0.74 and 0.65 for the mowed and grazed paddocks, respectively.

3.1.2. Soil Water Content and Evapotranspiration

Evapotranspiration (ET) measurements were also used for the parameterization and validation of the CenW model for both grassland sites. Soil water content observations were only available for the mowed paddock and were not used for the calibration of the model. Figure 4 shows the time series of daily observed and modeled soil water content (SWC) of the mowed paddock for three depths averaged over the entire soil profile.

Over the entire study period, the soil water content measured and modeled at different depth agree quite well (Figure 4a–c), confirming the correct set up of the soil water flows procedure in CenW. Averaged soil water content was generally well modeled, with an EF of 0.83. On average, over the study period and over the entire soil profile, modeled and measured SWC were 22.4% and 21.9%, respectively.

There was no systematic over- or underestimates of soil water content. Lower modeled SWC were found in spring 2006 and during summers of 2008 and 2010 (Figure 4) and were most likely due to higher CenW modeled water losses in spring and early summer than actual field conditions. Conversely, measured SWC was sometimes lower than modeled values. In 2007, observed soil water drawdown was faster than model simulation, but Figure 6d shows no discrepancies in ET, and so the problem seems to be linked to water drawdown. It could be that CenW extracted too much water from deeper layers and preserved it in the top layers. This situation was encountered following a water-limited period and could be due to cracks in the soil, causing preferential water flows not accounted for in the model or to the incomplete capture of vegetation dynamic in response to droughts.

Overall agreement for SWC is good, and remaining discrepancies could be due to measurement errors like the heavy rainfall in 2006 either not measured correctly, or not all water infiltrating but running off. Others could be due to shortcomings of the model, like not having soil cracks represented, or because of some measurement uncertainties reducing the model/data agreement. Because SWC and ET are tightly linked, achieving to get a good agreement between observed and modeled soil moisture is consistent with the good results reported for the modeling of daily and weekly evapotranspiration rates (Figure 5a and Table 1).

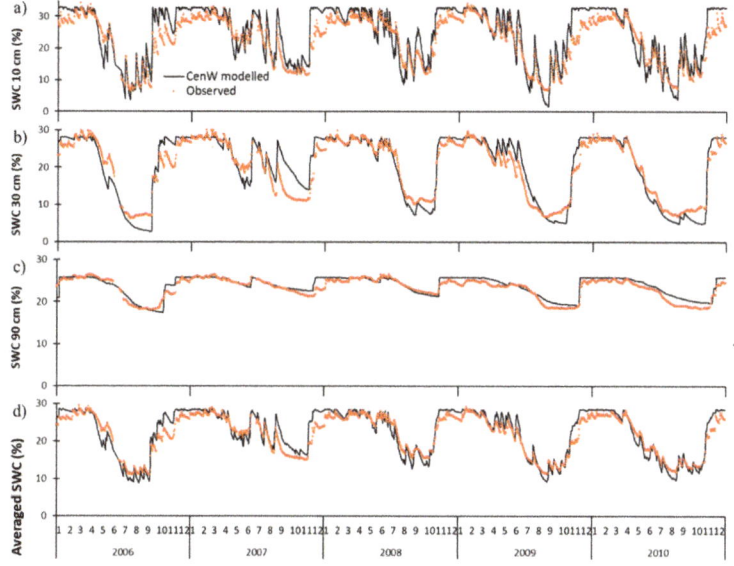

Figure 4. Time series of daily-modeled (black line) and observed (red symbols) soil water content at (**a**) 10 cm, (**b**) 30 cm, (**c**) 90 cm belowground, and (**d**) averaged over the whole soil profile (0–100 cm) for the mowed grassland.

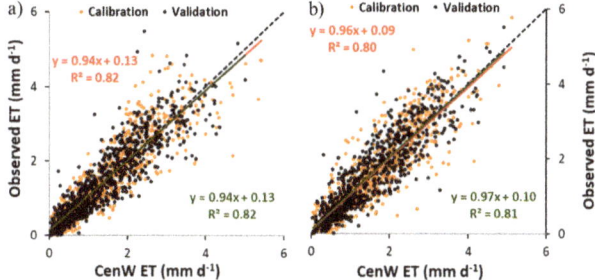

Figure 5. Scatter plots of observed vs. modeled evapotranspiration rates for the mown (**a**) and grazed (**b**) paddocks.

The CenW model explained more than 80% of the variation in daily ET for the two grassland sites. There was a close agreement between modeled and observed daily ET across the monitoring period (Figure 5), with R^2 of 0.82 and 0.81 and EF of 0.82 and 0.80 for mowed and grazed paddocks, respectively. The coefficients of the linear regression lines (Figure 5) show a tendency of the model to slightly underestimate low ET (positive intercepts) and overestimate high evapotranspiration rates (slopes lower than 1), however slopes and intercepts are very close to their optimal values, showing that there is no systematic differences between modeled and observed evapotranspiration rates.

Overall, very good agreements between the CenW modeled and observed daily CO_2 and water fluxes were achieved for both management practices (i.e., mowing and grazing). This indicated that the response of the model to climatic conditions and management practices were well captured in the simulations and that most of the processes encountered in the fields were properly implemented in

CenW. Student's *t*-tests were used to statistically test if slopes of linear regressions were significantly different from 1 and intercepts different than 0. Results showed that for water and all carbon fluxes, except NEE, for the two grasslands management, the slopes and intercepts were significant (p-values < 0.05). Even though GPP and ER were not used to parameterize CenW, there was nonetheless very good agreement between simulations and measurements (Figure 3a–d and Table 1). This indicate a high correlation between photosynthesis and ecosystem respiration rates derived from NEE data according to the Reichstein partitioning algorithm and fluxes modeled by the mechanistic CenW model.

3.2. Seasonal and Interannual Variabilities of Modeled and Observed Carbon Dioxide and Water Fluxes

3.2.1. Day-to-Day and Seasonal CO_2 and Water Fluxes Variability

For managed grassland ecosystems, important drivers of day-to-day and seasonal variabilities are management practices, particularly the timing and intensity of mowing and grazing that combine with the natural temporal climate variability to drive the behavior of ecosystems and strongly affect the CO_2 dynamic and C balance of managed grasslands [8].

Depending on a number of climatic factors (solar radiation, temperature, and precipitation), ecological factors (leaf area and water, nutrient, and temperature stresses), and management practices (nitrogen fertilization, mowing and grazing timing, duration, and intensity), modeled and observed CO_2 and water fluxes demonstrate pronounced temporal dynamics over several years [79], as exemplified by the time series presented in Figures 6 and 7.

The apparent day-to-day and seasonal variabilities of observation based GPP was well captured by the CenW model for both of the managed grassland sites (Figures 6a and 7a). The highest assimilation rates occurred during spring and summer, with GPP values up to 160 kgC ha^{-1} d^{-1} when growth conditions were the most favorable. Over the summer months, both modeled and observed GPP were reduced through water limitations, which occurred over most summer months but varied in intensity from year to year. Lower CO_2 assimilations rates were found during the winter months as temperature and radiation were low and limited photosynthesis and vegetation growth, but some gas exchange continued throughout even the coldest winters. After mowing events, during the peak growing season, GPP was strongly reduced down to wintertime levels. GPP typically dropped from preharvest values in the range of 120 to 160 kgC ha^{-1} d^{-1} to postharvest values between 20 and 50 kg C ha^{-1} d^{-1} (Figures 6a and 7a). These reductions of GPP by 2/3 are important and even if, on average, only three harvests were carried out each year, they have significant and long-lasting effects on ecosystem behavior and gas exchange. CenW managed to simulate accurately the recovery of the ecosystem gas exchanges rates (Figure 6, Figure 7, and Figure A1) after cutting and grazing events.

The day-to-day variability of observation-derived ER was also reasonably well captured in the model simulation and the seasonal pattern was well reproduced, in particular, displaying ongoing reasonably high respiration rates throughout the winter months (Figure 6b). Harvesting did not affect ER as strongly as GPP (Figure 6a,b) since autotrophic respiration from above ground vegetation is only part of the total ecosystem respiration, and (belowground) heterotrophic respiration was mostly unaffected by harvests. Because NEE is the difference between the two large fluxes of C assimilation through photosynthesis (GPP) and ecosystem respiration (ER), it was also affected by vegetation harvests (Figure 6c).

Mowing changed the pasture CO_2 status from a net carbon sink to a net source, and it usually took a few days to a week for NEE to become a sink again, and a few more weeks to return to preharvests carbon fixation levels. On average, over the five years of the study, the GPP recovery from mowing took 20 to 25 days. The seasonal and day-to-day variability of GPP and ER, controlled by the variability of climate conditions and timing of harvests causing the sharp reduction of photosynthesizing and respiring biomass, were well captured by the CenW model for the mowed grassland area.

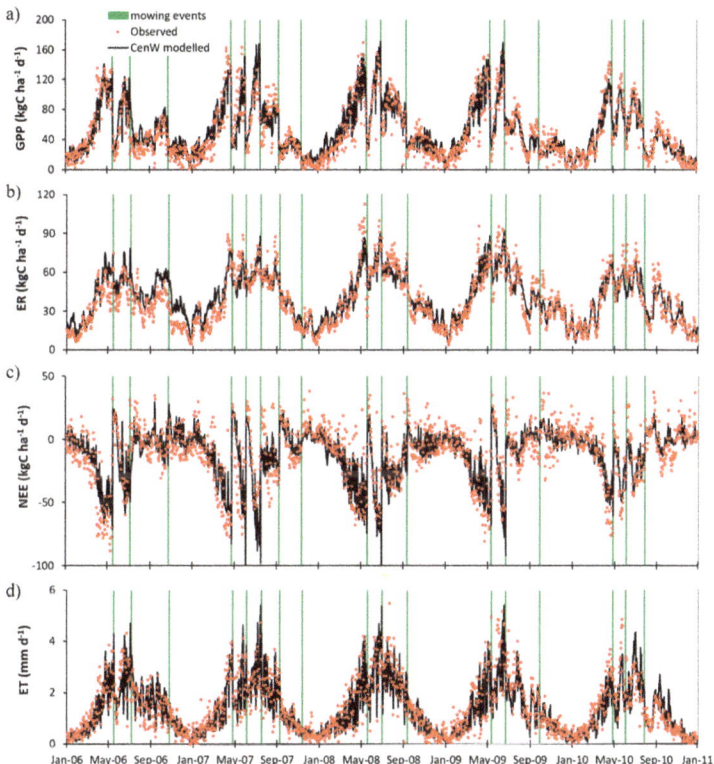

Figure 6. Time series of modeled (black line) and observed (red dots) daily carbon fluxes (**a**) GPP, (**b**) ER, and (**c**) NEE, and water fluxes (**d**) ET for the mowed grassland site. Vertical green lines represent mowing events.

Evapotranspiration flux (ET) was highest during spring and summer when climatic conditions were the most favorable for water losses and vegetation was the most active (Figure 6d). The removal of aboveground biomass by harvesting or grazing led to sharp reductions of the transpiration rates, partially compensated by the increase in soil transpiration caused by an increase of solar radiation reaching—and higher temperature at—the soil surface.

Grazing events greatly affected GPP and ET (Figure 7a,d) and caused massive spikes in modeled ER due to cattle respiration (Figure 7b). Similarly to the mowed paddock discussed above, the removal of photosynthesizing biomass by grazing animal caused an important subsequent reduction in carbon assimilation rates (GPP). However, in contrast to harvest events which were sudden and restricted to single days, the removal of biomass by dairy cows was progressive and spread over several consecutive days (on average five days). GPP reductions were therefore not as abrupt as for the mowed grassland (Figures 6a, 7a and A1) and generally, postgrazing daily modeled and observed GPP values agreed well and remained higher than postharvest values, which most likely resulted from the extent of biomass removal that differ between the two treatments.

Day to day variability and seasonality of ER and NEE of the grazed paddock were also reasonably well captured by CenW (Figure 7b,c), but there were large discrepancies between these two daily modeled and observed variables during most of the grazing events. These differences resulted from the fact that the CenW model specifically simulated grazers' respiration rates based on measured amounts of vegetation ingested, which caused the large pulses in modeled ER and NEE. According to

CenW simulations, the grazing animal respiration rate is 4.15 kg C head^{-1} d^{-1} (4 µmol CO$_2$ head^{-1} s^{-1}). In some cases, such pulses were visible but much smaller in the eddy covariance measurements than in the model. Measurements could only record what happened within the flux footprint, which varied with wind speed and direction while the CenW model simulated the whole paddock. If all dairy cows were not inside the footprint at any given time, it would have been impossible for the EC tower to measure total grazing animals' respiration while it was fully accounted for in the CenW model. At other times, a large number of cows might have been present within the flux footprint and their respiration would have been captured by the EC system. However, because this rate could have been an order of magnitude higher than the base respiratory carbon flux from the soil and pasture [56] the corresponding data could have been filtered out during the processing of EC fluxes. If the resultant data gaps during grazing events were filled using the traditional Reichstein gap-filling and partitioning algorithms it could have resulted in gaps being filled based on data collected during periods in the preceding and following week when there were no cows present within the flux footprint.

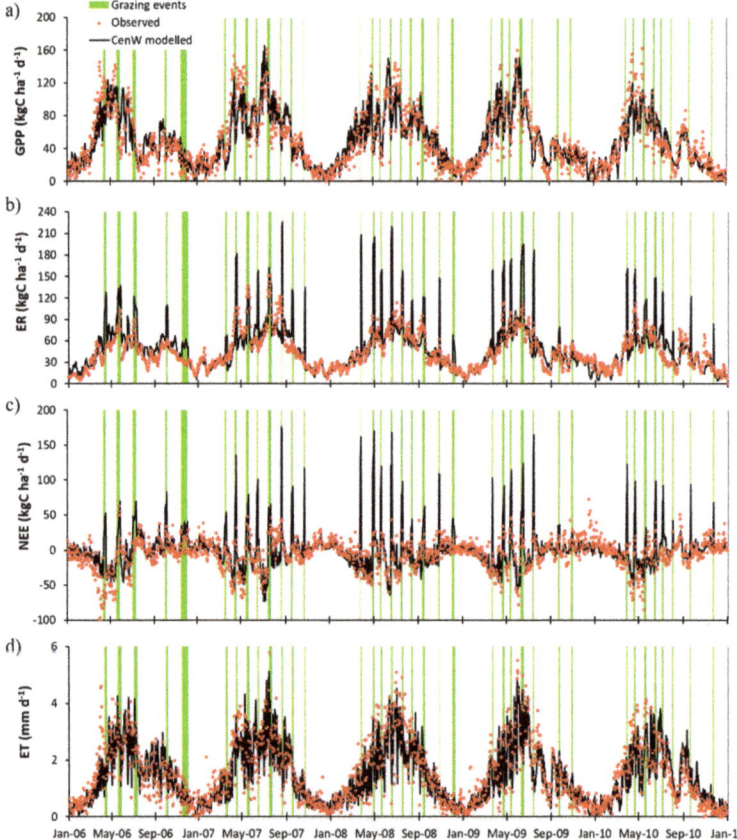

Figure 7. Time series of modeled (black line) and observed (red dots) daily carbon fluxes and water fluxes (**a**) GPP, (**b**) ER, (**c**) NEE, and (**d**) ET for the grazed grassland site. Vertical green lines represent grazing events.

3.2.2. Interannual Variability of CenW Modeled and EC Measurements of CO_2 and H_2O Fluxes

Interannual Variations in Mean Daily Fluxes

Generally, daily modeled and observed fluxes averaged over the five years of the study agreed very well for the mowed grassland site, as well as their interannual variations (+/− 1 SE from the 5-year daily averages). This is highlighted in Figure 8 by error bars (for EC observed fluxes) and yellow area (for CenW modeled fluxes) and confirms that the CenW model simulations captured well the fluxes variations due to differences in meteorological conditions and management for the mowed paddock.

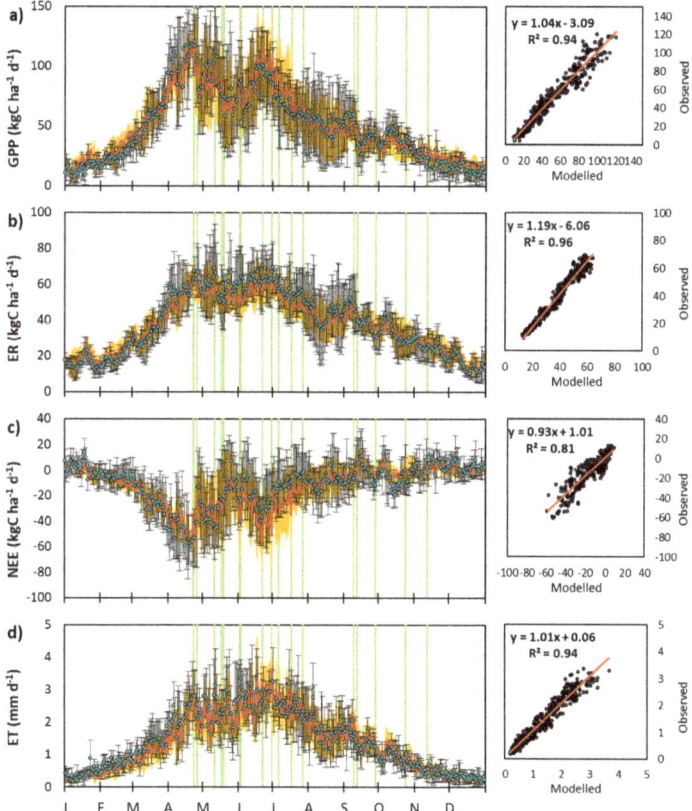

Figure 8. Daily carbon and water fluxes averaged for each day of the year over the study period (2006–2010) for the mowed pasture. The blue circles are used for observations, error bars represent one standard deviation around observed means; the red line is used for CenW modeled fluxes; and the yellow shaded areas represent one standard deviation around modeled means. GPP (**a**), ER (**b**), NEE (**c**), and ET (**d**).

Higher interannual variability was found during the most productive seasons (spring and summer), in which most of the harvest events occurred (on different days each year). GPP is more variable than ER because of the larger direct impact of harvest on photosynthesizing biomass that on total ER. Water limiting conditions and the onset of harvest events (end of April–early May) led to a substantial reduction of GPP and hence of the net ecosystem exchange rate with NEE averaged values during this period as low as wintertime fluxes.

For the grazed paddock, modeled and observed daily averages (over five years) agree very well for GPP (Figure 9a), as well as their interannual variability (error bars), confirming that the main biotic and abiotic factors controlling the dynamic of GPP were properly incorporated in the CenW ecosystem model. Weaker correlations were found between observed and modeled ER (Figure 9b) and NEE (Figure 9c) during grazing events while outside of grazing periods good agreements were retrieved. These large differences were likely caused by the noncapture of some or all of grazing animals' respiration by the EC system.

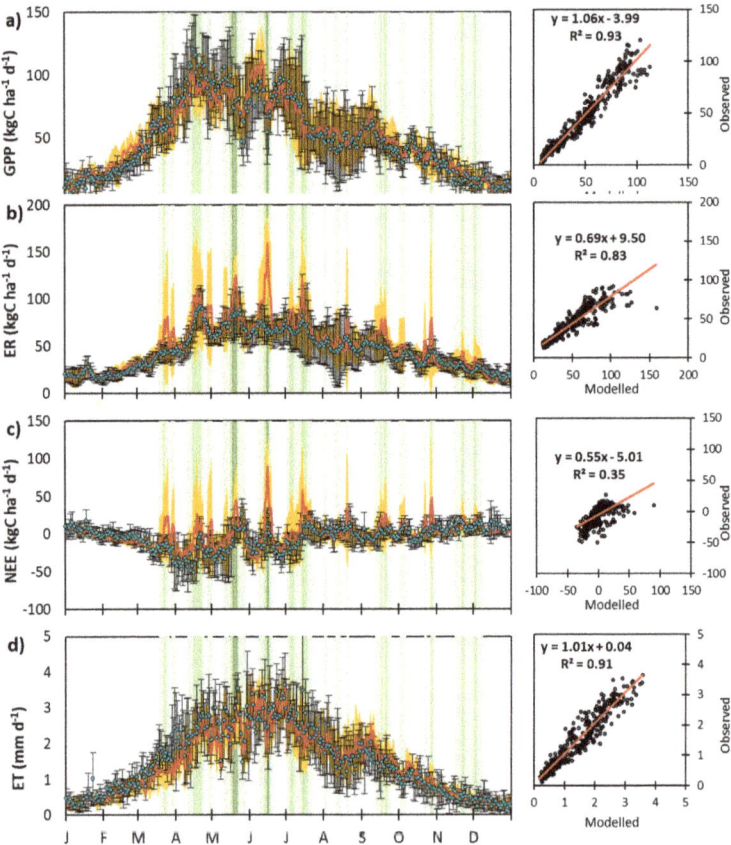

Figure 9. Daily carbon and water fluxes averaged for each day of the year over the study period (2006–2010) for the grazed pasture. Blue circles are used for observations, error bars represent one standard deviation around observed means; the red line is used for CenW modeled fluxes; and the yellow shaded areas represent one standard deviation around modeled means. GPP (**a**), ER (**b**), NEE (**c**), and ET (**d**).

Overall, model/data agreements are greatly variable and strongly depend of climate conditions and management practices (Figures 8 and 9).

Variability of Annual CO_2 and Water Fluxes

Correlations between climate, management practices and CO_2 and water fluxes are showed through a matrix plot (Figure 10). The different categories correspond to modeled and observed variables for the mowed and grazed paddocks. All points represent one year of either modeled or observed variables for the two sites summed/averaged over the summer half-year (15 April to 15 September), corresponding to the most productive time of the year. Lower panels show the scatter plots between the different selected variables and the upper panels give their correlation coefficients. For example, the pink-circled lower panel show the scatter plot of NEE and precipitation and the pink-circled upper panel give the corresponding correlation coefficients for the different categories (MG, MM, OG, and OM) and the overall correlation coefficient (Cor). The blue-circled area of the graph shows the selection of the most important relationships.

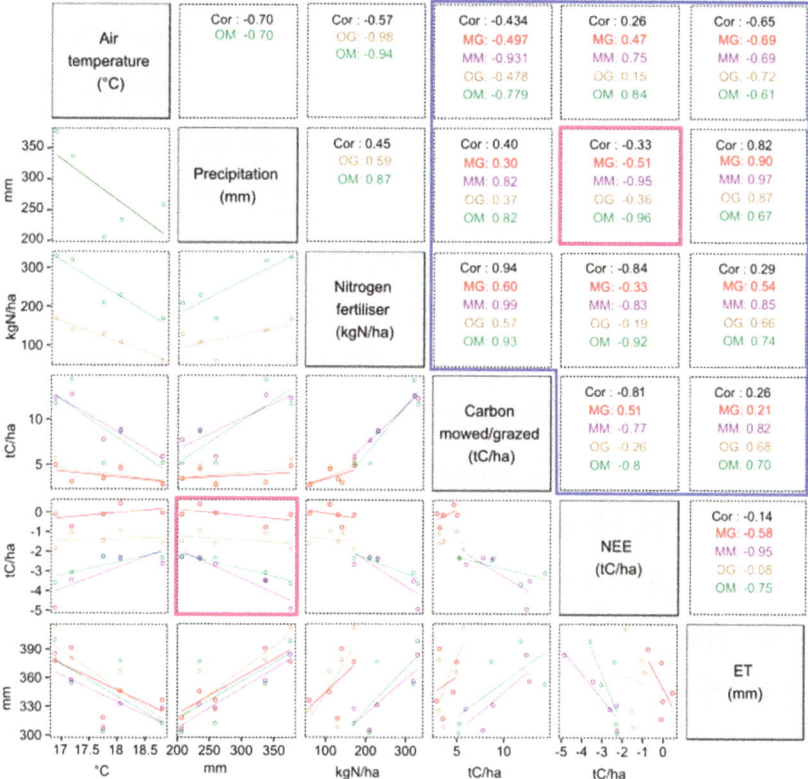

Figure 10. Matrix of paired plots showing the interannual variability of summer half-year (15th April–15th September) averaged climate drivers (air temperature and precipitation), management practices (N fertilizer application and amounts of C mowed/grazed), and observed and modeled CO_2 and H_2O fluxes for the mowed and grazed grassland sites (MG: modeled grazing; MM: modeled mowing; OG: observed grazing; OM: observed mowing). Variable names are given in the matrix diagonal. Paired scatterplots are in the lower triangle (below the diagonal in gray) with every point being the summer half-year of one year of the study period and colors are related to the different categories listed above. Their corresponding Pearson (linear) correlation coefficients are listed in the upper triangle (above the diagonal in gray). For example, the relationships between NEE and precipitation is shown in the pink circled scatter plot below the matrix diagonal and corresponding correlation coefficients for the different categories are given in the symmetric panel above the matrix diagonal (pink-circled). Important relationships are circled in blue on the upper panels.

First, it is striking (Figure 10) that annual CO_2 and H_2O fluxes were correlated with annual meteorological condition (air temperature and precipitation) and with management practices (C harvested/grazed and N fertilizer applications) but with marked differences across the two managed grassland sites.

For the mowed paddock, observed, and modeled amounts of carbon harvested are highly correlated with air temperature (OM: −0.78 and MM: −0.93), precipitation (OM and MM: 0.82), and N fertilizer (OM: 0.93 and MM: 0.99). On the contrary, correlations for the grazed paddock were lowest with N fertilizer (OG: 0.57 and MG: 0.60) and weak with climate (|OG| and |MG| <0.50).

The analysis also showed that, for the mowed paddock, the modeled and observed NEE were highly correlated with climate and management practices, but that for the grazed paddock NEE values were only weekly correlated with other variables. In this section and like for all this study, negative NEE represent a net gain, and a positive NEE is a net loss of CO_2 for the ecosystem. It is interesting that for the grazed paddock, modeled, and observed summer half-year NEE responded differently to the amount of vegetation grazed (i.e., CenW giving a positive moderate correlation of NEE with the amount of vegetation grazed while observations were giving a week negative correlation). This is due to the differences between modeled and observation-derived ER rates during grazing events and CenW simulating higher ER rates during grazing events: the more vegetation is eaten the more NEE increased (reduction of the sink strength of the pasture).

There were also high correlations between ET, climate and management practices for both grasslands, with a general upward trend of ET as precipitation, amounts of N fertilizer and C mowed/grazed increased and a downward trend with the increase of air temperature. More water vapor is returned to the atmosphere when there was more rainfall compared to dryer and hotter spring and summer periods.

The modeled and observed annual (full year average/sum) carbon and water balances for the mowed and grazed paddocks are shown in Figure 11. For the mowed paddock, observed annual GPP values ranged between 16 and 20.5 tC ha^{-1} yr^{-1} (five-year average: 18.2 tC ha^{-1} yr^{-1}) and modeled GPP values ranged between 15.3 and 22.7 tC ha^{-1} yr^{-1} (five-year average: 19.2 tC ha^{-1} yr^{-1}). For the grazed paddock, observed annual GPP values ranged between 15.9 and 20.3 tC ha^{-1} yr^{-1} (five-year average: 18.1 tC ha^{-1} yr^{-1}) and modeled GPP values ranged between 14.9 and 20.2 tC ha^{-1} yr^{-1} (five-year average: 18.2 tC ha^{-1} yr^{-1}).

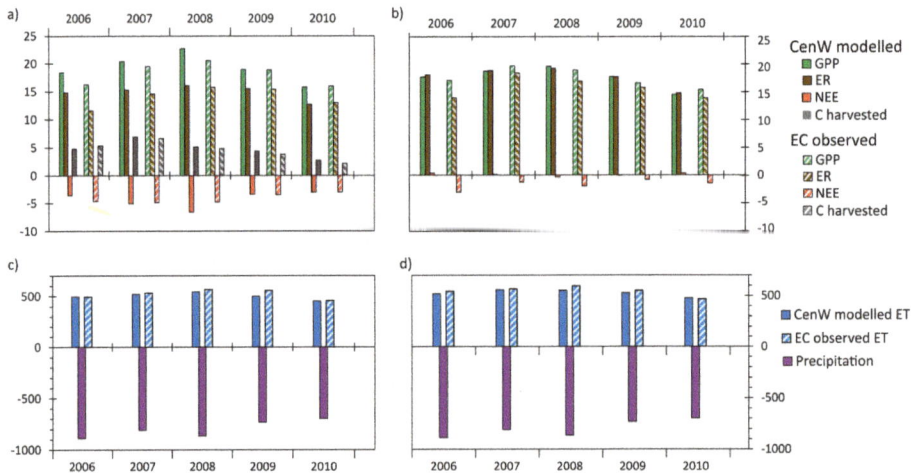

Figure 11. Bar plot showing modeled and observed annual CO_2 and water balances for the mowed ((**a**) and (**c**)) and grazed ((**b**) and (**d**)) paddocks. Carbon fluxes (NEE, GPP, ER, and C harvested are given in tC ha^{-1} and precipitation and ET are given in mm yr^{-1}).

On all years but 2006, CenW-modeled ER for the mowed paddock were lower than annual sums of EC-derived ER (Figure 11) with annual modeled ER values between 12.7 and 16.1 tC ha^{-1} yr^{-1} (five-year average: 14.9 tC ha^{-1} yr^{-1}), while observation-based ER values varied from 11.6 to 15.7 tC ha^{-1} yr^{-1} (five-year average: 14.1 tC ha^{-1} yr^{-1}). For the grazed paddock, model/data differences were even more important with annual EC-derived and modeled ER varying from 14.3 to 19.0 tC ha^{-1} yr^{-1} (five-year average: 16.3 tC ha^{-1} yr^{-1}) and from 15.2 to 19.9 tC ha^{-1} yr^{-1} (five-year average: 18.3 tC ha^{-1} yr^{-1}), respectively.

There is a good agreement between modeled and observed annual amounts of harvested C. Generally, the modeled and observed annual NEE agreed reasonably well for the mowed grassland sites but large differences were retrieved for the grazed paddock and might result from the miss or only partial capture of grazing animals' respiratory losses by the EC system. Five-year averages of observed NEE for the mowed and grazed grassland sites were −4.2 (−3.0 to −4.9) and −1.8 (−0.9 to −3.2) tC ha^{-1} yr^{-1}, respectively. CenW modeled NEE, averaged over 5 years, were −4.4 (−3.1 to −6.6) tC ha^{-1} yr^{-1} for the mowed paddock and 0.1 (−0.4 to 0.4) tC ha^{-1} yr^{-1} for the grazed pasture. Modeled annual net CO_2 fluxes, for the grazed paddock, were significantly lower than observed ones because, as we have seen, part of C harvested (grazed) is taken into account in observed NEE, while it is fully accounted for in the modeled NEE. Apparently, the mowed paddock fixed more CO_2 than the grazed paddock however, when harvested C is taken into account (Figure 11a) the mowed grassland C sink activity is drastically reduced.

Modeled and observed daily evapotranspiration rates generally agreed very well as we have seen in Section 3.1.2 with EF of 0.82 and 0.80 for the mowed and grazed paddocks, respectively. Both modeled and observed summer half-year evapotranspiration rates were higher for the grazed paddock because of the differences management practices (harvests and grazing) that affected vegetation dynamics. Even if there were fewer harvests than grazing events, the dramatic reduction of live foliage following grass cuttings affected water fluxes and reduced the annual amounts of evapotranspiration. Modeled summer half-year ET were also systematically lower than the observed results, which could result from (1) differences in modeled and observed roots dynamics (growth and senescence) affecting soil water extraction by plants and (2) the water returned directly to the field by cattle urinations which is not accounted for in the model. Observed five-year average annual ET were 521 and 544 mm yr^{-1} while modeled values were 503 to 527 mm yr^{-1} for the mowed and grazed paddocks, respectively.

It also as to be noted that the conventional gap-filling and partitioning tool [69] was not designed to deal neither with heterogeneities in ecosystems as it the case in intensively managed grasslands, like our study site, nor to take into account the varying magnitude of respiratory CO_2 losses from rotationally grazing animals that not depend of meteorological conditions. These conditions would add uncertainties in gap filled NEE fluxes and on their partitioning into GPP and ER [79].

4. Discussion

4.1. Performances of the CenW Model to Simulate Gas Exchanges of Mowed and Grazed Pastures

Mechanistic ecosystem models, like the CenW model, are useful tools to gain a better understanding of GHG emissions, yields, and carbon stock dynamics of managed grasslands as they can address, over long time periods, the complex interactions between climate, soil, vegetation and management practices [80–82]. Modeling studies have shown that models could achieve high accuracy in simulating greenhouse gas (CO_2 and H_2O) uptakes and emissions, yields, and carbon source/sink activity of managed grasslands for a wide range of climate and management conditions [17,56,83–86].

By using observation and models in conjunction, it is possible to improve our knowledge of the systems under study, identify weaknesses in datasets and models and to correct them. Over the study period, the site experienced large variations in temperature, moisture availability, and radiation, which are controlling factors of the exchange rates between the atmosphere and grasslands of CO_2 [87–90] and water fluxes [91]. In addition, grazing and cutting dramatically alter the way managed grassland ecosystems respond to climate drivers by the sharp removal of large amounts of live biomass [36,92,93] and thus strongly affect the seasonal and inter-annual variabilities of gas exchanges and the annual carbon budgets of the farm/paddock [22,79,94].

Modeling NEE accurately is generally difficult as NEE is calculated from the (relatively small) difference between the two largest carbon fluxes between the atmosphere and the ecosystem (vegetation + soil), i.e., GPP, which is the carbon gain through photosynthetic CO_2 assimilation and ER, which corresponds to total ecosystem respiratory losses by plant autotrophic, soil heterotrophic and also by grazing animals'. Accurate and reliable modeling of daily NEE fluxes requires that the CenW model properly incorporates and simulates the main processes driving the dynamics of both ecosystems (mowed and grazed grasslands). This study aimed to confirm the applicability of the CenW ecosystem model to simulate managed grassland systems.

Generally, better agreements between observed and modeled NEE and its components GPP and ER were found for the mowed paddock than for the grazed one. This could be because there were more frequent, longer and more spatially heterogeneous disturbances on the grazed than on the mowed paddock. In addition, uncertainties in the amounts of vegetation removal and respiration rates of grazing animals likely reduced the overall agreement between CenW outputs versus observations for the grazed paddock. Over the entire study period and for the two managed grasslands, CenW tends to slightly overestimate the lowest rates of C assimilation (Figure 3a,d) and ecosystem respiration (Figure 3b,e) and to underestimate large values of both uptake and emission of C fluxes as indicated by the slopes and intercepts of the linear regressions. After the careful parameterization of the CenW ecosystem model, very good agreements were found between simulated and observed carbon and water fluxes, highlighting that the model, parameterized with local data could appropriately be applied to intensively managed grasslands, as long as sufficient information on management practices are available.

4.2. Cow Respiration in Observed and Modeled CO_2 Fluxes

Discrepancies between modeled and observed ER and NEE fluxes, on the grazed paddock, could be caused by the uneven repartition of dairy cows on the field and the variability of the flux footprint: i.e. dairy cows were under-represented in the flux footprint, causing an under-estimation of animal respiratory losses of CO_2 in EC data. Some large grazer respiration fluxes could have also been excluded through the filtering of NEE data to remove outliers (condition 1, 2 and 3 in Section 2.1.2). Moreover, because the Reichstein gap-filling algorithm does not explicitly include respiration from grazing animals (e.g., animals disturbance) to either fill gaps in NEE time series or in the partitioning of NEE into GPP and ER, substantial bias could be added to the dataset [56,79]. For all measurements without grazing there was good agreement between the CenW model and EC observations, highlighting the correct parameterization of the model and confirming that most of the ecosystem processes were well embedded in the model. It also showed that large discrepancies were present during grazing periods because of the possible non or only partial capture of cows respiration in NEE data (Figures 7c, 9c and A1).

A recent EC study of CO_2 fluxes on two paired sites under rotational and continuous grazing management, found that ER for the rotationally grazed paddock was greatly affected by cattle respiration and that grazing animal's respiration was correctly accounted for in EC measurements [95]. In this study, we showed that there is, during grazing events, a possible underestimation of observation-derived ER from the EC system placed on the grazed paddock (Figure 12), which is consistent with other studies on managed grassland that found no effect of animals' respiration on EC-derived ER [92,93,96]. By comparing daily EC-derived ER with their CenW modeled counterparts (Figure 12), we showed that, on a few cases (red dots around the one to one line), modeled and observation-derived ER agree very well, showing that cattle respiration was properly modeled and measured by the flux tower. However, for most of the grazing events recorded, a large proportion or all of cattle respiration was not captured by EC (red dots strongly deviating from the one to one line). This was likely caused by the uneven spatial repartition of animals on the paddock, shifts in flux footprint, specific site conditions, stocking densities, and the processing of EC data [75,93].

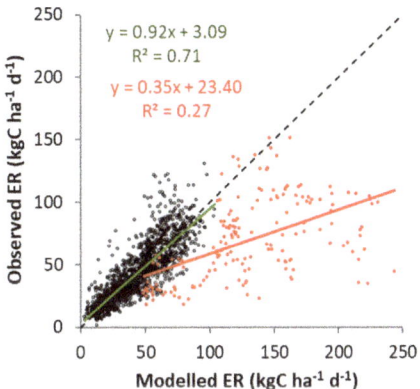

Figure 12. Scatter plot of observed vs modeled ecosystem respiration rates outside grazing periods (black) and during grazing periods (red) for the five years of the study.

In this study, we showed that the direct effect of grazing (i.e., the reduction of photosynthesizing biomass [22]) was properly captured in EC data but that the indirect effect (i.e., grazing animals' respiration [22]) was not captured by EC systems because of the stochastic position of cows and shifts in the flux footprint area. To fully understand why grazers' respiration is not accounted for and to find a way to correct EC data for this bias, using a biochemical, process-based ecosystem model incorporating management practices, like CenW, alongside dairy cows positioning devices and detailed flux footprint information, could be necessary [79,94].

Another problem is also related to the gap-filling process since algorithm used for this task did not use any information on cows' position and their respiration rates. If actual data affected by the presence of cows in the footprint were used to fill gaps outside of grazing periods, the large (depending of the number of cows) respiration would bias gap filled fluxes [79].

We advocate that a better way to process EC data on rotationally grazed pasture would be to exclude measurements taken during grazing periods and fill the gaps outside and inside these periods with data acquired when there was no cows in the paddocks [97]. This would insure to get the flux from the pasture only [79,94], and then calculate how much carbon is lost from the paddock due to grazers respiration based on stocking rates, grazing duration, and amount of ingested biomass. This imply that detailed information on cows movement and farming practices need to be recorded and used to process EC data from intensively manage grasslands, especially for systems with high stocking rates like rotational grazing.

The same dataset (we reprocessed meteorological and EC data for this study) was previously used to derive the net carbon storage (NCS) of the mowed and grazed grasslands [26]. They found that the grazed grassland have the potential to sequester more C than the mowed grassland but this implied that cows' respiration was accurately enough measured by EC. However, we showed by applying the CenW model that important respiratory losses (from grazing animals) were not accounted for in eddy covariance measurements taken on a intensively, rotationally grazed grassland (Figure 12). We showed that measured and modeled NEE fluxes agreed very well outside of grazing periods but strongly deviated during grazing events due to the noncapture of cows respiration by the EC system. CenW modeled annual NEE values (Figures 7c and A1) were found to be almost 10 times lower than the annual NEE values from EC data. As a result, the carbon storage capacity of this grazed grassland site would have been strongly reduced if cows respiration was adequately captured by the eddy covariance system.

4.3. Seasonal Variability of Observed and Modeled CO_2 and Water Fluxes

The natural variability of carbon and water fluxes, caused by short (day to day) and long term (seasonal) variations in climate conditions, management and vegetation dynamics is well captured in model runs. It is notable from time series (Figures 6–9) that grazing caused a greater interannual variability than mowing in all three carbon fluxes (i.e., GPP, ER, and NEE).

Unlike natural ecosystems, in intensively managed grasslands, the seasonal and interannual variability of carbon and water fluxes not only depends on the variability of the governing climate variables, but also on farming practices [36,98–100]. This is mostly a result of the rapid and sharp reduction of photosynthetically active biomass caused by mowing and grazing events that reduce GPP and switches the system from being a sink to a source of CO_2 [32,93,100]. However, because of the regular N fertilizer applications, and unlike for natural ecosystems, the N limitation is almost suppressed, promoting the rapid restart of GPP and vegetation growth.

Measured eddy covariance data showed a larger temporal variability than the modeled signals (Figures 8 and 9). This could be explained by the fact that EC data measurements contain random errors that add up to the "real" flux and that scale with the magnitude of the fluxes and hence vary diurnally and seasonally [101]. Therefore, this random error term could lead to an under or overestimation of the measured fluxes. Whereas CenW is always in the middle, where rates ought to be and the scatter in both directions cancelled out. It is also possible that these higher rates than the average are not artifacts and that the formalism of the model not allow to simulate these fluxes because we either do not know which processes are causing them or something not recorded happened on the farm. In the mowed paddock, ER was less affected by mowing events than GPP because only the autotrophic respiration of plant leaves term was affected by the sharp reduction of live foliage [88,95].

Seasonal dynamics of the soil moisture profile were controlled by temporal distributions of water gain from precipitation and ET losses, which were well captured by the model (for the mowed paddock). Soil water content observation and CenW showed larger variations in shallow soil layer than deeper in the soil profile and generally agree reasonably well.

5. Summary and Conclusions

This study investigated the performances of the CenW ecosystem model to simulate carbon and water fluxes from paired eddy covariance towers of two managed grassland systems in France with different management practices: mowing vs grazing. It showed that once parameterized, model/data agreement was very good for both sites and that CenW could adequately reproduce flux variability in response to management and climatic condition at daily, seasonal and interannual time scales. Model efficiencies for daily CO_2 fluxes were 0.65–0.80 and 0.73–0.85 for grazed and mowed paddocks, respectively. The mowed grassland ET, averaged SWC and harvested biomass were modeled with efficiencies of 0.82, 0.85, and 0.80, respectively. For the grazed paddock, model efficiency for daily ET was 0.80.

Our study showed that management practices highly determined the temporal dynamics and seasonal and interannual variabilities of CO_2 fluxes and the C status of the grazed pasture. In addition, most of previous studies which derived annual carbon budgets of managed pastures from EC measurements assumed or showed that grazing animal' respiratory losses were satisfactorily captured in NEE fluxes at annual time scales but showed weaknesses at smaller time scales [22,79].

It also highlighted that large discrepancies existed between measured and modeled net carbon exchange and ecosystem respiration rates during grazing events and that it is likely that large losses of CO_2 to the atmosphere were not fully captured by the eddy covariance system. The model/data comparison showed that flux processing and interpretation needed to be done carefully in grazed systems to account for the presence of dairy cows in the paddock. So far, only few studies have used eddy covariance measurement in combination with process-based models in grazed pastures. Here our results clearly demonstrate that grazing animals' respiratory flux is most often not captured by EC systems, which could lead to substantial bias in NEE data taken during grazing events.

In addition, because of the importance of CO_2 losses of carbon not accounted for in the annual C budget, large overestimations of the C status of the farm are likely to be made.

The capture or not by EC of cows respiration is site specific, using detailed ecosystem models incorporating farming practices and their effects on vegetation and C dynamics could help to identify and correct possible issues with EC data in intensively managed grasslands. Model/data agreements for the mowed paddock were higher than those obtained for the grazed paddock, certainly because mowed systems are less complex and disturbed than grazed ones.

Author Contributions: Conceptualization, A.C., N.P., and N.S.; Methodology, N.P., A.C., and M.K.; Software, M.K.; Validation, N.P., A.C., and M.K.; Formal Analysis, N.P.; Investigation, N.P. and A.C.; Resources, A.C.; Writing—Original Draft Preparation, N.P.; Writing—Review and Editing, A.C., N.S., C.F., K.K., M.K.; Supervision, A.C.; Project Administration, A.C.; Funding Acquisition, A.C.

Funding: The research leading to these results has received funding principally by the New Zealand Government to support the objectives of the "Livestock Research Group of the Global Research Alliance on Agricultural Greenhouse Gases under grants SOW12-GPLER-LCR-PM (Proposal ID 16949-15 LCR)", "The AnaEE France (ANR-11-INBS-0001)", "AllEnvi", and "CNRS-INSU".

Acknowledgments: We would like to thank the National Research Infrastructure "Agro-écosystèmes, Cycles Biogéochimique et Biodiversité" (ACBB) for their support in field experiment and the database. We are deeply indebted to Christophe de Berranger, Marie-Laure Decau and Nicolas Mascher for their substantial technical assistance. Last but not least we would like to thanks Dr. Gregory Starr (Department of Biological Sciences, University of Alabama, USA) for his technical support in refiltering the dataset.

Conflicts of Interest: The authors declare no conflict of interest.

Appendix A Management Records

Table A1. Management records for grasslands under mowing and grazing for the period 2006–2010 at the SOERE-ACBB, Lusignan, France.

	Mowed Paddock (P2)			Grazed Paddock (P4)				
Year	Date of Mowing	Date of N Fertilizer Application	Amount (kgN ha^{-1})	Starting Date of Grazing Event	Length of Grazing Period (Day)	Stocking Rate (Head ha^{-1})	Date of N Fertilizer Application	Amount (kgN ha^{-1})
2006	17-May	26-Feb	60	11-Apr	7	16.8	5-Apr	30
	6-Jun	24-May	60	19-May	10	16.8	24-May	30
	24-Oct	28-Sep	50	3-Jul	10	3.9		
				2-Oct	4.5	17.1		
				16-Nov	18	12.9		
2007	23-Apr	22-Feb	80	19-Mar	5	13.5	28-Mar	50
	5-Jun	27-Apr	60	16-Apr	4	21.3	19-Jun	30
	17-Jul	12-Jun	60	16-May	7	19.4	19-Sep	30
	10-Sep	26-Jul	60	14-Jun	3	16.1	20-Sep	30
	12-Nov	19-Sep	60	15-Jul	8	10.6		
		20-Sep	60	20-Aug	2	11		
				17-Sep	4	17.4		
				22-Oct	2	19.7		
2008	19-May	29-Jan	120	25-Mar	2	24.8	29-Jan	30
	30-Jun	22-May	90	28-Apr	4	22.9	22-May	30
	15-Sep	15-Jul	60	19-May	2.4	20.6	28-Jul	60
		17-Sep	60	16-Jun	4	20.6	17-Sep	50
				15-Jul	4	16.8		
				11-Aug	3.4	15		
				12-Sep	6	18.7		
				27-Oct	2	22.3		
				1-Dec	8.25	4.5		
2009	11-May	17-Feb	110	23-Mar	2	25.2	17-Feb	50
	22-Jun	19-May	60	20-Apr	4.5	24.5	19-May	60
	28-Sep	7-Oct	60	11-May	3.5	24.5		
				9-Jun	8	19		
				13-Jul	2.5	12.9		
				21-Sep	4	17		
				27-Oct	4	19.5		
2010	26-Apr	16-Mar	90	29-Mar	2.5	19.4	16-Mar	60
	2-Jun	29-Apr	70	19-Apr	3.5	19.4	29-Apr	50
	26-Jul	8-Jun	50	17-May	5.5	19		
				14-Jun	3.5	19		
				5-Jul	2.5	14.8		
				2-Aug	2	16.5		
				20-Sep	1.5	15.2		
				22-Nov	1.5	12.6		

Appendix B CenW Model Calibrated Parameters

Table A2. Main model parameters values used to simulate mowed and grazed grasslands after CenW calibration.

	Parameter Description	Lusignan Mowed	Lusignan Grazed	Units
	Minimum foliage turn-over	0.022	0.022	yr^{-1}
	Fine-root turn-over	2.49	2.49	yr^{-1}
	Low-light senescence limit	0.056	0.08	MJ m^{-2} d^{-1}
	Max daily low-light senescence	0.015	0.017	% d^{-1}
	Max drought foliage death rate	6.08	6.76	% d^{-1}
	Drought death of roots relative to foliage	0.062	0.066	–
	Mycorrhizal uptake	0.01	0.01	g kg^{-1} d^{-1}
	Soil water stress threshold (Wcrit)	0.60	0.60	–
	Respiration ratio per unit N	0.18	0.44	–
	beta parameter in T response of respiration	1.98	1.96	–
Stand	Temperature for maximum respiration	47	47	°C
	Growth respiration	0.29	0.32	–
	Time constant for acclimation response of respiration	364	247	d
	Water-logging threshold (Llog)	0.999	0.994	–
	Water-logging sensitivity (sL)	8.3	7.33	–
	Ratio of [N] in senescing and live foliage	0.99	0.99	–
	Ratio of [N] in average foliage to leaves at the top	0.83	0.78	–
	Biological N fixation	1.71	7.9	gN kgC^{-1}
	Growth Km for carbon	0.97	1.8	%
	Growth Km for nitrogen	1.94	3.7	%
	Drop of standing dead leaves	2.11	2.11	% d^{-1}
	Decomposability of standing dead relative to metabolic litter	0.7	0.7	–
	Specific leaf area	17.5	19.3	m^2 (kg DW)$^{-1}$
	Foliage albedo	6.77	6.75	%
	Transmissivity	1.57	1.56	%
	Loss as volatile organic carbon	0	0	%
	Threshold N concentrations (No)	6.33	5.76	gN (kg DW)$^{-1}$
	Non-limiting N concentration (Nsat)	41.6	42.4	gN (kg DW)$^{-1}$
	Light-saturated maximum photosynthetic rate (Amax)	45.7	47.2	μmol m^{-2} s^{-1}
	Maximum quantum yield	0.06	0.06	mol mol^{-1}
photosynthesis	Curvature in light response function	0.412	0.412	–
	Light extinction coefficient	0.86	0.86	–
	Ball–Berry stomatal parameter (unstressed) bb1	10.1	11.9	–
	Ball–Berry stomatal parameter (stressed) bb2	8	8	–
	Minimum temperature for photosynthesis (Tn)	-4.1	-4.1	°C
	Lower optimum temperature for photosynthesis (Topt, lower)	25.8	25.8	°C
	Upper optimum temperature for photosynthesis (Topt, upper)	30.06	30.06	°C
	Maximum temperature for photosynthesis (Tx)	38.8	38.8	°C
	Temperature damage sensitivity (sT)	0.04	0.04	–
	Threshold for frost damage	0.19	0.19	°C
	Allocation to reproductive organs	None	None	–
	Fine root: foliage target ratio (nitrogen-unstressed)	0.98	0.90	–
allocation	Fine root: foliage target ratio (nitrogen-stressed)	3.6	4.6	–
	Used target-oriented dynamic root-shoot allocation	Yes	Yes	–
	Fine root:foliage [N] ratio	0.82	0.82	–
	Relative temperature dependence of heterotrophic respn	0.49	0.75	–
	Foliar lignin concentration	11.9	12	%
	Root lignin concentration	14.6	14.6	%
	Organic matter transfer from surface to soil	90	90	% yr^{-1}
decomposition	Critical C:N ratio	8.03	8	–
	Ratio of C:N ratios in structural and metabolic pools	4.83	4.09	–
	Exponential term in lignin inhibition	5	5	–
	Water stress sens. of decomp. relative to plant processes	0.68	1.03	–
	Residual decomposition under dry conditions	0.05	0.05	–
	Mineral N immobilized	5.32	5.38	% d^{-1}
	Atmospheric N deposition	2	2	kgN ha^{-1} yr^{-1}
	Volatilization fraction	10.1	10.1	%
	Leaching fraction	0.46	0.46	–
site	Litter water-holding capacity	2	2	g gDW^{-1}
	Mulching effect of litter	2.8	2.8	% tDW^{-1}
	Canopy aerodynamic resistance	83	78.7	s m^{-1}
	Canopy rainfall interception	0.044	0.044	mm LAI^{-1}
	Maximum rate of soil evaporation	1.55	1.25	mm d^{-1}

Appendix C EC-Derived and Modeled GPP Time Series

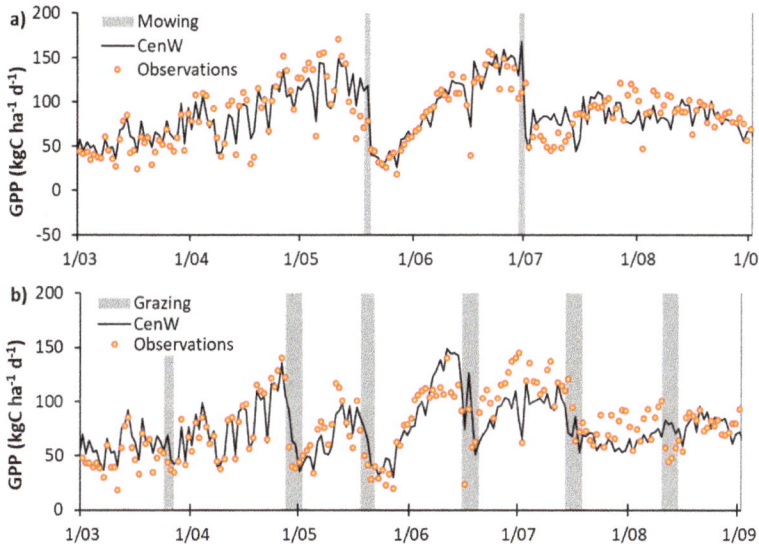

Figure A1. Daily time series of EC derived (dots) and modeled (black line) GPP for a) the mowed and b) the grazed grassland sites over 6 months (March to end August) of 2008. Vertical gray areas represent the recorded timing of harvests and grazing events for the two selected paddocks.

References

1. FAO and ITPS. *Status of the World's Soil Resources—Main Report*; FAO and ITPS: Rome, Italy, 2015; ISBN 978-92-5-109004-6.
2. FAOSTAT. Database Collection of the Food and Agriculture Organization of the United Nation. 2019. Available online: http://www.fao.org/faostat/en/#home (accessed on 4 January 2019).
3. Scurlock, J.M.O.; Hall, D.O. The global carbon sink: A grassland perspective. *Glob. Chang. Biol.* **1998**, *4*, 229–233. [CrossRef]
4. White, R.P.; Murray, S.; Rohweder, M. *Pilot Analysis of Global Ecosystems: Grassland Ecosystems*; World Resources Institute: Washington, DC, USA, 2000; ISBN 1-56973-461-5.
5. Conant, R.T.; Paustian, K.; Elliott, E.T. Grassland management and conversion into grassland: Effects on soil carbon. *Ecol. Appl.* **2001**, *11*, 343–355. [CrossRef]
6. Wang, W.; Fang, J. Soil respiration and human effects on global grasslands. *Glob. Planet. Chang.* **2009**, *67*, 20–28. [CrossRef]
7. Herrero, M.; Henderson, B.; Havlík, P.; Thornton, P.K.; Conant, R.T.; Smith, P.; Wirsenius, S.; Hristov, A.N.; Gerber, P.; Gill, M.; et al. Greenhouse gas mitigation potentials in the livestock sector. *Nat. Clim. Chang.* **2016**, *6*, 452–461. [CrossRef]
8. Soussana, J.F.; Allard, V.; Pilegaard, K.; Ambus, P.; Amman, C.; Campbell, C.; Ceschia, E.; Clifton-Brown, J.; Czobel, S.; Domingues, R.; et al. Full accounting of the greenhouse gas (CO_2, N_2O, CH_4) budget of nine European grassland sites. *Agric. Ecosyst. Environ.* **2007**, *121*, 121–134. [CrossRef]
9. Abberton, M.; Conant, R.; Batello, C. *Grassland Carbon Sequestration: Management, Policy and Economics*; Proceedings of the Workshop on the Role of Grassland Carbon Sequestration in the Mitigation of Climate Change; FAO: Rome, Italy, 2010; ISBN 978-92-5-106695-9.
10. Reid, R.S.; Thornton, P.K.; McCrabb, G.J.; Kruska, R.L.; Atieno, F.; Jones, P.G. Is it possible to mitigate greenhouse gas emissions in pastoral ecosystems of the tropics? *Environ. Dev. Sustain.* **2004**, *6*, 91–109. [CrossRef]

11. Allard, V.; Soussana, J.-F.; Falcimagne, R.; Berbigier, P.; Bonnefond, J.M.; Ceschia, E.; D'hour, P.; Hénault, C.; Laville, P.; Martin, C.; et al. The role of grazing management for the net biome productivity and greenhouse gas budget (CO_2, N_2O and CH_4) of semi-natural grassland. *Agric. Ecosyst. Environ.* **2007**, *121*, 47–58. [CrossRef]
12. Smith, P. Do grasslands act as a perpetual sink for carbon? *Glob. Chang. Biol.* **2014**, *20*, 2708–2711. [CrossRef] [PubMed]
13. Olff, H.; Ritchie, M.E.; Prins, H.H.T. Global environmental controls of diversity in large herbivores. *Nature* **2002**, *415*, 901–904. [CrossRef]
14. Jones, M.B.; Donnelly, A. Carbon sequestration in temperate grassland ecosystems and the influence of management, climate and elevated CO_2. *New Phytol.* **2004**, *164*, 423–439. [CrossRef]
15. Soussana, J.F.; Tallec, T.; Blanfort, V. Mitigating the greenhouse gas balance of ruminant production systems through carbon sequestration in grasslands. *Animal* **2010**, *4*, 334–350. [CrossRef]
16. McSherry, M.E.; Ritchie, M.E. Effects of grazing on grassland soil carbon: A global review. *Glob. Chang. Biol.* **2013**, *19*, 1347–1357. [CrossRef]
17. Senapati, N.; Jansson, P.-E.; Smith, P.; Chabbi, A. Modelling heat, water and carbon fluxes in mown grassland under multi-objective and multi-criteria constraints. *Environ. Model. Softw.* **2016**, *80*, 201–224. [CrossRef]
18. Ammann, C.; Flechard, C.R.; Leifeld, J.; Neftel, A.; Fuhrer, J. The carbon budget of newly established temperate grassland depends on management intensity. *Agric. Ecosyst. Environ.* **2007**, *121*, 5–20. [CrossRef]
19. Rumpel, C.; Crème, A.; Ngo, P.T.; Velásquez, G.; Mora, M.L.; Chabbi, A. The impact of grassland management on biogeochemical cycles involving carbon, nitrogen and phosphorus. *J. Soil Sci. Plant Nutr.* **2015**, *15*, 353–371. [CrossRef]
20. Fetzel, T.; Havlik, P.; Herrero, M.; Erb, K.-H. Seasonality constraints to livestock grazing intensity. *Glob. Chang. Biol.* **2017**, *23*, 1636–1647. [CrossRef] [PubMed]
21. Soussana, J.-F.; Loiseau, P.; Vuichard, N.; Ceschia, E.; Balesdent, J.; Chevallier, T.; Arrouays, D. Carbon cycling and sequestration opportunities in temperate grasslands. *Soil Use Manag.* **2004**, *20*, 219–230. [CrossRef]
22. Jérôme, E.; Beckers, Y.; Bodson, B.; Heinesch, B.; Moureaux, C.; Aubinet, M. Impact of grazing on carbon dioxide exchanges in an intensively managed Belgian grassland. *Agric. Ecosyst. Environ.* **2014**, *194*, 7–16. [CrossRef]
23. Oates, L.G.; Jackson, R.D. Livestock Management Strategy Affects Net Ecosystem Carbon Balance of Subhumid Pasture. *Rangel. Ecol. Manag.* **2014**, *67*, 19–29. [CrossRef]
24. Dlamini, P.; Chivenge, P.; Chaplot, V. Overgrazing decreases soil organic carbon stocks the most under dry climates and low soil pH: A meta-analysis shows. *Agric. Ecosyst. Environ.* **2016**, *221*, 258–269. [CrossRef]
25. Poeplau, C.; Marstorp, H.; Thored, K.; Kätterer, T. Effect of grassland cutting frequency on soil carbon storage—A case study on public lawns in three Swedish cities. *SOIL Discuss.* **2016**, *2*, 175–184. [CrossRef]
26. Senapati, N.; Chabbi, A.; Gastal, F.; Smith, P.; Mascher, N.; Loubet, B.; Cellier, P.; Naisse, C. Net carbon storage measured in a mowed and grazed temperate sown grassland shows potential for carbon sequestration under grazed system. *Carbon Manag.* **2014**, *5*, 131–144. [CrossRef]
27. Smith, P. How long before a change in soil organic carbon can be detected? *Glob. Chang. Biol.* **2004**, *10*, 1878–1883. [CrossRef]
28. Allen, D.E.; Pringle, M.J.; Page, K.L.; Dalal, R.C. A review of sampling designs for the measurement of soil organic carbon in Australian grazing lands. *Rangel. J.* **2010**, *32*, 227–246. [CrossRef]
29. Arrouays, D.; Marchant, B.P.; Saby, N.P.A.; Meersmans, J.; Orton, T.G.; Martin, M.P.; Bellamy, P.H.; Lark, R.M.; Kibblewhite, M. Generic Issues on Broad-Scale Soil Monitoring Schemes: A Review. *Pedosphere* **2012**, *22*, 456–469. [CrossRef]
30. Osborne, B.; Saunders, M.; Walmsley, D.; Jones, M.; Smith, P. Key questions and uncertainties associated with the assessment of the cropland greenhouse gas balance. *Agric. Ecosyst. Environ.* **2010**, *139*, 293–301. [CrossRef]
31. Mudge, P.L.; Wallace, D.F.; Rutledge, S.; Campbell, D.I.; Schipper, L.A.; Hosking, C.L. Carbon balance of an intensively grazed temperate pasture in two climatically contrasting years. *Agric. Ecosyst. Environ.* **2011**, *144*, 271–280. [CrossRef]
32. Rutledge, S.; Mudge, P.L.; Campbell, D.I.; Woodward, S.L.; Goodrich, J.P.; Wall, A.M.; Kirschbaum, M.U.F.; Schipper, L.A. Carbon balance of an intensively grazed temperate dairy pasture over four years. *Agric. Ecosyst. Environ.* **2015**, *206*, 10–20. [CrossRef]

33. Law, B.E.; Falge, E.; Gu, L.; Baldocchi, D.D.; Bakwin, P.; Berbigier, P.; Davis, K.; Dolman, A.J.; Falk, M.; Fuentes, J.D.; et al. Environmental controls over carbon dioxide and water vapor exchange of terrestrial vegetation. *Agric. For. Meteorol.* **2002**, *113*, 97–120. [CrossRef]
34. Chen, Z.; Yu, G.; Ge, J.; Wang, Q.; Zhu, X.; Xu, Z. Roles of Climate, Vegetation and Soil in Regulating the Spatial Variations in Ecosystem Carbon Dioxide Fluxes in the Northern Hemisphere. *PLoS ONE* **2015**, *10*, e0125265. [CrossRef]
35. Tian, H.; Lu, C.; Ciais, P.; Michalak, A.M.; Canadell, J.G.; Saikawa, E.; Huntzinger, D.N.; Gurney, K.R.; Sitch, S.; Zhang, B.; et al. The terrestrial biosphere as a net source of greenhouse gases to the atmosphere. *Nature* **2016**, *531*, 225–228. [CrossRef]
36. Wohlfahrt, G. Modelling Fluxes and Concentrations of CO_2, H_2O and Sensible Heat Within and Above a Mountain Meadow Canopy: A Comparison of Three Lagrangian Models and Three Parameterisation Options for the Lagrangian Time Scale. *Bound.-Layer Meteorol.* **2004**, *113*, 43–80. [CrossRef]
37. Jaksic, V.; Kiely, G.; Albertson, J.; Oren, R.; Katul, G.; Leahy, P.; Byrne, K.A. Net ecosystem exchange of grassland in contrasting wet and dry years. *Agric. For. Meteorol.* **2006**, *139*, 323–334. [CrossRef]
38. Keenan, T.F.; Baker, I.; Barr, A.; Ciais, P.; Davis, K.; Dietze, M.; Dragoni, D.; Gough, C.M.; Grant, R.; Hollinger, D.; et al. Terrestrial biosphere model performance for inter-annual variability of land-atmosphere CO_2 exchange. *Glob. Chang. Biol.* **2012**, *18*, 1971–1987. [CrossRef]
39. Fischer, E.M.; Sedláček, J.; Hawkins, E.; Knutti, R. Models agree on forced response pattern of precipitation and temperature extremes. *Geophys. Res. Lett.* **2014**, *41*, 8554–8562. [CrossRef]
40. Reyer, C. Forest Productivity Under Environmental Change—A Review of Stand-Scale Modeling Studies. *Curr. For. Rep.* **2015**, *1*, 53–68. [CrossRef]
41. Baldocchi, D.D.; Wilson, K.B. Modeling CO_2 and water vapor exchange of a temperate broadleaved forest across hourly to decadal time scales. *Ecol. Model.* **2001**, *142*, 155–184. [CrossRef]
42. Baldocchi, D.D. Assessing the eddy covariance technique for evaluating carbon dioxide exchange rates of ecosystems: Past, present and future. *Glob. Chang. Biol.* **2003**, *9*, 479–492. [CrossRef]
43. Baldocchi, D. Measuring and modelling carbon dioxide and water vapour exchange over a temperate broad-leaved forest during the 1995 summer drought. *Plant Cell Environ.* **1997**, *20*, 1108–1122. [CrossRef]
44. Zhu, Q.; Zhuang, Q. Parameterization and sensitivity analysis of a process-based terrestrial ecosystem model using adjoint method. *J. Adv. Model. Earth Syst.* **2014**, *6*, 315–331. [CrossRef]
45. Walker, W.E.; Harremoës, P.; Rotmans, J.; van der Sluijs, J.P.; van Asselt, M.B.A.; Janssen, P.; Krayer von Kraus, M.P. Defining Uncertainty: A Conceptual Basis for Uncertainty Management in Model-Based Decision Support. *Integr. Assess.* **2003**, *4*. [CrossRef]
46. Braswell, B.H.; Sacks, W.J.; Linder, E.; Schimel, D.S. Estimating diurnal to annual ecosystem parameters by synthesis of a carbon flux model with eddy covariance net ecosystem exchange observations. *Glob. Chang. Biol.* **2005**, *11*, 335–355. [CrossRef]
47. Stoy, P.C.; Katul, G.G.; Siqueira, M.B.S.; Juang, J.-Y.; McCarthy, H.R.; Kim, H.-S.; Oishi, A.C.; Oren, R. Variability in net ecosystem exchange from hourly to inter-annual time scales at adjacent pine and hardwood forests: A wavelet analysis. *Tree Physiol.* **2005**, *25*, 887–902. [CrossRef]
48. Siqueira, M.B.; Katul, G.G.; Sampson, D.A.; Stoy, P.C.; Juang, J.-Y.; Mccarthy, H.R.; Oren, R. Multiscale model intercomparisons of CO_2 and H_2O exchange rates in a maturing southeastern US pine forest. *Glob. Chang. Biol.* **2006**, *12*, 1189–1207. [CrossRef]
49. Hollinger, D.; Richardson, A. Uncertainty in eddy covariance measurements and its application to physiological models. *Tree Physiol.* **2005**, *25*, 873–885. [CrossRef]
50. Parsons, A.J.; Schwinning, S.; Carrère, P. Plant growth functions and possible spatial and temporal scaling errors in models of herbivory. *Grass Forage Sci.* **2001**, *56*, 21–34. [CrossRef]
51. Johnson, I.R.; Chapman, D.F.; Snow, V.; Eckard, R.; Parsons, A.; Lambert, M.G.; Cullen, B. DairyMod and EcoMod: Biophysical pasture-simulation models for Australia and New Zealand. *Aust. J. Exp. Agric.* **2008**, *48*, 621–631. [CrossRef]
52. Kirschbaum, M.U.F. CenW, a forest growth model with linked carbon, energy, nutrient and water cycles. *Ecol. Model.* **1999**, *118*, 17–59. [CrossRef]
53. Kirschbaum, M.U.F.; Keith, H.; Leuning, R.; Cleugh, H.A.; Jacobsen, K.L.; van Gorsel, E.; Raison, R.J. Modelling net ecosystem carbon and water exchange of a temperate Eucalyptus delegatensis forest using multiple constraints. *Agric. For. Meteorol.* **2007**, *145*, 48–68. [CrossRef]

54. Kirschbaum, M.U.F.; Watt, M.S. Use of a process-based model to describe spatial variation in Pinus radiata productivity in New Zealand. *For. Ecol. Manag.* **2011**, *262*, 1008–1019. [CrossRef]
55. Parton, W.J.; Schimel, D.S.; Cole, C.V.; Ojima, D.S. Analysis of Factors Controlling Soil Organic Matter Levels in Great Plains Grasslands1. *Soil Sci. Soc. Am. J.* **1987**, *51*, 1173–1179. [CrossRef]
56. Kirschbaum, M.U.F.; Rutledge, S.; Kuijper, I.A.; Mudge, P.L.; Puche, N.; Wall, A.M.; Roach, C.G.; Schipper, L.A.; Campbell, D.I. Modelling carbon and water exchange of a grazed pasture in New Zealand constrained by eddy covariance measurements. *Sci. Total Environ.* **2015**, *512–513*, 273–286. [CrossRef]
57. Kirschbaum, M.; Schipper, L.; Mudge, P.; Rutledge-Jonker, S.; Puche, N.J.; Campbell, D. The trade-offs between milk production and soil organic carbon storage in dairy systems under different management and environmental factors. *Sci. Total Environ.* **2016**, *577*, 61–72. [CrossRef]
58. Moni, C.; Chabbi, A.; Nunan, N.; Rumpel, C.; Chenu, C. Spatial dependance of organic carbon–metal relationships: A multi-scale statistical analysis, from horizon to field. *Geoderma* **2010**, *158*, 120–127. [CrossRef]
59. Chabbi, A.; Kögel-Knabner, I.; Rumpel, C. Stabilised carbon in subsoil horizons is located in spatially distinct parts of the soil profile. *Soil Biol. Biochem.* **2009**, *41*, 256–261. [CrossRef]
60. Kunrath, T.R.; de Berranger, C.; Charrier, X.; Gastal, F.; de Faccio Carvalho, P.C.; Lemaire, G.; Emile, J.-C.; Durand, J.-L. How much do sod-based rotations reduce nitrate leaching in a cereal cropping system? *Agric. Water Manag.* **2015**, *150*, 46–56. [CrossRef]
61. Senapati, N.; Chabbi, A.; Smith, P. Modelling daily to seasonal carbon fluxes and annual net ecosystem carbon balance of cereal grain-cropland using DailyDayCent: A model data comparison. *Agric. Ecosyst. Environ.* **2018**, *252*, 159–177. [CrossRef]
62. Fuchs, M.; Tanner, C.B. Calibration and Field Test of Soil Heat Flux Plates 1. *Soil Sci. Soc. Am. J.* **1968**, *32*, 326–328. [CrossRef]
63. Aubinet, M.; Grelle, A.; Ibrom, A.; Rannik, Ü.; Moncrieff, J.; Foken, T.; Kowalski, A.S.; Martin, P.; Berbigier, P.; Bernhofer, C.; et al. Estimates of the Annual Net Carbon and Water Exchange of Forests: The EUROFLUX Methodology. *Adv. Ecol. Res.* **2000**, *30*, 113–175.
64. Ferrara, R.M.; Loubet, B.; Di Tommasi, P.; Bertolini, T.; Magliulo, V.; Cellier, P.; Eugster, W.; Rana, G. Eddy covariance measurement of ammonia fluxes: Comparison of high frequency correction methodologies. *Agric. For. Meteorol.* **2012**, *158–159*, 30–42. [CrossRef]
65. Webb, E.K.; Pearman, G.I.; Leuning, L. Correction of Flux Measurements for Density Effects Due to Heat and Water-Vapor Transfer. *Quart. J. R. Meteorol. Soc.* **1980**, *106*, 85–100. [CrossRef]
66. Burba, G. *Eddy Covariance Method for Scientific, Industrial, Agricultural and Regulatory Applications: A Field Book on Measuring Ecosystem Gas Exchange and Areal Emission Rates*; LI-COR Biosciences: Lincoln, NE, USA, 2013; ISBN 978-0-615-76827-4.
67. Falge, E.; Baldocchi, D.; Olson, R.; Anthoni, P.; Aubinet, M.; Bernhofer, C.; Burba, G.; Ceulemans, R.; Clement, R.; Dolman, H.; et al. Gap filling strategies for long term energy flux data sets. *Agric. For. Meteorol.* **2001**, *107*, 71–77. [CrossRef]
68. Falge, E.; Baldocchi, D.; Olson, R.; Anthoni, P.; Aubinet, M.; Bernhofer, C.; Burba, G.; Ceulemans, R.; Clement, R.; Dolman, H.; et al. Gap filling strategies for defensible annual sums of net ecosystem exchange. *Agric. For. Meteorol.* **2001**, *107*, 43–69. [CrossRef]
69. Reichstein, M.; Falge, E.; Baldocchi, D.; Papale, D.; Aubinet, M.; Berbigier, P.; Bernhofer, C.; Buchmann, N.; Gilmanov, T.; Granier, A.; et al. On the separation of net ecosystem exchange into assimilation and ecosystem respiration: Review and improved algorithm. *Glob. Chang. Biol.* **2005**, *11*, 1424–1439. [CrossRef]
70. Moffat, A.M.; Papale, D.; Reichstein, M.; Hollinger, D.Y.; Richardson, A.D.; Barr, A.G.; Beckstein, C.; Braswell, B.H.; Churkina, G.; Desai, A.R.; et al. Comprehensive comparison of gap-filling techniques for eddy covariance net carbon fluxes. *Agric. For. Meteorol.* **2007**, *147*, 209–232. [CrossRef]
71. Lloyd, J.; Taylor, J.A. On the Temperature Dependence of Soil Respiration. *Funct. Ecol.* **1994**, *8*, 315–323. [CrossRef]
72. Sands, P. Modelling Canopy Production. I. Optimal Distribution of Photosynthetic Resources. *Funct. Plant Biol.* **1995**, *22*, 593–601. [CrossRef]
73. Baisden, W.T.; Amundson, R.; Brenner, D.L.; Cook, A.C.; Kendall, C.; Harden, J.W. A multiisotope C and N modeling analysis of soil organic matter turnover and transport as a function of soil depth in a California annual grassland soil chronosequence. *Glob. Biogeochem. Cycles* **2002**, *16*, 82-1–82-26. [CrossRef]

74. Pal, P.; Clough, T.; Kelliher, F.; van Koten, C.; Sherlock, R. Intensive Cattle Grazing Affects Pasture Litter-Fall: An Unrecognized Nitrous Oxide Source. *J. Environ. Qual.* **2012**, *41*, 444–448. [CrossRef]
75. Zeeman, M.J.; Hiller, R.; Gilgen, A.K.; Michna, P.; Plüss, P.; Buchmann, N.; Eugster, W. Management and climate impacts on net CO_2 fluxes and carbon budgets of three grasslands along an elevational gradient in Switzerland. *Agric. For. Meteorol.* **2010**, *150*, 519–530. [CrossRef]
76. Kelliher, F.M.; Clark, H. Chapter 9: Ruminants. In *Methane and Climate Change*; Reay, D., Smith, P., van Amstel, A., Eds.; Earthscan: London, UK, 2010; pp. 136–150.
77. Crush, J.R.; Waghorn, G.C.; Rolston, M.P. Greenhouse gas emissions from pasture and arable crops grown on a Kairanga soil in the Manawatu, North Island, New Zealand. *N. Z. J. Agric. Res.* **1992**, *35*, 253–257. [CrossRef]
78. Nash, J.E.; Sutcliffe, J.V. River flow forecasting through conceptual models part I—A discussion of principles. *J. Hydrol.* **1970**, *10*, 282–290. [CrossRef]
79. Felber, R.; Neftel, A.; Ammann, C. Discerning the cows from the pasture: Quantifying and partitioning the NEE of a grazed pasture using animal position data. *Agric. For. Meteorol.* **2016**, *216*, 37–47. [CrossRef]
80. Huntzinger, D.N.; Post, W.M.; Wei, Y.; Michalak, A.M.; West, T.O.; Jacobson, A.R.; Baker, I.T.; Chen, J.M.; Davis, K.J.; Hayes, D.J.; et al. North American Carbon Program (NACP) regional interim synthesis: Terrestrial biospheric model intercomparison. *Ecol. Model.* **2012**, *232*, 144–157. [CrossRef]
81. Warszawski, L.; Friend, A.; Ostberg, S.; Frieler, K.; Lucht, W.; Schaphoff, S.; Beerling, D.; Cadule, P.; Ciais, P.; Clark, D.B.; et al. A multi-model analysis of risk of ecosystem shifts under climate change. *Environ. Res. Lett.* **2013**, *8*, 044018. [CrossRef]
82. Chang, X.; Bao, X.; Wang, S.; Wilkes, A.; Erdenetsetseg, B.; Baival, B.; Avaadorj, D.; Maisaikhan, T.; Damdinsuren, B. Simulating effects of grazing on soil organic carbon stocks in Mongolian grasslands. *Agric. Ecosyst. Environ.* **2015**, *212*, 278–284. [CrossRef]
83. White, T.A.; Johnson, I.R.; Snow, V.O. Comparison of outputs of a biophysical simulation model for pasture growth and composition with measured data under dryland and irrigated conditions in New Zealand. *Grass Forage Sci.* **2008**, *63*, 339–349. [CrossRef]
84. Graux, A.-I.; Bellocchi, G.; Lardy, R.; Soussana, J.-F. Ensemble modelling of climate change risks and opportunities for managed grasslands in France. *Agric. For. Meteorol.* **2013**, *170*, 114–131. [CrossRef]
85. Ben Touhami, H.; Bellocchi, G. Bayesian calibration of the Pasture Simulation model (PaSim) to simulate European grasslands under water stress. *Ecol. Inform.* **2015**, *30*, 356–364. [CrossRef]
86. Ehrhardt, F.; Soussana, J.-F.; Bellocchi, G.; Grace, P.; McAuliffe, R.; Recous, S.; Sándor, R.; Smith, P.; Snow, V.; de Antoni Migliorati, M.; et al. Assessing uncertainties in crop and pasture ensemble model simulations of productivity and N_2O emissions. *Glob. Chang. Biol.* **2018**, *24*, 603–616. [CrossRef]
87. Baldocchi, D. "Breathing" of the terrestrial biosphere: Lessons learned from a global network of carbon dioxide flux measurement systems. *Aust. J. Bot.* **2008**, *56*, 1–26. [CrossRef]
88. Wohlfahrt, G.; Hammerle, A.; Haslwanter, A.; Bahn, M.; Tappeiner, U.; Cernusca, A. Seasonal and inter-annual variability of the net ecosystem CO_2 exchange of a temperate mountain grassland: Effects of weather and management. *J. Geophys. Res.* **2008**, *113*, 1–14. [CrossRef]
89. Yan, L.; Chen, S.; Huang, J.; Lin, G. Water regulated effects of photosynthetic substrate supply on soil respiration in a semiarid steppe. *Glob. Chang. Biol.* **2011**, *17*, 1990–2001. [CrossRef]
90. Cleverly, J.; Boulain, N.; Villalobos-Vega, R.; Grant, N.; Faux, R.; Wood, C.; Cook, P.G.; Yu, Q.; Leigh, A.; Eamus, D. Dynamics of component carbon fluxes in a semi-arid Acacia woodland, central Australia. *J. Geophys. Res. Biogeosci.* **2013**, *118*, 1168–1185. [CrossRef]
91. Eamus, D.; Cleverly, J.; Boulain, N.; Grant, N.; Faux, R.; Villalobos-Vega, R. Carbon and water fluxes in an arid-zone Acacia savanna woodland: An analyses of seasonal patterns and responses to rainfall events. *Agric. For. Meteorol.* **2013**, *182–183*, 225–238. [CrossRef]
92. Rogiers, N.; Eugster, W.; Furger, M.; Siegwolf, R. Effect of land management on ecosystem carbon fluxes at a subalpine grassland site in the Swiss Alps. *Theor. Appl. Climatol.* **2005**, *80*, 187–203. [CrossRef]
93. Peichl, M.; Carton, O.; Kiely, G. Management and climate effects on carbon dioxide and energy exchanges in a maritime grassland. *Agric. Ecosyst. Environ.* **2012**, *158*, 132–146. [CrossRef]
94. Felber, R.; Bretscher, D.; Münger, A.; Neftel, A.; Ammann, C. Determination of the carbon budget of a pasture: Effect of system boundaries and flux uncertainties. *Biogeosciences* **2016**, *13*, 2959–2969. [CrossRef]

95. Gourlez de la Motte, L.; Mamadou, O.; Beckers, Y.; Bodson, B.; Heinesch, B.; Aubinet, M. Rotational and continuous grazing does not affect the total net ecosystem exchange of a pasture grazed by cattle but modifies CO_2 exchange dynamics. *Agric. Ecosyst. Environ.* **2018**, *253*, 157–165. [CrossRef]
96. Lin, X.; Zhang, Z.; Wang, S.; Hu, Y.; Xu, G.; Luo, C.; Chang, X.; Duan, J.; Lin, Q.; Xu, B.; et al. Response of ecosystem respiration to warming and grazing during the growing seasons in the alpine meadow on the Tibetan plateau. *Agric. For. Meteorol.* **2011**, *151*, 792–802. [CrossRef]
97. Skinner, R.H. High Biomass Removal Limits Carbon Sequestration Potential of Mature Temperate Pastures. *J. Environ. Qual.* **2008**, *37*, 1319–1326. [CrossRef]
98. Falge, E.; Baldocchi, D.; Tenhunen, J.; Aubinet, M.; Bakwin, P.; Berbigier, P.; Bernhofer, C.; Burba, G.; Clement, R.; Davis, K.J.; et al. Seasonality of ecosystem respiration and gross primary production as derived from FLUXNET measurements. *Agric. For. Meteorol.* **2002**, *113*, 53–74. [CrossRef]
99. Falge, E.; Tenhunen, J.; Baldocchi, D.; Aubinet, M.; Bakwin, P.; Berbigier, P.; Bernhofer, C.; Bonnefond, J.-M.; Burba, G.; Clement, R.; et al. Phase and amplitude of ecosystem carbon release and uptake potentials as derived from FLUXNET measurements. *Agric. For. Meteorol.* **2002**, *113*, 75–95. [CrossRef]
100. Nieveen, J.P.; Campbell, D.I.; Schipper, L.A.; Blair, I.J. Carbon exchange of grazed pasture on a drained peat soil. *Glob. Chang. Biol.* **2005**, *11*, 607–618. [CrossRef]
101. Richardson, A.D.; Hollinger, D.Y.; Burba, G.G.; Davis, K.J.; Flanagan, L.B.; Katul, G.G.; William Munger, J.; Ricciuto, D.M.; Stoy, P.C.; Suyker, A.E.; et al. A multi-site analysis of random error in tower-based measurements of carbon and energy fluxes. *Agric. For. Meteorol.* **2006**, *136*, 1–18. [CrossRef]

© 2019 by the authors. Licensee MDPI, Basel, Switzerland. This article is an open access article distributed under the terms and conditions of the Creative Commons Attribution (CC BY) license (http://creativecommons.org/licenses/by/4.0/).

Article

Grassland Management Influences the Response of Soil Respiration to Drought

Gabriel Y. K. Moinet [1,*], Andrew J. Midwood [2], John E. Hunt [1], Cornelia Rumpel [3,4], Peter Millard [1] and Abad Chabbi [3,4,5]

1. Manaaki Whenua–Landcare Research, PO Box 69040, Lincoln 7640, New Zealand; huntj@landcareresearch.co.nz (J.E.H.); millardp@landcareresearch.co.nz (P.M.)
2. Department of Biology, University of British Columbia–Okanagan, Kelowna, BC V1V 1V7, Canada; andrew.midwood@ubc.ca
3. The French National Center for Scientific Research (CNRS), Institut d'Ecologie et des Sciences de l'Environnement Paris (IEES), UMR 7618, Batiment EGER, Aile B, F-78850 Thiverval Grignon, France; cornelia.rumpel@inra.fr (C.R.); abad.chabbi@inra.fr (A.C.)
4. AgroParisTech, French Natl Inst Agr Res INRA, UMR ECOSYS, F-78850 Thiverval Grignon, France
5. National Institute of Agricultural Research (INRA), Centre de recherché, Nouvelle-Aquitaine-Poitiers, URP3F, 86600 Lusignan, France
* Correspondence: moinetg@landcareresearch.co.nz; Tel.: +64-332-197-82

Received: 24 February 2019; Accepted: 3 March 2019; Published: 7 March 2019

Abstract: Increasing soil carbon stocks in agricultural grasslands has a strong potential to mitigate climate change. However, large uncertainties around the drivers of soil respiration hinder our ability to identify management practices that enhance soil carbon sequestration. In a context where more intense and prolonged droughts are predicted in many regions, it is critical to understand how different management practices will temper drought-induced carbon losses through soil respiration. In this study, we compared the impact of changing soil volumetric water content during a drought on soil respiration in permanent grasslands managed either as grazed by dairy cows or as a mowing regime. Across treatments, root biomass explained 43% of the variability in soil respiration ($p < 0.0001$). Moreover, analysis of the isotopic composition of CO_2 emitted from the soil, roots, and root-free soil suggested that the autotrophic component largely dominated soil respiration. Soil respiration was positively correlated with soil water content ($p = 0.03$) only for the grazed treatment. Our results suggest that the effect of soil water content on soil respiration was attributable mainly to an effect on root and rhizosphere activity in the grazed treatment. We conclude that farm management practices can alter the relationship between soil respiration and soil water content.

Keywords: soil respiration; soil volumetric water content; stable carbon isotopes; grassland; management practices

1. Introduction

Grasslands cover 26% of Earth's ice-free land surface area, representing 70% of the world's agricultural area [1] and containing 20% of global soil carbon stocks [2]. There is now a wide interest in improving carbon storage on land as a negative emission technology to stay below the 1.5 °C global warming limit [3]. Increasing soil carbon stocks in grasslands has a strong potential to contribute to this effort [4,5]. However, the impacts of different grassland management practices are not well understood [6], and recent reviews highlight a lack of data to clarify the mechanisms by which various management practices affect soil organic carbon (SOC) stocks [7,8]. Moreover, there are large uncertainties concerning the drivers of soil respiration (R_s), which is the second largest terrestrial carbon flux globally [9].

Global climate change is predicted to lead to an increase in the intensity and frequency of extreme events in large areas of the world. Droughts may cause intense water stress for plants and soil organisms [10]. In other areas, increasing rainfall together with evapotranspiration and changes in precipitation patterns could potentially lead to constrained water availability during the growing season [11]. In a context where the global water cycle is predicted to change, understanding the effect of soil water content on soil respiration is critical.

Soil respiration is the result of two processes: root and rhizosphere respiration (autotrophic component), and microbial decomposition of soil organic matter (heterotrophic component). Both plants [12] and micro-organisms [13] are sensitive to changes in soil water availability, resulting in soil respiration being sensitive to both soil water content [14,15] and the frequency and intensity of precipitation [16]. However, our understanding of the effect of drought on soil respiration remains relatively limited, notably due to differential responses of the autotrophic and heterotrophic components and the difficulties associated with quantifying them [17,18].

Usually, studies addressing the effects of soil water content on heterotrophic soil respiration are carried out in the laboratory in the absence of plants. Partitioning soil respiration into its autotrophic and heterotrophic components can be achieved with minimal soil disturbance in the field using the stable isotopes of carbon [17–21], although this can prove challenging in dry conditions where low CO_2 efflux may be below the detection limits of instrumentation [17]. Nonetheless, even when the rate of autotrophic and heterotrophic respiration cannot be determined, the ^{13}C isotopic signature ($\delta^{13}C$) of CO_2 emitted from the soil ($\delta^{13}CR_s$) can provide qualitative information on the dynamic soil respiration components. This is because the $\delta^{13}C$ of emitted CO_2 becomes more enriched as the contribution of microbially derived Soil Organic Matter (SOM) becomes an increasing part of total soil respiration [22,23].

Understanding the effect of management practices on the drought response of soil respiration and the plant and microbial processes driving it is critical to develop agricultural systems that are resistant to extreme climatic events. Here we compared the impact of changing soil water content during a severe drought on soil respiration and the $\delta^{13}C$ of CO_2 emitted from soils ($\delta^{13}CR_s$), roots, and rhizosphere ($\delta^{13}CR_a$), and root-free soils ($\delta^{13}CR_h$) in a permanent pasture grazed by dairy cows and a permanent pasture under a mowing regime with no grazing mammals. These two management systems are among the main types of grassland management in Europe [24] and have a strong potential to store carbon below ground and partly offset greenhouse gases emissions [25]. The main objective was to determine whether these two different grassland management practices influence the response of soil respiration and its components to drought.

2. Materials and Methods

2.1. Site Description

The site is located at the national long-term experimental observatory Système d'Observation et d'Expérimentation pour la Recherche en Environnement-Agroecosystems, Biogeochemical Cycles and Biodiversity (SOERE-ACBB) near Lusignan, western France (46°25′12.91″ N; 0°07′29.35″ E). The soil is classified as a Dystric Cambisol [26]. It developed from loamy parent material of unknown origin over red clay [27]. The soil profile can be divided into two main domains: Upper soil horizons are characterized by a loamy texture, classified as Cambisol, whereas lower soil horizons are clayey rubefied horizons, rich in kaolinite and iron oxides, classified as a Paleo-Ferralsol [28].

The two experimental paddocks of about 3 ha in surface area were converted to permanent grasslands in 2005. Since then, one paddock was grazed by a herd of dairy cows, and the other paddock was managed by periodically mowing and harvesting the biomass. The timing for harvest and mowing are made so as to maximize above-ground production and rarely happen at the same time. Following this guideline, harvest happened 2 days before our experiment started, while grazing had happened 2 weeks before the start of the experiment.

The plant community in the mown grassland was a mixture including *Dactylis glomerata* L. (cocksfoot) cultivar Ludac, *Festuca arundinacea* Schreb (tall fescue) cultivar Soni, and *Lolium perenne* L. (rye-grass) cultivar Milca. In the grazed grassland, *Trifolium repens* L. (white clover) cultivar Menna was added to the multispecies mixture. The mown grassland was cut four times a year with biomass exported, and nitrogen (N) fertilizer was applied at rates comprised between 120 and 310 kg N ha^{-1} year^{-1}. Fertilizer application rates were adjusted to maintain the nitrogen nutrition index between 0.9 and 1.0, that is, close to non-limiting nitrogen nutrition to near maximum plant production [28]. Grazing in the grazed paddock took place from March to December with 50 days per year using 15 to 20 livestock unit per hectare. Grazed grasslands did not receive nitrogen fertilization but nitrogen losses were returned by dung and urine and through the presence of leguminous species.

In 2011, six years after management conversion to permanent grasslands, bulk density was identical in the mown and grazed treatments, averaging 1.4 g cm^{-3} in the top 300 mm. Carbon stocks in 2014 ranged between 55 (mown grassland) and 64 (grazed grassland) t ha^{-1} in the first 300 mm. Nitrogen stocks were also similar in both grasslands, with 1.7 and 1.9 t ha^{-1} for the mown and grazed grasslands, respectively. No significant changes had occurred during the first 9 years (2014) after the conversion to permanent grasslands (A. Crème, personal communication).

2.2. Experimental Design

The experiment took place in early June 2017, at the beginning of a heat wave that affected large parts of Europe, including western France. At the beginning of a period of 10 days with no precipitation, we created a large range of soil water contents by applying a large artificial rainfall event. The responses of soil respiration (R_s) of the ^{13}C isotopic signature of CO_2 respired from the whole soil ($\delta^{13}CR_s$), and its autotrophic (live roots and rhizosphere, $\delta^{13}CR_a$) and heterotrophic (root and rhizosphere-free soil, $\delta^{13}CR_h$) components to changes in soil water content were compared for the grazing and mown systems.

One month before the start of the experiment, 10 rectangular plots of 1 × 2 m were positioned at random locations in each paddock, and four PVC collars (100 mm diameter, 30 mm depth) were fully inserted at random locations within each plot. On each day of measurement, a plot was randomly selected in each paddock (mown and grazed) and a set of measurements of R_s, $\delta^{13}CR_s$, $\delta^{13}CR_a$, $\delta^{13}CR_h$, soil volumetric water content (θ_s), and soil temperature (T_s) were taken from each collar.

A full set of measurements were taken on the four collars in one plot of each paddock before water was applied (day 0). Two hundred liters of water per plot was progressively applied over the course of 12 h to prevent run-off and pooling to the nine remaining plots in each paddock, equivalent to 100 mm of rainfall. Measurements were then made from the four collars in one randomly selected plot of each paddock every morning for the next 8 days, starting about 24 h after the end of watering to avoid the potential short-term burst of CO_2 emissions (the so-called "Birch effect") observed when dry soils are rewetted.

One plot in the grazed paddock was selected to characterize short-term changes in soil respiration due to the application of water. An additional set of four PVC collars were inserted in a 1 × 2 m area adjacent to the selected plot but received no water addition, constituting a control treatment. Measurements of R_s, θ_s, and T_s were taken 2 h and 1 h before water application on each of the eight collars. Water was then applied on one of the plots over the course of 1 h by spreading 100 mm evenly on the surface area using a hand watering can. Measurements of R_s, θ_s, and T_s were then carried out on the four watered and four dry collars every hour for 12 h.

2.3. Measurements of Soil Respiration and $\delta^{13}C$ of Respired CO_2

Except for the short-term watering experiment, where R_s was measured using a closed dynamic chamber system (EGM-4, PP systems, Amesbury, MA, USA), measurements of R_s were taken using two custom-built open chamber systems with four chambers each, adapted from Midwood et al. [29] and Midwood and Millard [30]. The chambers were placed on the collars set in the soil, and CO_2-free air was supplied to the chambers using mass flow controllers (model FMA5510, Omega Engineering Ltd., Stamford, CT, USA). The air was pumped out of the chambers using diaphragm pumps (TD−3, Brailsford and Co. Inc., Antrim, NH, USA), and the flow rate was controlled by mass flow controllers and adjusted to 15 mL min^{-1} lower than the inflow of CO_2-free air to avoid any air ingress from the atmosphere. The CO_2 concentration of the air leaving the chamber was measured using an infrared gas analyzer (IRGA) (Li840, LiCor Biosciences, Cambridge, UK), and the entire system was controlled by a datalogger (CR1000 and SDM-CV04, Campbell Scientific Ltd., Logan, UT, USA), allowing adjustment of the inflow of CO_2-free air to the measured CO_2 concentration to obtain a constant target concentration value in the chamber. After an equilibration period of 90 min and when the chamber's CO_2 concentrations were constant at 440 ppm, approximately 500 mL of respired air was collected in pre-evacuated air-tight bags (Tedlar® Keika Ventures, Chapel Hill, NC, USA) and the gas samples were analyzed for $\delta^{13}C$ values. All gas samples were analyzed for $\delta^{13}C$ values using a tuneable diode laser (TDL, TGA100A; Campbell Scientific Inc., Logan, UT, USA). The rate of R_s was calculated from the measured CO_2 concentration in the chamber and the flow rate of the CO_2-free air delivered to the chamber.

Measurements of the isotopic signatures of the CO_2 respired by the heterotrophic (root- and rhizosphere-free soil, $\delta^{13}CR_h$) and autotrophic (roots and rhizosphere, $\delta^{13}CR_a$) components of the soil were made by adapting the technique described by Snell et al. [23]. After R_S had been measured and the soil surface efflux sampled, roots and soils were collected. The collars were removed, and a soil core was extracted using a 100 mm diameter steel tube hammered into the soil to a depth of 250 mm. The soil from the core was broken up loosely and the roots removed by hand.

Shifts in $\delta^{13}CR_h$ have been shown to change exponentially with time after a soil core is extracted and broken up [19,23]. Thus, to obtain values of $\delta^{13}CR_h$ that are representative of the isotopic signature of the carbon pool used as a substrate before disturbance (time zero), one needs to work as rapidly as possible. To be as consistent as possible between samples, each operation was kept to a constant time. The process of breaking up the soil core in a tray and removing the roots by hand was kept close to 90 s. Subsequently, a subsample of root-free soil was placed in an air-tight bag, flushed quickly three times with nitrogen gas to purge atmospheric air from the soil, then filled with approximately 500 mL of CO_2-free air and allowed to incubate at the ambient temperature until the time from sampling the soil core reached 3 min. The air in the bag was then sampled and analyzed for $\delta^{13}CO_2$.

The roots were then cleaned of most of the remaining soil attached to them and placed in a separate air-tight bag. The bags were evacuated and then filled with approximately 500 mL of CO_2-free air and allowed to incubate at ambient temperature for between 20 and 60 min, after which the air in the bags was sampled for measurements of $\delta^{13}CO_2$. The roots were then dried at 105 °C for 24 h and weighted to give an estimate of the root biomass in each soil core.

2.4. Statistical Analyses

To test for the effect of soil volumetric water content (θ_s) on R_s, $\delta^{13}CR_s$, $\delta^{13}CR_a$, and $\delta^{13}CR_h$ for the paddocks managed as grazed and mown systems, a backwards stepwise regression approach was used separately on each explained variable [31]. Each measurement of R_s, $\delta^{13}CR_s$, $\delta^{13}CR_a$, and $\delta^{13}CR_h$ was treated as a sample. The full linear models included the three explanatory variables θ_s, root biomass, and paddock management (as a factor), as well as every two-way interaction between them. Model selection was based on a comparison of Akaike's Information Criterion (AIC), the model with the lowest AIC value being the most strongly supported. As a rule of thumb, when two models presented a $\Delta AIC < 2$, the simpler model was selected [32]. All statistical analyses were conducted using R version 3.4.2 [33] (R Development Core Team, 2017).

3. Results

3.1. Hourly Measurements: Short-Term Response of Soil Respiration to Water Addition

In the grazed paddock, soil volumetric water content (θ_s) greatly increased immediately after 100 mm of water was added (irrigated plot), approximately doubling from 0.2 to 0.4 m^3 m^{-3} and becoming significantly higher than in the control plot without water addition. θ_s then slowly decreased over the next 24 h in both plots but remained significantly higher in the irrigated plot (Figure 1a), with a difference between the irrigated and the control plot of over 0.11 m^3 m^{-3} throughout the 24 h.

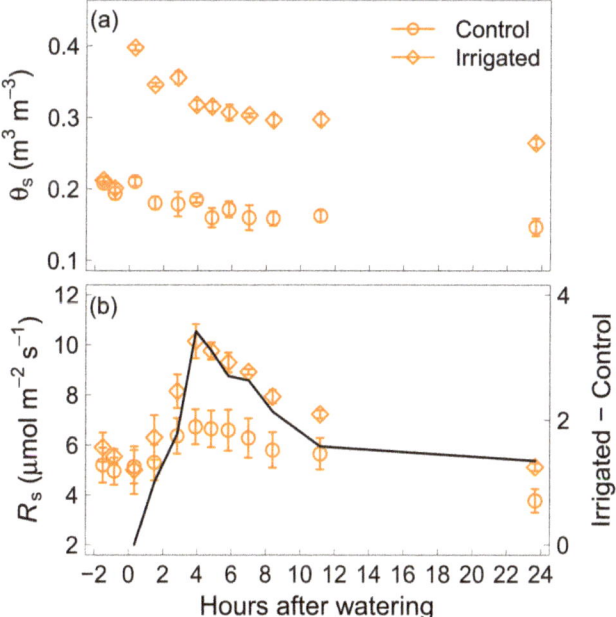

Figure 1. Soil volumetric water content (θ_S, panel **a**) and soil respiration (R_S, panel **b**) before and over 24 h after the addition of 100 mm of water in the grazed treatment for watered plots (diamonds) and control non-watered plots (circles). The line in panel b is the difference between mean R_s in the watered plots and mean R_s in the control plot for each hour after watering. Vertical bars represent one standard error of the mean ($n = 4$).

Soil respiration (R_s) started to increase 2 h after water was added to the irrigated plot, becoming significantly higher than in the adjacent control plot and remaining higher during the 24 h measuring period (Figure 1b). R_s also varied in the control plot due to diel variation. R_s peaked for both plots 4 h after water was added. The difference between mean R_s in the irrigated and the control plot was null before and just after water was added. This difference started to increase 1 h after water addition, was highest 4 h after water addition with a value of 3.4 µmolCO$_2$ m^{-2} s^{-1}, and then decreased and levelled off, being similar 11 h and 24 h after water addition (Figure 1b).

3.2. Daily Changes after Water Addition

To assess the effect of water addition on R_s, $\delta^{13}CR_s$, $\delta^{13}CR_a$, $\delta^{13}CR_h$, only the measurements taken from 24 h onward after water addition were included to exclude the pulse of respiration measured shortly after water addition.

Soil volumetric water content (θ_S) increased after the addition of 100 mm of water on the plots and remained higher than before watering for 3 days. After the initial increase, θ_S decreased steadily for the whole experiment period of 8 days (Figure 2a). Individual values of θ_S overall ranged from 0.05 to 0.31 m^3 m^{-3}.

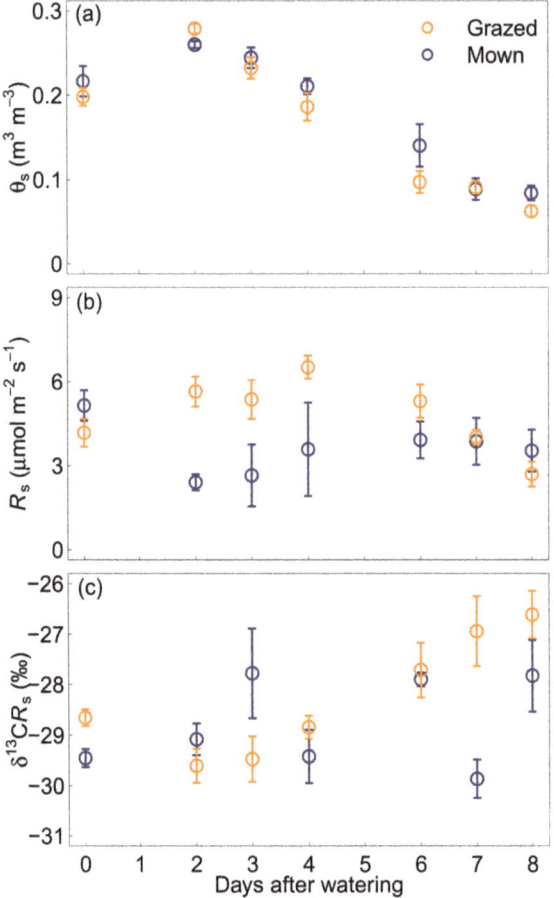

Figure 2. Soil volumetric water content (θ_s, panel **a**), soil respiration (R_s, panel **b**) and ^{13}C isotopic signature of soil respired CO$_2$ (δ^{13}CR$_s$, panel **c**) over the days after adding 100 mm of water in the grazed (orange circles) and mown (blue circles) grasslands. Vertical bars represent one standard error of the mean ($n = 4$).

Although the trend for θ_S was identical in both paddocks, R_s showed different patterns in the grazed and mown paddocks over the days after water was added (Figure 2b). In the grazed paddock, R_s roughly followed a similar pattern to θ_S, showing higher values after water addition for 3 days and decreasing steadily after that. R_s did not show any obvious pattern in the mown paddock and was more variable than in the grazed paddock. Similarly, the ^{13}C isotopic signature of soil-respired CO$_2$ (δ^{13}CR$_s$) presented a different pattern for the two paddocks (Figure 2c). Changes in δ^{13}CR$_s$ were tightly coupled with changes in θ_S in the grazed paddock, with more depleted values 2 days after water addition and a steady change to more enriched values over the measurement period. As for R_s, δ^{13}CR$_s$ did not show any obvious pattern over the days after water addition in the mown paddock.

The ^{13}C isotopic signature of CO$_2$ respired from the roots and from the root- and rhizosphere-free soil did not show clear patterns after water addition for either paddock (Figure S1).

3.3. Effect of Soil Volumetric Water Content

The model for R_s with the best fit selected from the backwards stepwise regression, explained 54% of the variability in R_s and incorporated the effect of root biomass ($F = 55.74$, $p < 0.0001$) and the interactive effect of soil water content and paddock management ($F = 4.99$, $p = 0.03$). The single-variate model including root biomass explained 43% of the variability in R_s (Figure 3a). Soil respiration was positively correlated with θ_s in the grazed paddock ($p = 0.03$) but was not significantly affected by θ_s in the mown paddock ($p = 0.2$, Figure 3b).

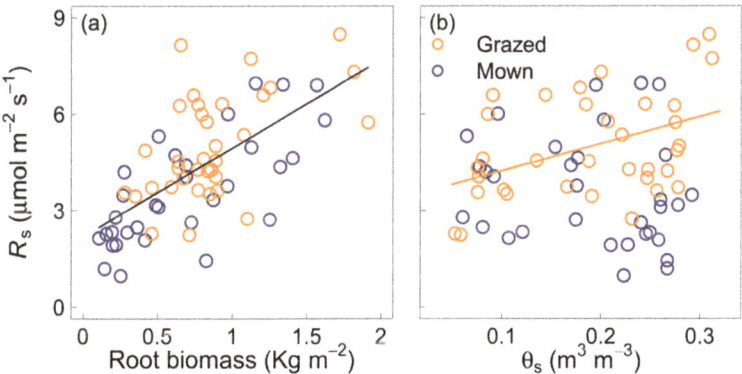

Figure 3. Soil respiration (R_S) as a function of root biomass (panel **a**) and soil volumetric water content (θ_s, panel **b**) for the grazed (orange circles) and mown (blue circles) grasslands. The lines represent the fit of linear regressions. The black line in panel a indicates identical fit for the two treatments. Only significant correlations appear.

As for R_s, the model for $\delta^{13}CR_s$ with the best fit incorporated the effect of root biomass ($F = 4.41$, $p = 0.04$) and the interactive effect of soil water content and paddock management ($F = 9.65$, $p < 0.01$). This model explained 39% of the variability in $\delta^{13}CR_s$. Soil volumetric water content had different effects for the two paddocks (Figure 4a), with a significant effect of θ_s on $\delta^{13}CR_s$ for the grazed paddock ($p < 0.01$), but no significant effect for the mown paddock ($p = 0.6$).

The model for $\delta^{13}CR_h$ with the best fit incorporated the interactive effect of root biomass and paddock management ($F = 7.63$, $p < 0.01$) and the interactive effect of soil water content and paddock management ($F = 6.44$, $p = 0.01$), and explained 39% of the variability in $\delta^{13}CR_h$. The effect of root biomass and of θ_s on $\delta^{13}CR_h$ was significant only for the grazed paddock ($p < 0.001$ for root biomass and $p = 0.04$ for θ_s). The model for $\delta^{13}CR_a$ with the best fit was a single-variate model incorporating only root biomass ($F = 7.23$, $p < 0.01$) and explained 10% of the variability in $\delta^{13}CR_a$. There was no significant effect of θ_s on $\delta^{13}CR_a$ or $\delta^{13}CR_h$ (Figure 4a,c).

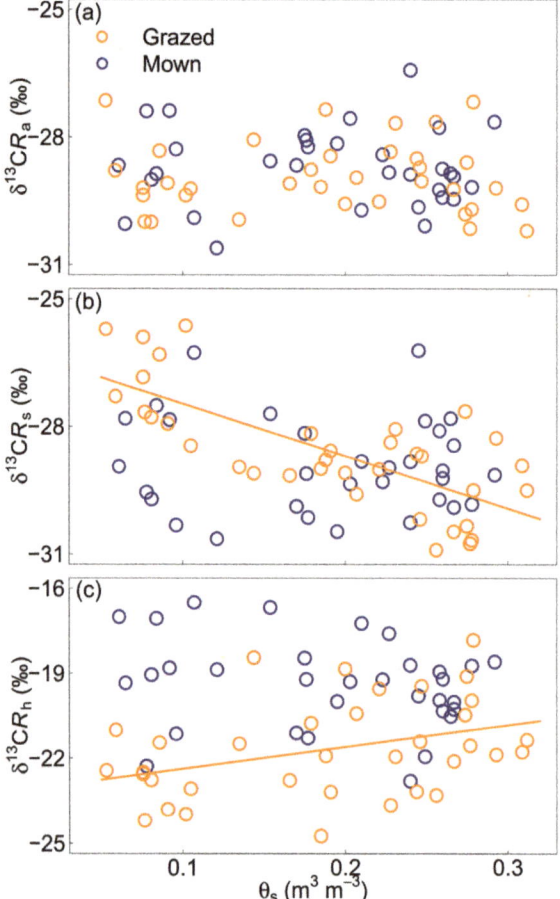

Figure 4. ^{13}C isotopic signature of CO_2 respired from the roots and rhizosphere ($\delta^{13}CR_a$, panel **a**), the whole soil ($\delta^{13}CR_s$, panel **b**) and the root- and rhizosphere-free soil ($\delta^{13}CR_h$, panel **c**) as functions of soil volumetric water content (θ_s) for the grazed (orange circles) and mown (blue circles) grasslands. The lines represent the fit of linear regressions. Only significant correlations appear.

It is noteworthy that soil temperature (T_s) ranged from 17.6 to 29.6 °C and was negatively correlated with θ_S ($F = 33.84$, $p < 0.0001$). Because of this correlation, the effect of T_s and θ_S was analyzed separately. In the model with the best fit for each backwards stepwise regression analysis, θ_S was replaced with T_s, and the effect was assessed by comparing the coefficient of determination (R^2) for the models including θ_s and T_s. Replacing θ_S by T_s in the best models resulted in a lower coefficient of determination for all the variables.

4. Discussion

4.1. Birch Effect

It is well documented that rewetting dry soils is usually followed by a burst of respiration, which can be very short-lived [34] or last for several days [35], after which respiration rates stabilize to that of a continuously wet soil. In our study, with a flush of CO_2 starting 4 h after watering and the difference between dry and wet soils coming to equilibrium over the few hours after that, the Birch effect was observed to last less than a day. Therefore, we considered measurements taken 24 h after watering for analyzing R_s and its components to exclude the influence of the Birch effect.

4.2. Soil Water Content and Isotopic Discrimination

Changes in $\delta^{13}CR_a$ and $\delta^{13}CR_h$ with soil water content were largely insignificant, in line with previous results where irrigated and non-irrigated plots were not found to differ significantly in their values of $\delta^{13}CR_a$ and to marginally affect $\delta^{13}CR_h$ [17]. This may appear surprising, particularly for the autotrophic component. Under conditions of limited water availability, plants close their stomata, improving their water use efficiency and usually leading to lower ^{13}C discrimination [36,37]. Moreover, substrates respired in the roots are supplied mainly by recent photo-assimilates [38,39]. We would, therefore, have expected more enriched values for $\delta^{13}CR_a$ with higher water stress and lower photosynthetic ^{13}C discrimination. Our result suggests that the substrates for respiration of the roots were not tightly coupled with photo-assimilates. Water stress may reduce the phloem diffusion rate, therefore, increasing the time lag between assimilation and utilization of compounds in the roots and rhizosphere and resulting in photosynthesis to decrease faster than respiration rates under water stress [38,39]. We conclude that carbohydrate reserves provided substrates for the autotrophic component of soil respiration and were not exhausted during the 8 days following watering.

4.3. Soil Water Content and Components of Soil Respiration

The ^{13}C signatures of CO_2 respired from the roots and from the whole soil were similar in value for high water content, ranging between −31 and −27‰. Probably due to the relatively low carbon content at this site, the contribution of SOM decomposition to total R_S was small and $\delta^{13}CR_s$ was largely dominated by the signature from the roots, also in line with the observation that roots can be a dominant contributor to total R_s in many ecosystems [40–42]. Not surprisingly, soil respiration was strongly positively correlated with root biomass for both treatments.

Soil water content is a major driver of root respiration and can be the limiting factor for plant activity in dry conditions [43–45]. When the soil was wet just after water addition, the contribution of roots and rhizosphere respiration to total soil respiration was probably at its maximum and was close to 100%, leading to values of $\delta^{13}CR_s$ and $\delta^{13}CR_a$ being indistinguishable. With such small changes in the components of soil respiration ($\delta^{13}CR_a$ and $\delta^{13}CR_h$), changes in $\delta^{13}CR_s$ must have been due to changes in the relative contribution of the autotrophic and heterotrophic components.

Because the autotrophic component dominated soil respiration, the enrichment of $\delta^{13}CR_s$ with decreasing water content in the grazed paddock was likely to be due to the decreasing contribution of the root and rhizosphere component rather than an increasing contribution of the heterotrophic component. This hypothesis is in line with the results from a laboratory study showing that microbial activity in the rhizosphere of grassland species was less affected than root activity [46]. This enrichment, therefore, suggests that in the grazed treatment, soil respiration decreased with water content due to a decreasing autotrophic component.

No changes were observed in the mown treatment. Carbohydrate reserves may vary with grass species and development stage [47]. It is possible that the plant community in the mown treatment presented a rooting system with more carbon reserves and increased resilience to drought. Another explanation could be found in the fact that grazing and mowing happened at different times. Due to a farm management decision, mowing had happened shortly before the beginning of our experiment. Grasses have been observed to mobilize their root reserves and allocate them above-ground for leaf regrowth after cutting or grazing [48,49]. It is, therefore, possible that root exudation and respiration were decreased due to mowing, thereby, cancelling the positive effect of adding water on root activity observed in the grazed treatment.

4.4. Consequences for Soil Carbon Sequestration

Agricultural management practices are known to influence the fate of soil carbon stocks, in interaction with climate and soil properties [5]. Although permanent grasslands are known to have a strong potential to store carbon below ground [24], soil carbon sequestration may be reversible under the effect of climate change and changes in management practices, particularly those involving soil disturbance [25]. Our results showed that different management practices and decisions are strong contributors to determine the response of soil carbon dynamics to soil water content. Although we were not able to conclude directly about which of mowing or grazing management will enhance drought resilience, we showed that the adaptation and resilience of agricultural grasslands to an intensified global water cycle will likely strongly depend on management practices and decisions. These decisions will, therefore, determine the magnitude of climate-induced changes in grasslands soil carbon stocks and should be the subject of careful evaluation.

5. Conclusions

Soil respiration in grazed and mown grasslands responded differently to changes in soil water content. It is possible that the unsynchronized cutting and grazing regimes partly explained this difference. Our results suggest that farm management practices and the timing of management decisions potentially alter the relationship between soil carbon dynamics and soil water content. Therefore, management choices and decisions will likely significantly contribute to driving changes in soil carbon stocks under the influence of climate change as the global water cycle changes and the intensity and frequency of droughts increase. More studies are needed to fully identify which management practices can lead to increased resilience and adaptation of grassland ecosystems to drought.

Supplementary Materials: The following are available online at http://www.mdpi.com/2073-4395/9/3/124/s1, Figure S1: ^{13}C isotopic signature of CO_2 respired from the roots ($\delta^{13}CR_a$, panel **a**) and of CO_2 respired from root and rhizosphere free soil ($\delta^{13}CR_h$, panel **b**) over the days after adding 100 mm of water in the grazed (orange circles) and mown (blue circles) grasslands. Vertical bars represent one standard error of the mean ($n = 4$).

Author Contributions: Conceptualization, P.M., A.C. and C.R.; Methodology, P.M., A.J.M. and J.E.H.; Formal Analysis, G.Y.K.M.; Investigation, G.Y.K.M., A.J.M., J.E.H. and P.M.; Resources, A.C. and P.M.; Writing—Original Draft Preparation, G.Y.K.M.; Writing—Review and Editing, C.R., A.C.; A.J.M. and P.M.; Project Administration, G.Y.K.M., P.M. and A.C.; Funding Acquisition, P.M & A.C.

Funding: This research was funded by ENVRIPlus transnational access grant—Grant Agreement number 654182-delivered to GYKM, and by the New Zealand Government to support the objectives of the Livestock Research Group of the Global Research Alliance on Agricultural Greenhouse Gases - Grants SOW12-GPLER LCR-PM (Proposal ID 16949-15 LCR).

Acknowledgments: We are grateful to Xavier Charrier, François Gastal, Jerome Chargelegue, Jean François Bouhiron, Christophe de Berranger, Christophe Huguet, Camille Bartolini, and Aicha Chabbi for help with site access, organization, and field measurements.

Conflicts of Interest: The authors declare no conflict of interest.

Abbreviations

R_s	Soil respiration
$\delta^{13}CR_s$	^{13}C isotopic signature of CO_2 respired from the whole soil
$\delta^{13}CR_a$	^{13}C isotopic signature of CO_2 respired from roots
$\delta^{13}CR_h$	^{13}C isotopic signature of CO_2 respired from root and rhizosphere free soil
θ_s	Soil volumetric soil water content
T_s	Soil temperature at 100 mm depth

References

1. Steinfeld, H.; Gerber, P.; Wassenaar, T.; Castel, V.; Rosales, M.; Haan, C. *Livestock's Long Shadow: Environmental Issues and Options*; Food & Agriculture Organization: Rome, Italy, 2006.
2. Stockmann, U.; Adams, M.A.; Crawford, J.W.; Field, D.J.; Henakaarchchi, N.; Jenkins, M.; Minasny, B.; McBratney, A.B.; de Courcelles, V.D.R.; Singh, K.; et al. The knowns, known unknowns and unknowns of sequestration of soil organic carbon. *Agric. Ecosyst. Environ.* **2013**, *164*, 80–99. [CrossRef]
3. De Coninck, H.; Revi, A.; Babiker, M.; Bertoldi, P.; Buckeridge, M.; Cartwright, A.; Dong, W.; Ford, J.; Fuss, S.; Hourcade, J.C.; et al. Strengthening and implementing the global response. In *Global Warming of 1.5 °C, An IPCC Special Report*; Masson-Delmotte, V., Zhai, P., Pörtner, H.O., Roberts, D., Skea, J., Shukla, P.R., Pirani, A., Eds.; Intergovernmental Panel on Climate Change: Geneva, Switzerland, 2018.
4. Paustian, K.; Lehmann, J.; Ogle, S.; Reay, D.; Robertson, G.P.; Smith, P. Climate-smart soils. *Nature* **2016**, *532*, 49–57. [CrossRef] [PubMed]
5. Chabbi, A.; Lehmann, L.; Ciais, P.; Loescher, H.L.; Cotrufo, M.F.; Don, A.; San-Clements, M.; Schipper, L.; Six, J.; Smith, P.; et al. Aligning agriculture and climate policy. *Nat. Clim. Chang.* **2017**, *7*, 307–309. [CrossRef]
6. Rumpel, C.; Crème, A.; Ngo, P.T.; Velásquez, G.; Mora, M.L.; Chabbi, A. The impact of grassland management on biogeochemical cycles involving carbon, nitrogen and phosphorus. *J. Soil Sci. Plant Nutr.* **2015**, *15*, 353–371. [CrossRef]
7. Dignac, M.-F.; Derrien, D.; Barré, P.; Barot, S.; Cécillon, L.; Chenu, C.; Chevallier, T.; Freschet, G.T.; Garnier, P.; Guenet, B.; et al. Increasing soil carbon storage: Mechanisms, effects of agricultural practices and proxies. A review. *Agron. Sustain. Dev.* **2017**, *37*, 14. [CrossRef]
8. Whitehead, D.; Schipper, L.A.; Pronger, J.; Moinet, G.Y.K.; Mudge, P.L.; Calvelo Pereira, R.; Kirschbaum, M.U.F.; McNally, S.R.; Beare, M.H.; Camps-Arbestain, M. Management practices to reduce losses or increase soil carbon stocks in temperate grazed grasslands: New Zealand as a case study. *Agric. Ecosyst. Environ.* **2018**, *265*, 432–443. [CrossRef]
9. Schlesinger, W.H.; Andrews, J.A. Soil respiration and the global carbon cycle. *Biogeochemistry* **2000**, *48*, 7–20. [CrossRef]
10. Field, C.B.; Intergovernmental Panel on Climate Change; Working Group II. *Climate Change 2014: Impacts, Adaptation, and Vulnerability. Part A*; Intergovernmental Panel on Climate Change: Geneva, Switzerland, 2014.
11. Borken, W.; Matzner, E. Reappraisal of drying and wetting effects on C and N mineralization and fluxes in soils. *Glob. Chang. Biol.* **2009**, *15*, 808–824. [CrossRef]
12. Tezara, W.; Mitchell, V.J.; Driscoll, S.D.; Lawlor, D.W. Water stress inhibits plant photosynthesis by decreasing coupling factor and ATP. *Nature* **1999**, *401*, 914–917. [CrossRef]
13. Moyano, F.E.; Manzoni, S.; Chenu, C. Responses of soil heterotrophic respiration to moisture availability: An exploration of processes and models. *Soil Biol. Biochem.* **2013**, *59*, 72–85. [CrossRef]
14. Davidson, E.A.; Verchot, L.V.; Cattânio, J.H.; Ackerman, I.L.; Carvalho, J.E.M. Effects of soil water content on soil respiration. *Biogeochemistry* **2000**, *48*, 53–69. [CrossRef]
15. Moinet, G.Y.K.; Cieraad, E.; Turnbull, M.H.; Whitehead, D. Effects of irrigation and addition of nitrogen fertiliser on net ecosystem carbon balance for a grassland. *Sci. Total Environ.* **2017**, *579*, 1715–1725. [CrossRef] [PubMed]
16. Canarini, A.; Dijkstra, F.A. Dry-rewetting cycles regulate wheat carbon rhizodeposition, stabilization and nitrogen cycling. *Soil Biol. Biochem.* **2015**, *81*, 195–203. [CrossRef]
17. Moinet, G.Y.K.; Cieraad, E.; Hunt, J.E.; Fraser, A.; Turnbull, M.H.; Whitehead, D. Soil heterotrophic respiration is insensitive to changes in soil water content but related to microbial access to organic matter. *Geoderma* **2016**, *274*, 68–78. [CrossRef]
18. Huang, S.; Ye, G.; Lin, J.; Chen, K.; Xu, X.; Ruan, H.; Tan, F.; Chen, H.Y.H. Autotrophic and heterotrophic soil respiration responds asymmetrically to drought in a subtropical forest in the Southeast China. *Soil Biol. Biochem.* **2018**, *123*, 242–249. [CrossRef]
19. Millard, P.; Midwood, A.J.; Hunt, J.E.; Barbour, M.M.; Whitehead, D. Quantifying the contribution of soil organic matter turnover to forest soil respiration, using natural abundance δ^{13}C. *Soil Biol. Biochem.* **2010**, *42*, 935–943. [CrossRef]

20. Moinet, G.Y.K.; Hunt, J.E.; Kirschbaum, M.U.F.; Morcom, C.P.; Midwood, A.J.; Millard, P. The temperature sensitivity of soil organic matter decomposition is constrained by microbial access to substrates. *Soil Biol. Biochem.* **2018**, *116*, 333–339. [CrossRef]
21. Moinet, G.Y.K.; Midwood, A.J.; Hunt, J.E.; Whitehead, D.; Hannam, K.D.; Jenkins, M.; Brewer, M.J.; Adams, M.A.; Millard, P. Estimates of rhizosphere priming effects are affected by soil disturbance. *Geoderma* **2018**, *313*, 1–6. [CrossRef]
22. Boström, B.; Comstedt, D.; Ekblad, A. Isotope fractionation and ^{13}C enrichment in soil profiles during the decomposition of soil organic matter. *Oecologia* **2007**, *153*, 89–98. [CrossRef] [PubMed]
23. Snell, H.S.K.; Robinson, D.; Midwood, A.J. Minimising methodological biases to improve the accuracy of partitioning soil respiration using natural abundance ^{13}C. *Rapid Commun. Mass Spectrom.* **2014**, *28*, 2341–2351. [CrossRef] [PubMed]
24. Soussana, J.F.; Allard, V.; Pilegaard, K.; Ambus, P.; Amman, C.; Campbell, C.; Ceschia, E.; Clifton-Brown, J.; Czobel, S.; Domingues, R.; et al. Full accounting of the greenhouse gas (CO_2, N_2O, CH_4) budget of nine European grassland sites. *Agric. Ecosyst. Environ.* **2007**, *121*, 121–134. [CrossRef]
25. Soussana, J.F.; Tallec, T.; Blanfort, V. Mitigating the greenhouse gas balance of ruminant production systems through carbon sequestration in grasslands. *Animal* **2010**, *4*, 334–350. [CrossRef] [PubMed]
26. FAO-ISRIC. *Guidelines for Soil Description*, revised 3rd ed.; Food and Agricultural Organisation: Rome, Italy, 1990.
27. Chabbi, A.; Kögel-Knabner, I.; Rumpel, C. Stabilised carbon in subsoil horizons is located in spatially distinct parts of the soil profile. *Soil Biol. Biochem* **2009**, *41*, 256–261. [CrossRef]
28. Senapati, N.; Chabbi, A.; Giostri, A.F.; Yeluripati, J.B.; Smith, P. Modelling nitrous oxide emissions from mown-grass and grain-cropping systems: Testing and sensitivity analysis of DailyDayCent using high frequency measurements. *Sci. Total Environ.* **2016**, *572*, 955–977. [CrossRef] [PubMed]
29. Midwood, A.J.; Thornton, B.; Millard, P. Measuring the ^{13}C content of soil-respired CO_2 using a novel open chamber system. *Rapid Commun. Mass Spectrom.* **2008**, *22*, 2073–2081. [CrossRef] [PubMed]
30. Midwood, A.J.; Millard, P. Challenges in measuring the δ^{13}C of the soil surface CO_2 efflux. *Rapid Commun. Mass Spectrom.* **2011**, *25*, 232–242. [CrossRef] [PubMed]
31. Zuur, A.F.; Ieno, E.N.; Walker, N.J.; Saveliev, A.A.; Smith, G.M. *Mixed Effects Models and Extensions in Ecology with R*; Gail, M., Krickeberg, K., Samet, J.M., Tsiatis, A., Wong, W., Eds.; Spring Science and Business Media: New York, NY, USA, 2009.
32. Burnham, K.P.; Anderson, D.R. *Model Selection and Multimodel Inference: A Practical Information-Theoretic Approach*; Springer: New York, NY, USA, 2003.
33. R Core Team. *R: A Language and Environment for Statistical Computing*; R Foundation for Statistical Computing: Vienna, Austria, 2018.
34. Powers, H.H.; Hunt, J.E.; Hanson, D.T.; McDowell, N.G. A dynamic soil chamber system coupled with a tunable diode laser for online measurements of δ^{13}C, δ^{18}O, and efflux rate of soil-respired CO_2: Measurements of δ^{13}C, δ^{18}O, and efflux rate of soil-respired CO_2. *Rapid Commun. Mass Spectrom.* **2010**, *24*, 243–253. [CrossRef] [PubMed]
35. Fierer, N.; Schimel, J.P. Effects of drying-rewetting frequency on soil carbon and nitrogen transformations. *Soil Biol. Biochem.* **2002**, *34*, 777–787. [CrossRef]
36. Read, J.; Farquhar, G. Comparative studies in Nothofagus (Fagaceae). I. Leaf carbon isotope discrimination. *Funct. Ecol.* **1991**, *5*, 684–695. [CrossRef]
37. Schulze, E.-D.; Williams, R.J.; Farquhar, G.D.; Schulze, W.; Langridge, J.; Miller, J.M.; Walker, B.H. Carbon and nitrogen isotope discrimination and nitrogen nutrition of trees along a rainfall gradient in northern Australia. *Funct. Plant Biol.* **1998**, *25*, 413–425. [CrossRef]
38. Kuzyakov, Y.; Gavrichkova, O. Review: Time lag between photosynthesis and carbon dioxide efflux from soil: A review of mechanisms and controls. *Glob. Chang. Biol.* **2010**, *16*, 3386–3406. [CrossRef]
39. Barthel, M.; Cieraad, E.; Zakharova, A.; Hunt, J.E. Sudden cold temperature delays plant carbon transport and shifts allocation from growth to respiratory demand. *Biogeosciences* **2014**, *11*, 1425–1433. [CrossRef]
40. Raich, J.W.; Schlesinger, W.H. The global carbon dioxide flux in soil respiration and its relationship to vegetation and climate. *Tellus B* **1992**, *44*, 81–99. [CrossRef]
41. Hanson, P.J.; Edwards, N.T.; Garten, C.T.; Andrews, J.A. Separating root and soil microbial contributions to soil respiration: A review of methods and observations. *Biogeochemistry* **2000**, *48*, 115–146. [CrossRef]

42. Subke, J.-A.; Inglima, I.; Francesca Cotrufo, M. Trends and methodological impacts in soil CO_2 efflux partitioning: A metaanalytical review. *Glob. Chang. Biol.* **2006**, *12*, 921–943. [CrossRef]
43. Flanagan, L.B.; Wever, L.A.; Carlson, P.J. Seasonal and interannual variation in carbon dioxide exchange and carbon balance in a northern temperate grassland. *Glob. Chang. Biol.* **2002**, *8*, 599–615. [CrossRef]
44. Hunt, J.E.; Kelliher, F.M.; McSeveny, T.M.; Ross, D.J.; Whitehead, D. Long-term carbon exchange in a sparse, seasonally dry tussock grassland. *Glob. Chang. Biol.* **2004**, *10*, 1785–1800. [CrossRef]
45. Zhang, L.; Guo, H.; Jia, G.; Wylie, B.; Gilmanov, T.; Howard, D.; Ji, L.; Xiao, J.; Li, J.; Yuan, W.; et al. Net ecosystem productivity of temperate grasslands in northern China: An upscaling study. *Agric. For. Meteorol.* **2014**, *184*, 71–81. [CrossRef]
46. Sanaullah, M.; Chabbi, A.; Rumpel, C.; Kuzyakov, Y. Carbon allocation in grassland communities under drought stress followed by ^{14}C pulse labeling. *Soil Biol. Biochem.* **2012**, *55*, 132–139. [CrossRef]
47. White, L.M. Carbohydrate reserves of grasses: A review. *J. Range Manag.* **1973**, *26*, 13–18. [CrossRef]
48. Steen, E.; Larsson, K. Carbohydrates in roots and rhizomes of perennial grasses. *New Phytol.* **1986**, *104*, 339–346. [CrossRef]
49. Donaghy, D.J.; Fulkerson, W.J. The importance of water-soluble carbohydrate reserves on regrowth and root growth of *Lolium perenne* (L.). *Grass Forage Sci.* **1997**, *52*, 401–407. [CrossRef]

© 2019 by the authors. Licensee MDPI, Basel, Switzerland. This article is an open access article distributed under the terms and conditions of the Creative Commons Attribution (CC BY) license (http://creativecommons.org/licenses/by/4.0/).

Review

Drivers, Process, and Consequences of Native Grassland Degradation: Insights from a Literature Review and a Survey in Río de la Plata Grasslands

Guadalupe Tiscornia [1],*, Martín Jaurena [2] and Walter Baethgen [3]

[1] Agro-Climate and Information System Unit (GRAS), National Institute of Agricultural Research (INIA Uruguay), Ruta 48 KM.10, Canelones 90200, Uruguay
[2] Pastures and Forages National Research Program, National Institute of Agricultural Research (INIA Uruguay), Ruta 5 KM.386, Tacuarembó 45000, Uruguay; mjaurena@inia.org.uy
[3] International Research Institute for Climate and Society (IRI), Columbia University, 61 Route 9W, Palisades, NY 10964, USA; baethgen@iri.columbia.edu
* Correspondence: gtiscornia@inia.org.uy; Tel.: +598-2367-7641

Received: 23 March 2019; Accepted: 5 May 2019; Published: 10 May 2019

Abstract: Natural grasslands are being progressively degraded around the world due to human-induced action (e.g., overgrazing), but there is neither a widely accepted conceptual framework to approach degradation studies nor a clear definition of what "grassland degradation" is. Most of the drivers, processes, and consequences related to grassland degradation are widespread and are usually separately quoted in the literature. In this paper, we propose a comprehensive framework with different conceptual categories, for monitoring grassland degradation, and a new definition based on current ones. We provide a conceptual update of grassland degradation based on a literature review and an expert survey, focused on the Río de la Plata grasslands (RPG). We identified "drivers" as external forces or changes that cause degradation; "processes" as measurable changes in grasslands conditions that can be evaluated using indicators; and "consequences" as the impacts or results of the process of grassland degradation. We expect that this conceptual framework will contribute to monitoring programs, to support management decisions, to design conservation measures, and to communicate the importance of grasslands conservation and the different concepts involved. Particularly for RPG, we expect that this paper will contribute to promote sustainable management practices in this important and often neglected ecosystem.

Keywords: productivity; forage; biomass; species diversity; indicators; grazing

1. Introduction

Grasslands cover an estimated area of 40% of the land surface [1]. There are different types of native grasslands, with communities determined mainly by climate and soil conditions, by grazing animals and by fire [2]. Globally, grasslands make a significant contribution to food security, providing forage for ruminants used for meat and milk production [3,4] and by the provision of many other important ecosystem services for human well-being [5,6]. However, the status of grasslands conservation varies around the world, but it is generally far from satisfactory [7]: they have shown signs of degradation, due primarily to overgrazing [2,8–12], other improper management practices [13–20], and climate change [12–14,16–18,20]. In fact, grasslands are among the ecosystems with the highest species richness in the world [21]. A global estimation of Gang et al. [17] showed that almost half of grassland ecosystems were degraded and almost 5% of these grasslands experiences strong to extreme levels of degradation.

Given its relevance, grassland degradation has become an emerging topic in the fields of grassland management and environmental protection. Even though many authors and international organizations call for preservation of natural grasslands [13,22,23], there is not a consensual definition of what natural grassland degradation is. The conceptual definitions of degradation are very broad, involving the deterioration of grassland quality, productivity, economic potential, service function, recovery ability, and diversity [24]. Others focus only on the decline of forage production or grassland productivity [9,25–28], and others incorporate invasion of non-native plant species [29,30], dominance of non-palatable species [31] or functions related to biogeochemical and water cycle [12,32–34]. In summary, the term "degradation" referring to the condition of a natural grassland, is widely used in different circumstances by different stakeholders, from researchers and rural extension agents, to policy makers, without a common agreed definition. These extremely variable definitions of grassland degradation reflect the wide variety of services expected from this ecosystem, with each definition corresponding to one or a few services.

Although definitions differ across human and environmental contexts, we argue that all of them can be grouped in conceptual categories related to degradation: e.g., drivers, process, or consequences. Many research reports focus on the process indicators of grassland degradation: e.g., Cao et al. [35] define the degradation as "an adverse reduction in biodiversity and biomass production, increased soil erosion, and nutrient loss". Wick et al. [15] in contrast, focus on consequences and define grassland degradation as the reduction in their ability to provide ecosystem goods and services. Other studies combine different conceptual categories, such as consequences and process indicators: "degradation (...) leads to desertification, reduces grassland productivity and biodiversity" [36]; others combine drivers and process indicators: overgrazing, woody-plant encroachment, and invasion by non-native plant species, led to the reduction in the quantity or nutritional quality of the vegetation available for grazing" [29]. Clarifying these conceptual categories is crucial to build conceptual models that can explain the dynamics of grassland degradation and restoration.

Understanding the forces that drive the state of grasslands ecosystems, is an important first step for preventing their degradation. For this task, the State and Transition Model (STM) [37], simple diagrams that conceptualize the complex dynamic of the vegetation, which has been proposed as a useful tool to assess grassland management and conservation [38]. According to these models, there are "normal" or non-degraded vegetation states that are known to be very stable, since they rapidly return to its original condition after a small perturbation. However, discontinuous and irreversible changes (at least in the medium term) occur after a major disturbance, when the thresholds are exceeded and, a "normal" stable state is replaced by an alternative stable but unwanted (degraded) one [15,39]. It is frequently reported in the literature that primary driving forces of grassland degradation, which can be grouped into natural events (e.g., the increase in the frequency of extreme drought) and anthropogenic (management) factors (e.g., overgrazing or grazing abandonment) or by the interaction of both factors, can cause discontinuous changes in vegetation compositions that are non-reversible. These transitions to degraded communities are triggered when the driving forces overcome the resilience of the reference community [40,41]. However, the dividing line between what is considered "normal" and what is considered degraded is far from absolute.

Much has been studied about degradation in water-limited grasslands [8,12,42–44], which are ecosystems very prone to soil desertification. In these grasslands, prolonged droughts and overgrazing are commonly recognized as the two common underlying drivers that lead to degradation. However, in mesic and humid grasslands, which are alternatively limited by temperature, water and nutrients, an in-depth synthesis of knowledge is needed to understand driving forces, process, and consequences of grassland degradation.

Given the importance and the need for preserving natural grasslands for livestock production and diversity, it is essential to reach a conceptual framework to study the process of grassland degradation. Considering de Quiroz [45] statement that degradation is intrinsically related to human management objectives, we believe that there is a need for a review of the studies conducted in grasslands used for livestock production. In this context, important questions arise: what is the condition of a degraded natural grassland? Which are the main drivers that influence grassland degradation? How ecological process are affected by grassland degradation? What are the main consequences of grassland degradation? The goal of this paper is to provide a conceptual update of grassland degradation based on an extensive worldwide literature review and an expert survey focused on the Río de la Plata region.

2. Methodology

2.1. Literature Review

A literature search was conducted in December 2016 with the main objectives of identifying the drivers, process indicators, and consequences related to degradation. We first used the free search engine Google Scholar (https://scholar.google.com/) to search across a wide range of academic sources and not only peer-reviewed literature. We searched by the terms "natural grassland degradation" and "rangeland degradation" related exclusively to livestock production; a complementary search in Spanish was conducted with the terms: "degradación", "campo natural" and "pastizales". Search in Portuguese was omitted, even though it is a very important language and the "Campos" region includes Brazil, due to the fact that in general the Brazilian journals have abstracts in English. Although there are many uses of natural grasslands such as recreation, hunting, and conservation, we address the issue of degradation of natural grassland, not of those intensively farmed.

A total of 5910 results were obtained: 5345 from the English-language search and another 565 from the Spanish-language search. Since we wanted to conduct an in-depth analysis of the concepts and indicators used to describe grassland degradation around the world, we limited our analysis to the most-cited 100 papers, though substitutions were made to ensure that at least one paper from each of the major grassland regions was represented (South American, North America, East Europe, Asia, Australia, and South Africa). Out of these 100 papers, we identified and systematized 27 papers that provided an explicit definition of grasslands degradation (see Supporting material in supplementary file).

Additionally, and to enrich and reinforce drivers, process indicators, and consequences and the conceptual framework, we updated the literature review by specifically searching in different scientific databases (Scopus, Science Direct, Springer, JSTOR, among others). As a result of this new search, an additional 40 papers were also systematized.

2.2. Expert Survey

To adapt our propose conceptual framework to the focus region, we center the expert survey on a regional level. We then conducted a regional survey focused on the Río de la Plata Grasslands (RPG).

This survey was first sent to the regional authors with more citations on "Río de la Plata Grasslands" research, and to take advantage of the expert network, we applied a "snowball sampling" technique [46]: we asked the experts to recommend three more people to whom the survey should be sent. This second group of experts was added to the expert list and surveys were sent to them as well. The survey was first conducted via email or phone, between July 2014 and March 2016.

The survey target population [47] consisted of 39 experts from different institutions: 54% from universities (Departments of Agronomy or Natural Sciences/Ecology), 13% from the private sector (agronomic advisors and/or farmers), 33% from other institutions (research institutes, extension institutes or the government, and non-profit organizations). The respondent's disciplinary backgrounds were 85% Grassland management scientists and 15% Botany/Ecology. Only 8% were women. In terms of nationality, 22 were Uruguayans, 11 Argentineans, and 6 Brazilians.

Five purposely open-ended questions were included in the survey: 1. What do you consider to be a degraded natural grassland and how can you define it? 2. Mention three relevant indicators that you would use to characterize a degraded natural grassland. 3. How do you define a non-degraded natural grassland? 4. Mention three relevant indicators that you would use to characterize a non-degraded natural grassland. 5. Identify two or three of the most relevant characteristics that could be used to differentiate the two conditions.

3. Conceptual Framework for Grassland Degradation

In the present study, we established a framework that identified three distinct conceptual categories: drivers, processes, and consequences of natural grasslands degradation (Figure 1). For this work, we define "drivers" as the external forces or changes that can cause degradation, such as overgrazing or land use change. "Processes" are the conditions that create a sequence of changes in the properties of the grassland ecosystem, and can be evaluated using indicators, such as reduction in plant growth. We use the term "process indicator" as a variable used to measure and infer a conclusion from the phenomenon of degradation [48], such as vegetation height or bare ground cover. Finally, we consider "consequences" as the impacts or results of the process of grassland degradation, such as reduction in the Aboveground Net Primary Productivity (ANPP).

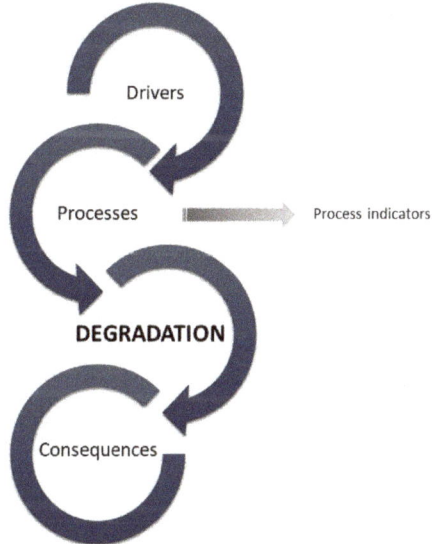

Figure 1. Conceptual framework of drivers, processes, and consequences of natural grasslands degradation.

3.1. Which Are the Main Drivers that Influence Grassland Degradation?

In Akiyama and Kawamura [9] review, they discussed several mechanisms of grassland degradation highlighting climate change and anthropogenic disturbances as the two main underlying factors that lead to degradation. Globally, the factors of climate change most related to grassland degradation are warming and the increase in the frequency of prolonged droughts [13,16,17,33,35,49]. Liu et al. [18], Gang et al. [17], Yang et al. [50], and Zhou et al. [20,51] presented quantitative assessments of the relative roles of climate change and anthropogenic perturbations, revealing that any of them can be the dominant driver of degradation depending on the study context. However, most of these studies did not consider the interactions between drivers. In this sense, Sala et al. [52] shows that there is clear evidence to support nonlinear responses of ecosystems and synergistic interactions among many global drivers of biodiversity change (e.g., invasions of alien species that are promoted by both human disturbances and climatic changes). On the other hand, Klein et al. [53] evince a decrease on ANPP and forage quality (for a of 1.0–2.0 °C warming) but also showed an interaction with grazing that could mitigate that decline. Additionally, Carlyle et al. [54] analyzed different grasslands types and found, by manipulating temperature, water, and grazing conditions, that biomass production depends on the interaction between the three variables under site-specific conditions.

Overgrazing, which occurs when stocking rates exceed the carrying capacity of grasslands, is the most widespread anthropogenic driver of degradation [9,55,56]. In fact, stocking rates are less variable than the carrying capacity, and this is ultimately related to primary production and, therefore, to the climatic conditions. In this way, overgrazing clearly interacts with climate variability. In this regard, Liu et al. [18] report that Mongolian steppe degradation can be attributed both to climate change (decreasing precipitation and increasing temperature) and to the increase in the stocking rate. On the other hand, extreme under-grazing (or grazing abandonment) could also be a source of degradation affecting productivity and species composition. Although long term livestock exclosures increased standing biomass and vegetation cover, and they may be a useful management tool to restore the vegetation in degraded grassland of arid regions [57,58], they sometimes decreased grassland biodiversity [59,60] and ANPP [59].

Regarding grassland degradation, there is a long-standing controversy about the use of fire. On one hand, Snyman [61] suggested that fire may potentially degrade semiarid grassland based on the fact that the recovery of the above and below ground biomass production and water use efficiency, took at least two years after burning. On the other hand, the use of fire appears to be relevant in fire-prone communities (e.g., mesic grasslands) where fire is an evolutionary force that has been used for hundreds of years to maintain grassland structure and function [62].

Finally, the improper fertilization and/or exotic species introduction [63–65] and invasive alien species [66] are also other important factor influencing grassland degradation.

3.2. How Ecological Process Are Affected by Grassland Degradation?

3.2.1. Energy Flux

The most direct signs of grassland degradation are the change in vegetation parameters such as the decrease in litter mass, vegetation cover and height [12,31,67,68]. At the same time, the decline in vegetation vigor in degraded grasslands is also reflected in the decrease of the canopy leaf area index. This is highly relevant, since the leaf area index is one of the main controllers of the primary production through its role in the photosynthesis [69–71]. On the other hand, the decrease in vigor of the vegetation could be accompanied by the substitution of high forage quality species for low forage quality ones [8,31,35,67]. This species turnover could lead to a lower leaf area efficiency than non-degraded grasslands and, therefore, to a greater reduction in plant growth.

3.2.2. Biogeochemical Cycles

Grasslands have a high potential for carbon storage; however, this capacity strongly depends on how grasslands are managed [72]. The decrease in litter and plant cover in degraded grasslands can lead to a slowdown of the nutrient cycling [67,73,74] and to a significant reduction of soil organic carbon [12,74–79] and nitrogen [80]. Additionally, the reduction in plant cover also results in a reduced infiltration, which in turn, increases soil erosion [81] and reduced soil water content [82]. This process can also slow down even more the nutrient cycling process via feedback cycles.

3.2.3. Water Cycle

The hydrological functioning of grasslands could change dramatically by the decrease in plant and litter cover. The decline in plant cover can lead to an increase in soil bulk density and a decline in the soil structural stability [67,83,84]. The degradation of grasslands increases evaporation rates [85,86] and surface runoff [83,87], reducing soil-water infiltration capacity [67,83,84]. This in turn reduces the effective soil water content [43] and increases soil loss due to erosion [88], leading to greater risks of degradation [89]. This process is highly relevant in semiarid regions, where plant and litter cover are scare per se.

3.3. What are the Main Consequences of Grassland Degradation?

In countries with most of the territory occupied by grasslands and where extensive livestock production is an important economic activity, grassland degradation can cause significant economic and environmental problems. Some economic consequences arise, such as the decreased capacity to produce commodities [24,45,90] (e.g., meat and wool). This occur because degraded grasslands have a decreased capacity to provide forage for livestock (e.g., decreased ANPP) [8,18,24,26,35,55,91,92]. In addition, in degraded grasslands, high forage quality species could be replaced by low productive species [8,31,35,67], with the consequent loss of forage nutritional value [31]. Such changes imply a reduction in the carrying capacity and, therefore, a reduction in livestock production [9,56].

The degradation of the ecosystem services (benefits that society receives from ecosystems) involving water and nutrient cycle, energy flow, climate regulation, biodiversity and erosion control among others [6], has been repeatedly reported as one of the main consequences. Wen et al. [93] reported that primary production and other ecosystem services (carbon storage, nitrogen recycling, and plant diversity) of degraded grassland were always lower than those provided by non-degraded ones.

Changes in surface properties of the ecosystems, could severely affect the energy balance, which may affect the regional climate [94–96]. The decrease in litter quantity, vegetation height and biomass [12,31,67,68]; the replacement of high forage quality species by low quality ones [8,31,35,67]; the increase in runoff and soil losses [88]; and the decrease in litter [73], are key processes that, separately or together, can lead to a reduced plant growth which results in a lower secondary production.

Grassland degradation decrease soil organic carbon and nitrogen stocks and promote the emission of greenhouse gases into the atmosphere [12,80,93]. According to the two-year study of Zhang et al. [80], total carbon and nitrogen stored in a semiarid grassland ecosystem was reduced, under severe degradation conditions, by 16% and 10% respectively. Indeed, in a comprehensive meta-analysis Dlamini et al. [79] showed that grassland degradation reduced soil organic carbon by 16% and 8% in dry and wet climates, respectively. In some regions, degradation goes beyond its effects on ecosystem services, as it can lead to the completely loss of grassland habitat or to desertification [97].

Other important consequence of grassland degradation is the loss of animal and/or plant biodiversity [24,35,65]. Grazing usually increases species richness in mesic grasslands, while it generally reduces species richness in semiarid and arid grassland [98–100]. Even though there are some analysis as the one reported by Eldridge et al. [101], where grazing reduces ecosystem structure, function, and composition on different bioclimatic conditions (arid to humid and sub- humid) in Australia, the negative effects of grazing were greater in drier environments. Nevertheless, the way in which species decrease in overgrazed grasslands is mostly dependent on the grazing history and the position on the moisture gradient [102]. In mesic grasslands with long history of grazing, the peak of species richness generally occurs under moderate grazing intensities [102–104], meanwhile in sites with short history of grazing this peak happen at light grazing intensities. On the other hand, in semiarid grasslands with a long history of grazing, this peak take place at light grazing intensities, and then, as grazing intensity increases, species richness should decline slightly; meanwhile in sites with short history of grazing the species richness decrease linearly with the increase of grazing from non-grazed to severe grazing intensities [102,103].

4. Conceptual Framework Adapted to RPG

The South American RPG are one of the largest temperate and subtropical grasslands regions in the world. The RPG cover the central-eastern part of Argentina, most of Uruguay and southern Brazil [105] (Figure 2).

In the RPG, extensive livestock production has taken place for more than 300 years and therefore stocking rate management is a key factor for grassland conservation [106]. Specifically, the RPG have a predominance of C4 grasses, and C3 grasses to a lesser extent, and are the habitat of 4864 plant species [19]; 385 bird species and 90 mammal species [107].

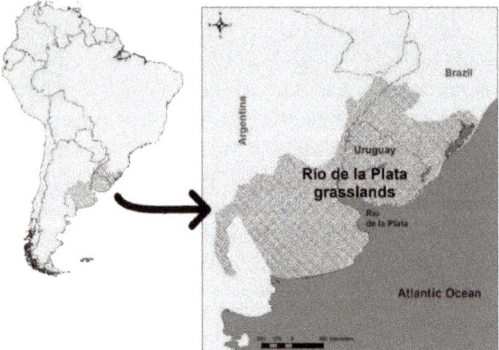

Figure 2. Map of South America highlighting the Río de la Plata grasslands region. Adapted from Miñarro and Bilenca [43].

Causes for degradation in RPG extensive livestock-based systems have been reported by many researchers, such as: (i) overgrazing by animals [56]; (ii) invasive alien species [108]; and (iii) nutrient addition and/or the introduction of exotic forage species into native grassland [63–65].

In RPG, animals graze all year round at relatively constant stocking rates [31], while ANPP of these grasslands shows large seasonal and inter-annual variations [109]. Under these conditions, grasslands can become recurrently overgrazed, mainly when periods of low forage production (e.g., severe droughts) coincide with high stocking rates. Indeed, the decreased forage biomass and the substitution of palatable species by unpalatable ones that can lead to a reduced plant growth and forage quality [8,31] are some of the processes that most concern in RPG. Besides that, weed invasions [108,110] and degradation promoted by fertilization and exotic forage species over-seeded in native grasslands are also concern. In these cases, the invasive weeds and the over-seeded species often weaken native species as they compete them for resources, such as space and light. However, as the over-seeded species do not persist, they end up facilitating the colonization and dominance of invasive species [65].

The consequences of overgrazing by livestock in RPG are strongly dependent on grassland type. Studies of Altesor et al. [59] in Southern Campos in Uruguay, Jaurena et al. [111] in northern Campos in Uruguay, and Fedrigo et al. [56] in Campos in Río Grande do Sul-Brazil indicate that these grasslands are particularly resistant to overgrazing. Meanwhile, in the Flooding Pampa in Argentina, the increase of grazing intensity induces a quick replacement of C3 and C4 native grasses by exotic annual forbs and grasses [112]. In addition, the most problematic invasive weed in the RPG is *Eragrostis plana* Nees, which has already invaded 20% of the native grassland in Río Grande do Sul, decreasing native plant diversity and livestock production [113].

4.1. How Did Experts Percived the Degradation of Natural Grassland?

According to the expert survey, in most cases, the experts recognized that the grassland degradation is strongly dependent on grassland community, related mainly to different soil types. They referred to degradation as a complex process that have diverse dynamics on each specific community or specific region. Most of the experts focused on the importance of the primary (plant) and secondary (animal) productivity. Some of grassland management specialist mentioned that the dominance of some exotic forage species, if they are palatable and nutritious for livestock, they did not consider the presence of these species (e.g., ryegrass *Lolium multiflorum* Lam.) to be an indicator of degradation. On the other hand, others defined the degradation problem related to the abiotic conditions that cause a change in the community state which is not reversible after the disturbance occurs.

4.2. Which Indicators Are the Most Important to Characterize Degradation?

According to the conceptual framework previously described (Figure 1), we identified the drivers, process indicators, and consequences related to degradation from the literature review and experts survey. It is noteworthy that 63% of papers alluded to process indicators, 50% mentioned at least one driver, and 36% referred to consequences. The most commonly cited drivers (46% of the papers) were related to human-induced processes (mainly overgrazing). Loss of ecosystem processes, service and function (24%) and soil erosion (10%) were the most frequently declared consequences.

Half of the experts identified overgrazing in specific communities as the most mentioned driver. The most commonly cited consequences increased soil erosion (41%) followed by reduction of resilience (including drought resistance) and the reduction of livestock production (28% each).

The most frequently mentioned indicators were vegetation or bare soil cover, productivity (ANPP related to the potential of the community), plant species or functional groups diversity, and species and functional type composition.

It is worth noticing that climate change was not considered to be an important driver for the experts but is was cited on the 15% pf the papers. Although drought resistance and resilience were mention as important consequences in the expert survey.

A comparison of the most relevant indicators extracted from the two different scales addressed in the study (regional and global) is shown in Table 1.

Table 1. Frequency (%) of the most relevant indicator mentioned in the literature review (from a total of 67 analyzed papers) and expert survey (from 29 experts who answered the survey). Some papers or experts mention more than one indicator. The "+" and "−" signs indicate the effect of the natural grassland's degradation on that indicator: a decrease (−) or an increase (+). Indicators are grouped following the conceptual framework present in Figure 1. *1 (overgrazing, under-grazing, fragmentation); *2 (high forage value: quality and stability in time); *3 (related to livestock preference); *4 (mainly livestock production); *5 (water cycle, nutrient cycle, energy flow, species dynamic).

Conceptual Category	Subtopic	Indicator	Effect	% of Papers (Literature)	% of Experts (Survey)
Drivers		Human-induced processes *1 [6,8–10,13–16,25,29,34,35,52,55,56,63–65,91,92,106,114–122]		46	52
		Result of natural events (drought, climate change) [13,14,16,17,25,33,35,49,55,115]		15	
		Productivity (ANPP related to the potential in a specific soil type) [12,15,24,25,29,32,34,36,45,55,90–92,115,123,124]	−	24	55
	Productivity	Vegetation and/or bare soil cover [12,24,25,32,34,55,108,110,114,116,125–127]	− or +	19	72
		Productivity indicator species % *2 [4,123,124]	−	4	45
		Plant height [12,92,125]	−	4	31
		Biomass or Forage availability [9,10,14,35,55,92,115,126]	−	12	7
		Quality indicator species (%) *3 [14,24,25,29,31,128]	−	10	10
Process indicators		Plant species richness, plant or functional groups diversity [8,12,24,26,29,34,35,65,123,125]	−	15	55
		Species and functional type composition [8,29,115,123,126]	− or +	7	55
	Biodiversity	Weeds (%) [8]	+	1	41
		Structural heterogeneity [29]	−	1	48
		Non-native plant species % [65,108,110,127]	+	6	41
		Key, endemic or rare species [32]	−	1	28
		Soil bulk density [32,87]	+	3	17
	Soil process	Soil organic matter [32,74–78,92]	−	10	14
		Soil nutrients [29,35,116]	+	4	21
		Litter presence [67,68,73]	−	4	10

Table 1. Cont.

Conceptual Category	Subtopic	Indicator	Effect	% of Papers (Literature)	% of Experts (Survey)
Consequences		Soil erosion [29,34,35,67,87,88,116]	+	10	41
		Drought resistance and resilience [18,24,76,129]	−	6	28
		Secondary production *4 [9,24,34,45,56,90,123,129]	−	12	28
		Ecosystem processes, services and function *5 [12,15,24,29,32,45,56,67,74,75,83,84,87,88,123,124]	−	24	23
		Plant health [15]	−	1	21
		Soil seed bank [130,131]	−	3	3

4.3. Conceptual Framework Proposed of Monitoring Río de la Plata Grassland Degradation

According to the information discussed above, a complete conceptual framework diagram presenting the relation between those concepts is shown in Figure 3.

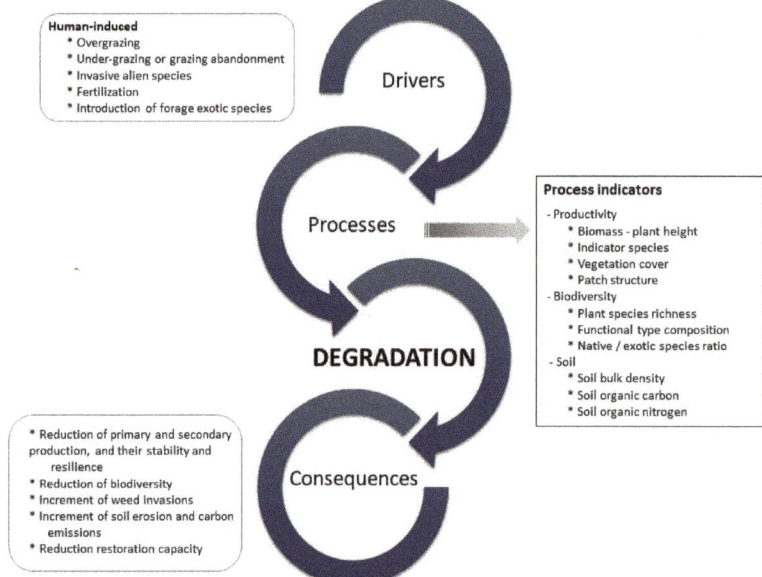

Figure 3. Proposed conceptual framework of RPG degradation with drivers, process indicators, and consequences based on literature review and expert survey.

5. What is a Degraded Native Grassland?

The term "degradation" referring to the condition of a natural grassland, is widely used in different circumstances by different stakeholders (from researchers and rural extension agents to policy makers), without a common agreed definition. Based on our review and considering that the interaction of grassland and herbivorous is the central key of the analysis, we propose as common definition of grassland degradation, a retro-progressive process in which a grassland community changes to a lower quality state, losing its capacity to be grazed by herbivores. Specifically, grassland degradation happens when you have a general decrease in productivity, soil properties, and diversity due to human activities and natural processes. This new definition is based on the fact that as plant communities degrade, above ground standing biomass and plant production decrease [12,18,29,67,68], and soil-vegetation cover and litter amount decrease too [67,74]. These effects could occur in conjunction with changes in species composition [8,29,35,67], reductions in species diversity [24,35,65], increased soil erosion [35,67,87,88] and augmented soil nutrient loss [67,74]; among other important effects. In all cases, different stakeholders agree that the degradation process leads to a lower quality of grassland ecosystem.

6. Discussion

The importance of natural grasslands to food security, to the provision of ecosystem services, and to the economy of many developing countries such as Uruguay, Argentina, and Brazil, imply that their degradation is a major political, economic, and environmental issue. This importance is reinforced by the experts' perception that degradation can be increased by the effects of livestock production intensification and climate change. Although it is widely known and repeatedly reported in the literature that overgrazing is one of the main global drivers of degradation, the experimental evidence of irreversible changes to degraded states in RPG is limited and it is highly dependent on the community type. The STMs developed in the RPG region by Andrade et al. [132] provide a conceptual framework of stages and thresholds that help researchers to study degradation and restoration. In the present review paper, we advance in the theoretical framework detailing the main drivers, process indicators, and consequences. Despite some differences evidence on the global literature review and the regional expert survey, some of the most frequently mentioned topics were the same, and they have some common process indicators, suggesting that the problem of degradation may have a common interpretation. Taking this into consideration, grassland degradation for livestock production could be monitored assessing drivers, process indicators, and consequences as in the propose framework.

From our perspective, to further advance in the knowledge of grassland degradation process in RPG, we represent the interactive effects of rainfall variability and overgrazing, in new conceptual STMs (Figure 4A–D). Specifically, to support the decisions of ecosystem management, these models try to link (i) multiple drivers operating simultaneously, in different site-specific conditions (e.g., soil types bioclimatic zones and evolutionary history of grazing), (ii) the main process affected (e.g., decreased on primary production by a reduced of the leaf area index, or the reduction on plant species diversity by an increment on interspecific competitive exclusion for light) and its indicators, and (iii) the expected consequences of management practices in different climatic scenarios.

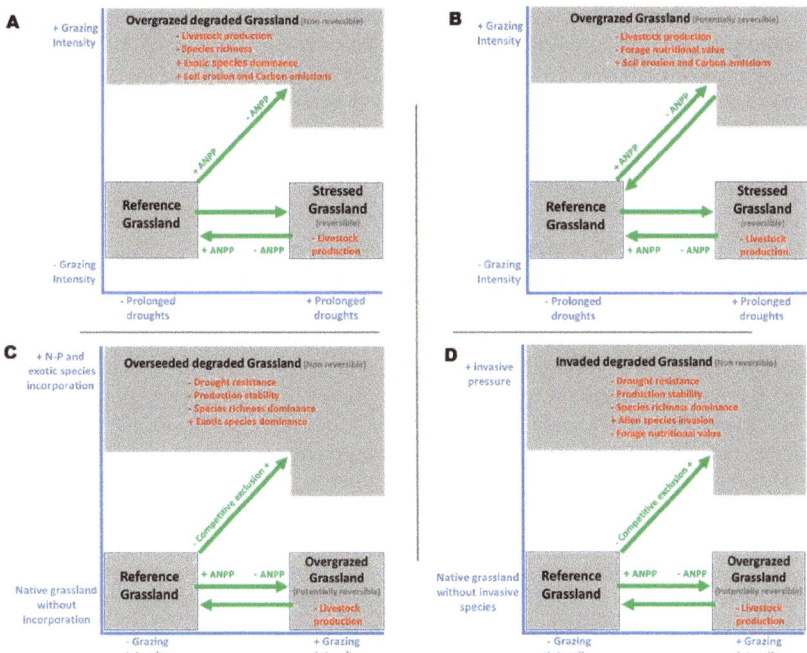

Figure 4. Conceptual framework proposed for the study of grassland degradation in the RPG represented by four state and transition models: (**A**) Grassland degradation by overgrazing and increased rainfall variability in sites with short evolutionary history of grazing; (**B**) Grassland degradation by overgrazing and increased rainfall variability in sites with long evolutionary history of grazing; (**C**) Grassland degradation by intensification (nitrogen and phosphorous incorporation combined with over-seeding of exotic forage species) with different grazing intensities; and (**D**) Grassland degradation by invasive alien species with different grazing intensities. Drivers are remarked in blue, process in green and consequences in red. ANPP (above ground net primary production). The two-sided arrows represent potentially reversible processes when the disturbance is removed, while the one-sided arrows represent an irreversible shift to a degraded state.

Some studies conducted in the Flooding Pampas in Argentina [112,133] reveal that grazing would induce rapid changes to alternative degraded states (Figure 4A). However, on the other hand, studies carried out on Campos grasslands [56,134] suggest that grazing intensity and climate variability would induce phase changes within states (slow, continuous and reversible community changes), but would not cause change into alternative degraded stable states (Figure 4B). Similar results are reported by Porensky et al., [135] in northern mixed-grass prairie in the USA. The differences in responses of plant community to overgrazing, which experts mentioned repeatedly, could be related with the interaction between grassland bioclimatic conditions and the evolutive history of grazing [103]. However, there is a lack of knowledge in that history on a specific region with an adequate spatial-temporal resolution, this has repercussions in being able to capture grassland dynamics (essential to understand short term evolution). To overcome this limitation, Oesterheld and Semmartin [136] propose to consider the regional set of species, the covariation of production and grazing intensity, and the positive biotic interactions that protect plants from herbivores, for a better understanding of why different grasslands communities respond differently to grazing. Additionally, invasive alien species [108] and nutrient addition and/or the introduction of exotic forage species into native grasslands [63–65] in interaction with grazing intensity would induce changes to alternative degraded states (Figure 4C,D).

As we already mentioned, grassland degradation has become a widespread and common problem in different regions of the world and in this review paper we have found that different regions and grassland types have some common drivers, process, and consequences related to degradation and thus, have the potential to be analyzed and monitored by a common framework. However, the reviewed articles also revealed that other drivers, processes and consequences were common within a similar rainfall variability conditions, and different according to a bioclimatic gradient and evolutive history of grazing. Climate change appears to be a significant driver on arid grasslands [20,50,51,137] while it is not mentioned in humidity grasslands analysis. On the other hand, light grazing intensity or grazing exclusion, have different effects depending on the bioclimatic region, with a positive effect on arid environments [57,58], and a negative one on humidity regions [59,60]. Regarding consequences, desertification is exclusively reported on arid and semiarid areas [9].

Although it seems that the responses to the underlying drivers of degradation are different between semiarid, mesic, and humid grasslands, we suggest that this bioclimatic gradient mask the true effect of rainfall variability. The results of Davidowitz [138] showed that more drier grasslands generally are climatically more variable than the humid ones, so inter-annual rainfall variability is inversely related to its mean. For many processes, inter-annual rainfall variability can be a more important climatic parameter than the average precipitations [139]. For instance, where rainfall variability is high in relation to its average, species need to adapt to first survive prolonged drought periods and then, to take advantage of long rainfall periods [140]. Additionally, in arid and semiarid rangelands, the variability in ANPP could be higher than variability in rainfall [141,142].

Although several studies quantified the effects of climate change and anthropogenic management on grassland degradation, simulating actual vs. potential net primary production effects [17,20,50,51,135], their interactions have not been addressed and the effect of frequency and intensity increment of extreme precipitation events on grassland degradation, remains almost unexplored.

In this review, we found that most of grassland degradation studies were conducted in water-limited grasslands. In these grasslands, prolonged droughts and overgrazing are the two common underlying drivers that lead to degradation. Sloat et al. [143] found that grasslands regions with high rainfall variability support lower livestock densities than less-variable areas. Considering that stocking rates are less variable than the carrying capacity and therefore than climatic conditions, overgrazing clearly interacts with the increased rainfall variability. From these non-equilibrium concept of grassland dynamics proposed by Ellis and Swift [144], it has been predicted that the potential degradation by overgrazing is low in environments with relatively high rainfall variability. These studies argue that in periods of drought, herbivores population is reduced, thereby decreasing their potential to degrade grasslands. We consider that this application of the non-equilibrium concept could contribute to understand the dynamic of wildlife-based ecosystems where animals self-regulate their populations during prolonged droughts. However, this is not the general case in extensive livestock-based systems where farmers seek to maximize profit. In this context, farmers have two contrasting scenarios for the application of management practices to mitigate drought effects. On one hand, they could provide supplemental emergency feed (e.g., crop residues or grain by-products), use stockpile forage and facilitate the access to water sources to minimize animal mortality and to avoid weight losses in drought periods, maintaining the stocking rates. This managements practices could lead to a positive feedback that cause further degradation [145]. As a second option, if these same management options are combined with the reduction of the stocking rate to match availability of lower forage and to minimizes the effect of grazing on vegetation, a negative feedback could probably occur that will prevent degradation.

The diversity of grasslands, the differential impacts of climate and management on them, and the variety of uses and human dimensions throughout the world, are the main reasons for the coexistence of different conceptual significances of degradation. de Queiroz [45] state that definitions of degradation are different depending on the human management objectives. Given that these objectives can differ, the definitions of what constitutes degradation also differ. We argue that definitions can also differ due to emphasis on different conceptual categories related to degradation and that is why it is important of clarify these conceptual categories. Moreover, the concept of degradation does not denote the same set of conditions for different stakeholders. Bedunah and Angerer [5] highlighted the complexity of the term degradation by claiming that rangeland scientists need to have a key role in collecting, understanding, and commenting on degradation definitions used by stakeholders. Agreeing on a definition of a complex concept such as grassland degradation is also important for research purposes, since it contributes to identify what processes need to be described and which variables need to be measured to systematically assess the degree of degradation.

7. Conclusions

Grassland degradation has multicausal drivers; however, the complexity of this process remains mostly unexplored, since most of the studies focus on single drivers of vegetation change. From our study, we proposed a new conceptual model which consider the multiple drivers that operate simultaneously and their interactions in different site-specific conditions.

Despite the difference in scale (global and regional), some of the mentioned topics were the same and they have some common process indicators suggesting that the problem of degradation, may have a common interpretation and can be analyzed with a common conceptual framework.

We believe that the proposed conceptual framework for the degradation of grasslands used for livestock production is a valuable contribution to monitoring programs and to support grassland management decisions, by clarifying the different concepts involved in the process (drivers, process indicators, and consequences) and structuring the degradation analysis allowing identifying key aspects. In addition, the framework could be a useful tool to communicate the importance of grasslands conservation, both to a general audience and to specialists in the field.

Despite this, we also found that some drivers, process, and consequences were common within a bioclimatic zone and could be different along a bioclimatic gradient. This gives some particularities to the systems that needs to be abroad and could be also analyzed according to the proposed conceptual framework.

Considering the RPG region, we expect that this paper will contribute to reach a conceptual update of grassland degradation based on the mentioned concepts that can help to promote a dual goal of production and conservation in this important and often neglected ecosystem.

Supplementary Materials: The following are available online at http://www.mdpi.com/2073-4395/9/5/239/s1, supplementary file: Supporting Literature review.pdf.

Funding: This research has been conducted as part of a Ph.D. program (Facultad de Agronomía, Universidad de la República, Uruguay) supported by the Agro-climate and Information System Unit (GRAS) of the National Agricultural Research Institute (INIA Uruguay) and a partial scholarship provided by the National Agency for Research and Innovation (ANII Uruguay).

Acknowledgments: We want to especially thank the experts who took the time to answer the questions or discussed the topic. We also want to thank Cathy Vaughan from IRI (International Research Institute for Climate and Society Columbia University) and Andrea Ruggia from INIA (National Institute of Agricultural Research, Uruguay) for their comments, reviews, and contributions. Finally, a special mention to Valentín Picasso from the Agronomy Department, University of Wisconsin for his great contribution to this paper.

Conflicts of Interest: The authors declare no conflict of interest.

References

1. Gibson, D.J. *Grasses and Grassland Ecology*; Oxford University Press: Oxford, UK, 2009; p. 313.
2. Allen, V.G.; Batello, C.; Berretta, E.J.; Hodgson, J.; Kothmann, M.; Li, X.; McIvor, J.; Milne, J.; Morris, C.; Peeters, A.; et al. An international terminology for grazing lands and grazing animals. *Grass Forage Sci.* **2011**, *66*, 2–28. [CrossRef]
3. Lund, H.G. Accounting for the world's rangelands. *Rangelands* **2007**, *29*, 3–10. [CrossRef]
4. O'Mara, F.P. The role of grasslands in food security and climate change. Review: Part of a highlight on breeding strategies for forage and grass improvement. *Ann. Bot.* **2012**, *110*, 1263–1270. [CrossRef] [PubMed]
5. Bedunah, D.J.; Angerer, J.P. Rangeland degradation, poverty, and conflict: How can rangeland scientists contribute to effective responses and solutions? *Rangel. Ecol. Manag.* **2012**, *65*, 606–612. [CrossRef]
6. Modernel, P.; Rossing, W.A.H.; Corbeels, M.; Dogliotti, S.; Picasso, V.; Tittonell, P. Land use change and ecosystem service provision in Pampas and Campos grasslands of southern South America. *Environ. Res. Lett.* **2016**, *11*, 113002. [CrossRef]
7. Suttie, J.M.; Reynolds, S.G.; Batello, C. *Grasslands of the World*; Food and Agriculture Organization of the United Nations: Rome, Italy, 2005; p. 535.
8. Berretta, E.J.; Risso, D.F.; Montossi, F.; Pigurina, G. Campos in Uruguay. In *Grassland Ecophysiology and Grazing Ecology*; Lemaire, G., Hodgson, J., de Moraes, A., Nabinger, C., Carvalho, P.C., Eds.; CABI Publishing: New York, NY, USA, 2000; Volume 19, pp. 377–394.
9. Akiyama, T.; Kawamura, K. Grassland degradation in China: Methods of monitoring, management and restoration. *Jpn. Soc. Grassl. Sci.* **2007**, *53*, 1–17. [CrossRef]
10. Wessels, K.J.; Prince, S.D.; Carroll, M.; Malherbe, J. Relevance of rangeland degradation in semiarid northeastern South Africa to the non-equilibrium theory. *Ecol. Appl.* **2007**, *17*, 815–827. [CrossRef]
11. Harris, R.B. Rangeland degradation on the Qinghai-Tibetan plateau: A review of the evidence of its magnitude and causes. *J. Arid Environ.* **2010**, *74*, 1–12. [CrossRef]
12. Han, J.G.; Zhang, Y.J.; Wang, C.J.; Bai, W.M.; Wang, Y.R.; Han, G.D.; Li, L.H. Rangeland degradation and restoration management in China. *Rangel. J.* **2008**, *30*, 233–239. [CrossRef]
13. Veldman, J.W.; Buisson, E.; Durigan, G.; Fernandes, G.W.; Le Stradic, S.; Mahy, G.; Negreiros, D.; Overbeck, G.E.; Veldman, R.G.; Zaloumis, N.P.; et al. Toward an old-growth concept for grasslands, savannas, and woodlands. *Front. Ecol. Environ.* **2015**, *13*, 154–162. [CrossRef]
14. Mansour, K.; Mutanga, O.; Everson, T. Remote sensing based indicators of vegetation species for assessing rangeland degradation: Opportunities and challenges. *Afr. J. Agric. Res.* **2012**, *7*, 3261–3270.
15. Wick, A.F.; Geaumont, B.A.; Sedivec, K.; Hendrickson, J. Grassland degradation. In *Biological and Environmental Hazards, Risks and Disasters*; Shroder, J.F., Sivanpillai, R., Eds.; Elsevier: New York, NY, USA, 2016; Volume 8, pp. 257–276. ISBN 9780123964717.
16. Zhou, H.; Zhao, X.; Tang, Y.; Gu, S.; Zhou, L. Alpine grassland degradation and its control in the source region of the Yangtze and Yellow Rivers, China. *Grassl. Sci.* **2005**, *51*, 191–203. [CrossRef]
17. Gang, C.; Zhou, W.; Chen, Y.; Wang, Z.; Sun, Z.; Li, J.; Qi, J.; Odeh, I. Quantitative assessment of the contributions of climate change and human activities on global grassland degradation. *Environ. Earth Sci.* **2014**, *72*, 4273–4282. [CrossRef]
18. Liu, Y.Y.; Evans, J.P.; McCabe, M.F.; de Jeu, R.A.M.; van Dijk, A.I.J.M.; Dolman, A.J.; Saizen, I. Changing Climate and Overgrazing Are Decimating Mongolian Steppes. *PLoS ONE* **2013**, *8*, e57599. [CrossRef]

19. Andrade, B.O.; Marchesi, E.; Burkart, S.; Setubal, R.B.; Lezama, F.; Perelman, S.; Schneider, A.A.; Trevisan, R.; Overbeck, G.E.; Boldrini, I.I. Vascular plant species richness and distribution in the Río de la Plata grasslands. *Bot. J. Linn. Soc.* **2018**, *188*, 250–256. [CrossRef]
20. Zhou, W.; Gang, C.; Zhou, L.; Chen, Y.; Li, J.; Ju, W.; Odeh, I. Dynamic of grassland vegetation degradation and its quantitative assessment in the northwest China. *Acta Oecol.* **2014**, *55*, 86–96. [CrossRef]
21. Wilson, J.B.; Peet, R.K.; Dengler, J.; Pärtel, M. Plant species richness: The world records. *J. Veg. Sci.* **2012**, *23*, 796–802. [CrossRef]
22. Henwood, W.D. Editorial–the world's temperate grasslands: A beleaguered biome. *Parks* **1998**, *8*, 1–2.
23. Carvalho, P.C.; Batello, C. Access to land, livestock production and ecosystem conservation in the Brazilian Campos biome: The natural grasslands dilemma. *Livest. Sci.* **2009**, *120*, 158–162. [CrossRef]
24. Xu, X.; Liu, J.; Shao, Q. Spatial and temporal characteristics of grassland degradation in the riverhead area of the Yellow River. In *Second International Conference on Earth Observation for Global Changes*; International Society for Optics and Photonics: Bellingham, WA, USA, 2009; Volume 7471, pp. 74710R-1–74710R-7.
25. Manssour, K. Rangeland Degradation Assessment Using Remote Sensing and Vegetation Species. Ph.D. Thesis, Faculty of Science and Agriculture, University of KwaZulu-Natal, Durban, South Africa, 2011.
26. Liu, Y.; Zha, Y.; Gao, J.; Ni, S. Assessment of grassland degradation near Lake Qinghai, West Chine, using Landsat TM and in situ reflectance spectra data. *Int. J. Remote Sens.* **2004**, *25*, 4177–4189. [CrossRef]
27. Rosengurtt, B. *Estudios Sobre Praderas Naturales del Uruguay 3ª Contribución*; Barreiro y Ramos: Montevideo, Uruguay, 1943; p. 281.
28. Berretta, E.J.; do Nascimento, D., Jr. *Glosario Estructurado de Términos Sobre Pasturas y Producción Animal*; Diálogo 32; IICA–PROCISUR: Montevideo, Uruguay, 1991; p. 127. ISBN 92-9039-180-4.
29. Behmanesh, B.; Barani, H.; Sarvestani, A.A.; Shahraki, M.R.; Sharafatmandrad, M. Rangeland degradation assessment: A new strategy based on indigenous ecological knowledge of pastoralists. *Solid Earth Discuss.* **2015**, *7*, 2999–3019. [CrossRef]
30. Millot, J.C.; Methol, R.; Risso, D. *Relevamiento de Pasturas Naturales y Mejoramientos Extensivos en Áreas Ganaderas del Uruguay: Informe Técnico de la Comisión Honoraria del Plan Agropecuario*; Consultora F.U.C.R.E.A; MGAP: Montevideo, Uruguay, 1987; p. 199.
31. Pallarés, O.R.; Berretta, E.J.; Maraschin, G.E. The South American campos ecosystem. In *Grasslands of the World*; Suttie, J., Reynolds, S.G., Batello, C., Eds.; FAO: Roman, Italy, 2005; pp. 171–219.
32. Tácuna, R.E.; Aguirre, L.; Flores, E.R. Influencia de la revegetación con especies nativas y la incorporación de materia orgánica en la recuperación de pastizales degradados. *Ecol. Appl.* **2015**, *14*, 191–200. [CrossRef]
33. Wang, Z.; Deng, X.; Song, W.; Li, Z.; Chen, J. What is the main cause of grassland degradation? A case study of grassland ecosystem service in the middle-south Inner Mongolia. *Catena* **2017**, *150*, 100–107. [CrossRef]
34. Abdalla, K.; Mutema, M.; Chivenge, P.; Everson, C.; Chaplot, V. Grassland degradation significantly enhances soil CO_2 emission. *Catena* **2018**, *167*, 284–292. [CrossRef]
35. Cao, J.; Yeh, E.T.; Holden, N.M.; Qin, Y.; Ren, Z. The roles of overgrazing, climate change and policy as drivers of degradation of China's grasslands. *Nomadic Peoples* **2013**, *17*, 82–101. [CrossRef]
36. Li, S.; Verburg, P.H.; Lv, S.; Wu, J.; Li, X. Spatial analysis of the driving factors of grassland degradation under conditions of climate change and intensive use in Inner Mongolia, China. *Reg. Environ. Chang.* **2012**, *12*, 461–474.
37. Westoby, M.; Walker, B.; Noy-Meir, I. Opportunistic management for rangelands not at equilibrium. *J. Range Manag.* **1989**, *42*, 266–274. [CrossRef]
38. Briske, D.D.; Fuhlendorf, S.D.; Smeins, F.E. State-and-transition models, thresholds, and rangeland health: A synthesis of ecological concepts and perspectives. *Rangel. Ecol. Manag.* **2005**, *58*, 1–10. [CrossRef]
39. Lopez, D.R.; Cavallero, L.; Brizuela, M.A.; Aguiar, M.R. Ecosystemic structural–functional approach of the state and transition model. *Appl. Veg. Sci.* **2011**, *14*, 6–16. [CrossRef]
40. Briske, D.D.; Fuhlendorf, S.D.; Smeins, F.E. A unified framework for assessment and application of ecological thresholds. *Rangel. Ecol. Manag.* **2006**, *59*, 225–236. [CrossRef]
41. Briske, D.D.; Bestelmeyer, B.T.; Stringham, T.K.; Shaver, P.L. Recommendations for development of resilience-based state-and-transition models. *Rangel. Ecol. Manag.* **2008**, *61*, 359–367. [CrossRef]
42. Snyman, H.A. Rangeland degradation in a semi-arid South Africa—I: Influence on seasonal root distribution, root/shoot ratios and water-use efficiency. *J. Arid Environ.* **2005**, *60*, 457–481. [CrossRef]

43. Snyman, H.A.; Du Preez, C.C. Rangeland degradation in a semi-arid South Africa—II: Influence on soil quality. *J. Arid Environ.* **2005**, *60*, 483–507. [CrossRef]
44. Cai, H.; Yang, X.; Xu, X. Human-induced grassland degradation/restoration in the central Tibetan Plateau: The effects of ecological protection and restoration projects. *Ecol. Eng.* **2015**, *83*, 112–119. [CrossRef]
45. de Queiroz, J.S. *Range Degradation in Botswana*; Pastoral Development Network Paper, No. 35b; Overseas Development Institute (ODI): London, UK, 1993; p. 19.
46. Atkinson, R.; Flint, J. Accessing hidden and hard-to-reach populations: Snowball research strategies. *Soc. Res. Update* **2001**, *33*, 1–4.
47. Vogt, W.P.; Johnson, R.B. *Dictionary of Statistics & Methodology: A Nontechnical Guide for the Social Sciences: A Nontechnical Guide for the Social Sciences*; Sage Publications: Thousand Oaks, CA, USA, 2011; p. 437.
48. Heink, U.; Kowarik, I. What are indicators? On the definition of indicators in ecology and environmental planning. *Ecol. Indic.* **2010**, *10*, 584–593. [CrossRef]
49. Chen, B.; Zhang, X.; Tao, J.; Wu, J.; Wang, J.; Shi, P.; Zhang, Y.; Yu, C. The impact of climate change and anthropogenic activities on alpine grassland over the Qinghai-Tibet Plateau. *Agric. For. Meteorol.* **2014**, *189*, 11–18. [CrossRef]
50. Yang, Y.; Wang, Z.; Li, J.; Gang, C.; Zhang, Y.; Zhang, Y.; Oden, I.; Qi, J. Comparative assessment of grassland degradation dynamics in response to climate variation and human activities in China, Mongolia, Pakistan and Uzbekistan from 2000 to 2013. *J. Arid Environ.* **2016**, *135*, 164–172. [CrossRef]
51. Zhou, W.; Yang, H.; Huang, L.; Chen, C.; Lin, X.; Hu, Z.; Li, J. Grassland degradation remote sensing monitoring and driving factors quantitative assessment in China from 1982 to 2010. *Ecol. Indic.* **2017**, *83*, 303–313. [CrossRef]
52. Sala, O.E.; Chapin, F.S.; Armesto, J.J.; Berlow, E.; Bloomfield, J.; Dirzo, R.; Huber-Sanwald, E.; Huenneke, L.F.; Jackson, R.B.; Kinzig, A.; et al. Global biodiversity scenarios for the year 2100. *Science* **2000**, *287*, 1770–1774. [CrossRef]
53. Klein, J.A.; Harte, J.; Zhao, X.Q. Experimental warming, not grazing, decreases rangeland quality on the Tibetan Plateau. *Ecol. Appl.* **2007**, *17*, 541–557. [CrossRef]
54. Carlyle, C.N.; Fraser, L.H.; Turkington, R. Response of grassland biomass production to simulated climate change and clipping along an elevation gradient. *Oecologia* **2014**, *174*, 1065–1073. [CrossRef] [PubMed]
55. Yao, Z.; Zhao, C.; Yang, K.; Liu, W.; Li, Y.; You, J.; Xiao, J. Alpine grassland degradation in the Qilian Mountains, China—A case study in Damaying Grassland. *Catena* **2016**, *137*, 494–500. [CrossRef]
56. Fedrigo, J.K.; Ataide, P.F.; Filho, J.A.; Oliveira, L.V.; Jaurena, M.; Laca, E.A.; Overbeck, G.E.; Nabinger, C. Temporary grazing exclusion promotes rapid recovery of species richness and productivity in a long-term overgrazed Campos grassland. *Restor. Ecol.* **2017**, *26*, 677–685. [CrossRef]
57. Al-Rowaily, S.L.; El-Bana, M.I.; Al-Bakre, D.A.; Assaeed, A.M.; Hegazy, A.K.; Ali, M.B. Effects of open grazing and livestock exclusion on floristic composition and diversity in natural ecosystem of Western Saudi Arabia. *Saudi J. Biol. Sci.* **2015**, *22*, 430–437. [CrossRef] [PubMed]
58. Qasim, S.; Gul, S.; Shah, M.H.; Hussain, F.; Ahmad, S.; Islam, M.; Rehman, G.; Yaqoob, M.; Shah, S.Q. Influence of grazing exclosure on vegetation biomass and soil quality. *Int. Soil Water Conserv. Res.* **2017**, *5*, 62–68. [CrossRef]
59. Altesor, A.; Oesterheld, M.; Leoni, E.; Lezama, F.; Rodríguez, C. Effect of grazing on community structure and productivity of a Uruguayan grassland. *Plant Ecol.* **2005**, *179*, 83–91. [CrossRef]
60. Yao, X.; Wu, J.; Gong, X.; Lang, X.; Wang, C. Grazing exclosures solely are not the best methods for sustaining alpine grasslands. *PeerJ* **2019**, *7*, e6462. [CrossRef]
61. Snyman, H.A. Short-term responses of southern African semi-arid rangelands to fire: A review of impact on soils. *Arid Land Res. Manag.* **2015**, *29*, 222–236. [CrossRef]
62. Bond, W.J.; Keeley, J.E. Fire as a global 'herbivore': The ecology and evolution of flammable ecosystems. *Trends Ecol. Evol.* **2005**, *20*, 387–394. [CrossRef] [PubMed]
63. Brambilla, D.M.; Nabinger, C.; Kunrath, T.R.; Carvalho, P.C.D.F.; Carassai, I.J.; Cadenazzi, M. Impact of nitrogen fertilization on the forage characteristics and beef calf performance on native pasture overseeded with ryegrass. *Rev. Bras. Zootec.* **2012**, *41*, 528–536. [CrossRef]
64. de Ávila, M.R.; Nabinger, C.; Brambilla, D.M.; Carassai, I.J.; Kunrath, T.R. The effects of nitrogen enrichment on tiller population density and demographics of annual ryegrass overseeded on natural pastures South of Brazil. *Afr. J. Agric. Res.* **2013**, *8*, 3013–3018.

65. Jaurena, M.; Lezama, F.; Salvo, L.; Cardozo, G.; Ayala, W.; Terra, J.; Nabinger, C. The dilemma of improving native grasslands by overseeding legumes: Production intensification or diversity conservation. *Rangel. Ecol. Manag.* **2016**, *69*, 35–42. [CrossRef]
66. Stohlgren, T.J.; Binkley, D.; Chong, G.W.; Kalkhan, M.A.; Schell, L.D.; Bull, K.A.; Otsuki, Y.; Newman, G.; Bashkin, M.; Son, Y. Exotic plant species invade hot spots of native plant diversity. *Ecol. Monogr.* **1999**, *69*, 25–46. [CrossRef]
67. Li, C.; Hao, X.; Zhao, M.; Han, G.; Willms, W.D. Influence of historic sheep grazing on vegetation and soil properties of a Desert Steppe in Inner Mongolia. *Agric. Ecosyst. Environ.* **2008**, *128*, 109–116. [CrossRef]
68. Yan, L.; Zhou, G.; Zhang, F. Effects of Different Grazing Intensities on Grassland Production in China: A Meta-Analysis. *PLoS ONE* **2013**, *8*, e81466. [CrossRef]
69. Bircham, J.S.; Hodgson, J. The influence of sward condition on rates of herbage growth and senescence in mixed swards under continuous stocking management. *Grass Forage Sci.* **1983**, *38*, 323–331. [CrossRef]
70. Asner, G.P.; Scurlock, J.M.O.; Hicke, J.A. Global synthesis of leaf area index observations: Implications for ecological and remote sensing studies. *Glob. Ecol. Biogeogr.* **2003**, *12*, 191–205. [CrossRef]
71. Sbrissia, A.F.; Duchini, P.G.; Zanini, G.D.; Santos, G.T.; Padilha, D.A.; Schmitt, D. Defoliation Strategies in Pastures Submitted to Intermittent Stocking Method: Underlying Mechanisms Buffering Forage Accumulation over a Range of Grazing Heights. *Crop Sci.* **2018**, *58*, 945–954. [CrossRef]
72. McSherry, M.E.; Ritchie, M.E. Effects of grazing on grassland soil carbon: A global review. *Glob. Chang. Biol.* **2013**, *19*, 1347–1357.
73. Willms, W.D.; Smoliak, S.; Bailey, A.W. Herbage production following litter removal on Alberta native grasslands. *J. Range Manag.* **1986**, *39*, 536–540. [CrossRef]
74. Liu, N.; Zhang, Y.; Chang, S.; Kan, H.; Lin, L. Impact of Grazing on Soil Carbon and Microbial Biomass in Typical Steppe and Desert Steppe of Inner Mongolia. *PLoS ONE* **2012**, *7*, e36434. [CrossRef]
75. Zhang, R.; Bai, Y.; Zhang, T.; Henkin, Z.; Degen, A.A.; Jia, T.; Guo, C.; Long, R.; Shang, Z. Driving Factors That Reduce Soil Carbon, Sugar, and Microbial Biomass in Degraded Alpine Grasslands. *Rangel. Ecol. Manag.* **2019**, *72*, 396–404. [CrossRef]
76. Steffens, M.; Kölbl, A.; Totsche, K.U.; Kögel-Knabner, I. Grazing effects on soil chemical and physical properties in a semiarid steppe of Inner Mongolia. *Geoderma* **2008**, *143*, 63–72. [CrossRef]
77. Zhao, F.; Ren, C.; Shelton, S.; Wang, Z.; Pang, G.; Chen, J.; Wang, J. Grazing intensity influence soil microbial communities and their implications for soil respiration. *Agric. Ecosyst. Environ.* **2017**, *249*, 50–56. [CrossRef]
78. Zhou, G.; Zhou, X.; He, Y.; Shao, J.; Hu, Z.; Liu, R.; Zhou, H.; Hosseinibai, S. Grazing intensity significantly affects belowground carbon and nitrogen cycling in grassland ecosystems: A meta-analysis. *Glob. Chang. Biol.* **2017**, *23*, 1167–1179. [CrossRef]
79. Dlamini, P.; Chivenge, P.; Chaplot, V. Overgrazing decreases soil organic carbon stocks the most under dry climates and low soil pH: A meta-analysis shows. *Agric. Ecosyst. Environ.* **2016**, *221*, 258–269. [CrossRef]
80. Zhang, G.; Kang, Y.; Han, G.; Mei, H.; Sakurai, K. Grassland degradation reduces the carbon sequestration capacity of the vegetation and enhances the soil carbon and nitrogen loss. *Acta Agric. Scand. Sect. B-Soil Plant Sci.* **2011**, *61*, 356–364. [CrossRef]
81. Mchunu, C.; Chaplot, V. Land degradation impact on soil carbon losses through water erosion and CO_2 emissions. *Geoderma* **2012**, *177*, 72–79. [CrossRef]
82. Yi, X.S.; Li, G.S.; Yin, Y.Y. The impacts of grassland vegetation degradation on soil hydrological and ecological effects in the source region of the Yellow River—A case study in Junmuchang region of Maqin country. *Procedia Environ. Sci.* **2012**, *13*, 967–981. [CrossRef]
83. Bertol, I.; Gomes, K.E.; Denardin, R.B.N.; Machado, L.A.Z.; Maraschin, G.E. Propriedades físicas do solo relacionadas a diferentes níveis de oferta de forragem numa pastagem natural. *Pesqui. Agropecuária Bras.* **1998**, *33*, 779–786.
84. Willatt, S.T.; Pullar, D.M. Changes in soil physical-properties under grazed pastures. *Aust. J. Soil Res.* **1984**, *22*, 343–348. [CrossRef]
85. Babel, W.; Biermann, T.; Coners, H.; Falge, E.; Seeber, E.; Ingrisch, J.; Willinghöfer, S. Pasture degradation modifies the water and carbon cycles of the Tibetan highlands. *Biogeosciences* **2014**, *11*, 6633–6656. [CrossRef]
86. Li, Z.; Wu, W.; Liu, X.; Fath, B.D.; Sun, H.; Liu, X.; Cao, J. Land use/cover change and regional climate change in an arid grassland ecosystem of Inner Mongolia, China. *Ecol. Model.* **2017**, *353*, 86–94. [CrossRef]

87. Xie, Y.; Sha, Z. Quantitative analysis of driving factors of grassland degradation: A case study in Xilin River Basin, Inner Mongolia. *Sci. World J.* **2012**, *2012*, 14. [CrossRef]
88. Ludwig, J.A.; Wilcox, B.P.; Breshears, D.D.; Tongway, D.J.; Imeson, A.C. Vegetation patches and runoff–erosion as interacting ecohydrological processes in semiarid landscapes. *Ecology* **2005**, *86*, 288–297. [CrossRef]
89. Sun, L.; Yang, L.; Hao, L.; Fang, D.; Jin, K.; Huang, X. Hydrological effects of vegetation cover degradation and environmental implications in a semiarid temperate Steppe, China. *Sustainability* **2017**, *9*, 281. [CrossRef]
90. Mansour, K.; Mutanga, O.; Everson, T.; Adam, E. Discriminating indicator grass species for rangeland degradation assessment using hyperspectral data resampled to AISA Eagle resolution. *ISPRS J. Photogramm. Remote Sens.* **2012**, *70*, 56–65. [CrossRef]
91. Paudel, K.P.; Andersen, P. Assessing rangeland degradation using multi temporal satellite images and grazing pressure surface model in Upper Mustang, Trans Himalaya, Nepal. *Remote Sens. Environ.* **2010**, *114*, 1845–1855. [CrossRef]
92. Hoppe, F.; Kyzy, T.Z.; Usupbaev, A.; Schickhoff, U. Rangeland degradation assessment in Kyrgyzstan: Vegetation and soils as indicators of grazing pressure in Naryn Oblast. *J. Mt. Sci.* **2016**, *13*, 1567–1583. [CrossRef]
93. Wen, L.; Dong, S.; Li, Y.; Li, X.; Shi, J.; Wang, Y.; Liu, D.; Ma, Y. Effect of degradation intensity on grassland ecosystem services in the alpine region of Qinghai-Tibetan Plateau, China. *PLoS ONE* **2013**, *8*, e58432. [CrossRef]
94. Pielke, R.A.; Avissar, R.; Raupach, M.; Dolman, A.J.; Zeng, X.; Denning, A.S. Interactions between the atmosphere and terrestrial ecosystems: Influence on weather and climate. *Glob. Chang. Biol.* **1998**, *4*, 461–475. [CrossRef]
95. Pielke, R.A., Sr.; Adegoke, J.; BeltraáN-Przekurat, A.; Hiemstra, C.A.; Lin, J.; Nair, U.S.; Niyogi, D.; Nobis, T.E. An overview of regional land-use and land-cover impacts on rainfall. *Tellus B Chem. Phys. Meteorol.* **2007**, *59*, 587–601. [CrossRef]
96. Foley, J.A.; Costa, M.H.; Delire, C.; Ramankutty, N.; Snyder, P. Green surprise? How terrestrial ecosystems could affect earth's climate. *Front. Ecol. Environ.* **2003**, *1*, 38–44.
97. Feng, Q.; Ma, H.; Jiang, X.; Wang, X.; Cao, S. What has caused desertification in China? *Sci. Rep.* **2015**, *5*, 15998. [CrossRef]
98. Bakker, E.S.; Ritchie, M.E.; Olff, H.; Milchunas, D.G.; Knops, J.M. Herbivore impact on grassland plant diversity depends on habitat productivity and herbivore size. *Ecol. Lett.* **2006**, *9*, 780–788. [CrossRef]
99. Altesor, A.; Piñeiro, G.; Lezama, F.; Jackson, R.B.; Sarasola, M.; Paruelo, J.M. Ecosystem changes associated with grazing in subhumid South American grasslands. *J. Veg. Sci.* **2006**, *17*, 323–332. [CrossRef]
100. Lezama, F.; Baeza, S.; Altesor, A.; Cesa, A.; Chaneton, E.J.; Paruelo, J.M. Variation of grazing-induced vegetation changes across a large-scale productivity gradient. *J. Veg. Sci.* **2014**, *25*, 8–21. [CrossRef]
101. Eldridge, D.J.; Poore, A.G.; Ruiz-Colmenero, M.; Letnic, M.; Soliveres, S. Ecosystem structure, function, and composition in rangelands are negatively affected by livestock grazing. *Ecol. Appl.* **2016**, *26*, 1273–1283. [CrossRef]
102. Cingolani, A.M.; Noy-Meir, I.; Díaz, S. Grazing effects on rangeland diversity: A synthesis of contemporary models. *Ecol. Appl.* **2005**, *15*, 757–773. [CrossRef]
103. Milchunas, D.G.; Sala, O.E.; Lauenroth, W. A generalized model of the effects of grazing by large herbivores on grassland community structure. *Am. Nat.* **1988**, *132*, 87–106. [CrossRef]
104. Deng, L.; Shangguan, Z.P. Species composition, richness and aboveground biomass of natural grassland in Hilly-Gully Regions of the Loess Plateau. *China J. Integr. Agric.* **2014**, *13*, 2527–2536. [CrossRef]
105. Soriano, A.; León, R.J.C.; Sala, O.E.; Lavado, R.S.; Deregibus, V.A.; Cahuepé, M.A.; Scaglia, O.A.; Velazquez, C.A.; Lemcoff, J.H. Río de la Plata grasslands. In *Ecosystems of the World 8A. Natural Grasslands. Introduction and Western Hemisphere*; Coupland, R.T., Ed.; Elsevier: New York, NY, USA, 1992; pp. 367–407.
106. Overbeck, G.E.; Müller, S.C.; Fidelis, A.; Pfadenhauer, J.; Pillar, V.D.; Blanco, C.C.; Boldrini, I.I.; Both, R.; Forneck, E.D. Brazil's neglected biome: The South Brazilian Campos. *Perspect. Plant Ecol. Evol. Syst.* **2007**, *9*, 101–116. [CrossRef]
107. Bilenca, D.; Miñarro, F. *Identificación de Áreas Valiosas de Pastizal (AVPs) en las Pampas y Cam-pos de Argentina, Uruguay y sur de Brasil*; Fundación Vida Silvestre Argentina: Buenos Aires, Argentina, 2004; p. 352.
108. Fonseca, C.R.; Guadagnin, D.L.; Emer, C.; Masciadri, S.; Germain, P.; Zalba, S.M. Invasive alien plants in the Pampas grasslands: A tri-national cooperation challenge. *Biol. Invasions* **2013**, *15*, 1751–1763. [CrossRef]
109. Guido, A.; Varela, R.D.; Baldassini, P.; Paruelo, J. Spatial and temporal variability in aboveground net primary production of Uruguayan grasslands. *Rangel. Ecol. Manag.* **2014**, *67*, 30–38. [CrossRef]

110. Medeiros, R.B.; Pillar, V.D.; Reis, J.C.L. Expansão de Eragrostis plana Ness (capim-annoni-2), no Rio Grande do Sul e indicativos de controle. In *Reunión del grupo técnico regional del Cono Sur en mejoramiento y utilización de los recursos forrajeros del área tropical y subtropical, Grupo Campos, 20. Salto, Uruguay, 2004*; Regional Norte de la Universidad de la República: Salto, Uruguay, 2004; pp. 208–211.
111. Jaurena, M.; Bentancur, O.; Ayala, W.; Rivas, M. Especies indicadoras y estructura de praderas naturales de basalto con cargas contrastantes de ovinos. *Agrociencia Urug.* **2011**, *15*, 103–114.
112. Sala, O.E.; Oesterheld, M.; León, R.J.C.; Soriano, A. Grazing effects upon plant community structure in subhumid grasslands of Argentina. *Vegetatio* **1986**, *67*, 27–32.
113. Focht, T.; Medeiros, R.B.D. Prevention of natural grassland invasion by Eragrostis plana Nees using ecological management practices. *Rev. Bras. Zootec.* **2012**, *41*, 1816–1823. [CrossRef]
114. Lin, L.; Li, Y.K.; Xu, X.L.; Zhang, F.W.; Du, Y.G.; Liu, S.L.; Gou, X.W.; Cao, G.M. Predicting parameters of degradation succession processes of Tibetan Kobresia grasslands. *Solid Earth* **2015**, *6*, 1237–1246. [CrossRef]
115. Liu, B.; You, G.; Li, R.; Shen, W.; Yue, Y.; Lin, N. Spectral characteristics of alpine grassland and their changes responding to grassland degradation on the Tibetan Plateau. *Environ. Earth Sci.* **2015**, *74*, 2115–2123. [CrossRef]
116. Geerken, R.; Ilaiwi, M. Assessment of rangeland degradation and development of a strategy for rehabilitation. *Remote Sens. Environ.* **2004**, *90*, 490–504. [CrossRef]
117. Hao, C.; Wu, S. The effects of land-use types and conversions on desertification in Mu Us Sandy Land of China. *J. Geogr. Sci.* **2006**, *16*, 57–68. [CrossRef]
118. Baldi, G.; Paruelo, J.M. Land-use and land cover dynamics in South American temperate grasslands. *Ecol. Soc.* **2008**, *13*. [CrossRef]
119. Vega, E.; Baldi, G.; Jobbagy, E.G.; Paruelo, J. Land use change patterns in the Río de la Plata grasslands: The influence of phytogeographic and political boundaries. *Agric. Ecosyst. Environ.* **2009**, *134*, 287–292. [CrossRef]
120. Mugerwa, S.; Emmanuel, Z. Drivers of grassland ecosystems' deterioration in Uganda. *Appl. Sci. Rep.* **2014**, *2*, 103–111.
121. Li, C.; de Jong, R.; Schmid, B.; Wulf, H.; Schaepman, M.E. Spatial variation of human influences on grassland biomass on the Qinghai-Tibetan plateau. *Sci. Total Environ.* **2019**, *665*, 678–689. [CrossRef] [PubMed]
122. McGranahan, D.A.; Engle, D.M.; Wilsey, B.J.; Fuhlendorf, S.D.; Miller, J.R.; Debinski, D.M. Grazing and an invasive grass confound spatial pattern of exotic and native grassland plant species richness. *Basic Appl. Ecol.* **2012**, *13*, 654–662. [CrossRef]
123. Liu, J.; Xu, X.; Shao, Q. Grassland degradation in the "Three-River Headwaters" region, Qinghai Province. *J. Geogr. Sci.* **2008**, *18*, 259–273. [CrossRef]
124. Zhao, Y.; He, C.; Zhang, Q. Monitoring vegetation dynamics by coupling linear trend analysis with change vector analysis: A case study in the Xilingol steppe in northern China. *Int. J. Remote Sens.* **2012**, *33*, 287–308. [CrossRef]
125. Ho, P.; Azadi, H. Rangeland degradation in North China: Perceptions of pastoralists. *Environ. Res.* **2010**, *110*, 302–307.
126. Miehe, S.; Kluge, J.; Von Wehrden, H.; Retzer, V. Long-term degradation of Sahelian rangeland detected by 27 years of field study in Senegal. *J. Appl. Ecol.* **2010**, *47*, 692–700. [CrossRef]
127. Guido, A.; Vélez-Martin, E.; Overbeck, G.E.; Pillar, V.D. Landscape structure and climate affect plant invasion in subtropical grasslands. *Appl. Veg. Sci.* **2016**, *19*, 600–610. [CrossRef]
128. Semmartin, M.; Di Bella, C.; de Salamone, I.G. Grazing-induced changes in plant species composition affect plant and soil properties of grassland mesocosms. *Plant Soil* **2010**, *328*, 471–481. [CrossRef]
129. Pickup, G. Estimating the effects of land degradation and rainfall variation on productivity in rangelands: An approach using remote sensing and models of grazing and herbage dynamics. *J. Appl. Ecol.* **1996**, *33*, 819–832. [CrossRef]
130. Snyman, H.A. Soil seed bank evaluation and seedling establishment along a degradation gradient in a semi-arid rangeland. *Afr. J. Range Forage Sci.* **2004**, *21*, 37–47. [CrossRef]
131. Vieira, M.D.S.; Bonilha, C.L.; Boldrini, I.I.; Overbeck, G.E. The seed bank of subtropical grasslands with contrasting land-use history in southern Brazil. *Acta Bot. Braz.* **2015**, *29*, 543–552. [CrossRef]

132. Andrade, B.O.; Koch, C.; Boldrini, I.I.; Vélez-Martin, E.; Hasenack, H.; Hermann, J.M.; Kollman, J.; Pillar, V.D.; Overbeck, G.E. Grassland degradation and restoration: A conceptual framework of stages and thresholds illustrated by southern Brazilian grasslands. *Nat. Conserv.* **2015**, *13*, 95–104. [CrossRef]
133. Chaneton, E.J.; Perelman, S.B.; Omacini, M.; Leon, R.J.C. Grazing, environmental heterogeneity, and alien plant invasions in temperate Pampa grasslands. *Biol. Invasions* **2002**, *4*, 7–24. [CrossRef]
134. Altesor, A.; López-Marsico, L.; Paruelo, J. (Eds.) *Bases Ecológicas y Tecnológicas Para el Manejo de Pastizales II*; Serie FPTA-INIA (69); INIA: Montevideo, Uruguay, 2019; p. 167.
135. Porensky, L.M.; Mueller, K.E.; Augustine, D.J.; Derner, J.D. Thresholds and gradients in a semi-arid grassland: Long-term grazing treatments induce slow, continuous and reversible vegetation change. *J. Appl. Ecol.* **2016**, *53*, 1013–1022. [CrossRef]
136. Oesterheld, M.; Semmartin, M. Impact of grazing on species composition: Adding complexity to a generalized model. *Austral Ecol.* **2011**, *36*, 881–890. [CrossRef]
137. Wang, Z.; Zhang, Y.; Yang, Y.; Zhou, W.; Gang, C.; Zhang, Y.; Li, J.; An, R.; Wang, K.; Odeh, I.; et al. Quantitative assess the driving forces on the grassland degradation in the Qinghai–Tibet Plateau, in China. *Ecol. Inform.* **2016**, *33*, 32–44. [CrossRef]
138. Davidowitz, G. Does precipitation variability increase from mesic to xeric biomes? *Glob. Ecol. Biogeogr.* **2002**, *11*, 143–154. [CrossRef]
139. Whitford, W.G. Ecology of Desert Systems. *J. Mammal.* **2002**, *84*, 1122–1124.
140. Van Etten, E.J. Inter-annual rainfall variability of arid Australia: Greater than elsewhere? *Aust. Geogr.* **2009**, *40*, 109–120. [CrossRef]
141. Le Houérou, H.N. A probabilistic approach to assessing arid rangelands' productivity, carrying capacity and stocking rates. In *Drylands: Sustainable Use of Rangelands into the Twenty First Century*; Squires, V.R., Sidahmed, A.E., Eds.; IFAD: Rome, Italy, 1998; pp. 159–172.
142. Wiegand, T.; Snyman, H.A.; Kellner, K.; Paruelo, J.M. Do grasslands have a memory: Modeling phytomass production of a semiarid South African grassland. *Ecosystems* **2004**, *7*, 243–258. [CrossRef]
143. Sloat, L.L.; Gerber, J.S.; Samberg, L.H.; Smith, W.K.; Herrero, M.; Ferreira, L.G.; Godde, C.M.; West, P.C. Increasing importance of precipitation variability on global livestock grazing lands. *Nat. Clim. Chang.* **2018**, *8*, 214. [CrossRef]
144. Ellis, J.E.; Swift, D.M. Stability of African pastoral ecosystems: Alternate paradigms and implications for development. *Rangel. Ecol. Manag. J. Range Manag. Arch.* **1998**, *41*, 450–459. [CrossRef]
145. Briske, D.D.; Zhao, M.; Han, G.; Xiu, C.; Kemp, D.R.; Willms, W.; Havstad, K.; Kang, L.; Wang, Z.; Wu, J.; et al. Strategies to alleviate poverty and grassland degradation in Inner Mongolia: Intensification vs. production efficiency of livestock systems. *J. Environ. Manag.* **2015**, *152*, 177–182. [CrossRef]

© 2019 by the authors. Licensee MDPI, Basel, Switzerland. This article is an open access article distributed under the terms and conditions of the Creative Commons Attribution (CC BY) license (http://creativecommons.org/licenses/by/4.0/).

Review

Tallgrass Prairie Responses to Management Practices and Disturbances: A Review

Pradeep Wagle * and Prasanna H. Gowda

USDA, Agricultural Research Service, Grazinglands Research Laboratory, El Reno, OK 73036, USA; prasanna.gowda@ars.usda.gov
* Correspondence: pradeep.wagle@ars.usda.gov; Tel.: +1-405-262-5291; Fax: +1-405-262-0133

Received: 1 November 2018; Accepted: 9 December 2018; Published: 12 December 2018

Abstract: Adoption of better management practices is crucial to lessen the impact of anthropogenic disturbances on tallgrass prairie systems that contribute heavily for livestock production in several states of the United States. This article reviews the impacts of different common management practices and disturbances (e.g., fertilization, grazing, burning) and tallgrass prairie restoration on plant growth and development, plant species composition, water and nutrient cycles, and microbial activities in tallgrass prairie. Although nitrogen (N) fertilization increases aboveground productivity of prairie systems, several factors greatly influence the range of stimulation across sites. For example, response to N fertilization was more evident on frequently or annually burnt sites (N limiting) than infrequently burnt and unburnt sites (light limiting). Frequent burning increased density of C_4 grasses and decreased plant species richness and diversity, while plant diversity was maximized under infrequent burning and grazing. Grazing increased diversity and richness of native plant species by reducing aboveground biomass of dominant grasses and increasing light availability for other species. Restored prairies showed lower levels of species richness and soil quality compared to native remnants. Infrequent burning, regular grazing, and additional inputs can promote species richness and soil quality in restored prairies. However, this literature review indicated that all prairie systems might not show similar responses to treatments as the response might be influenced by another treatment, timing of treatments, and duration of treatments (i.e., short-term vs. long-term). Thus, it is necessary to examine the long-term responses of tallgrass prairie systems to main and interacting effects of combination of management practices under diverse plant community and climatic conditions for a holistic assessment.

Keywords: biomass; fertilization; grazing; burning; restoration

1. Introduction

Tallgrass prairie, North America's most endangered ecosystem, had occupied more than 68×10^6 ha of the North American Great Plains before European settlement and the acreage of native prairie has declined as high as 99.9% by now [1]. Tallgrass prairie grasslands still contribute heavily for livestock production in several states of the United States. These grasslands can range from unmanaged low productive systems to highly managed high productive systems. Overall, grasslands are usually managed less effectively than croplands. In addition, tallgrass prairie grasslands frequently experience several disturbances such as grazing, fire, and drought. Thus, understanding the consequences of different management practices and disturbances on the plant community composition and production, nutrient cycling, and microbial activities in tallgrass prairie is of great importance for both economic and conservation purposes. However, the response of tallgrass prairie systems to management practices and disturbances is complicated as the prairie system consists of both C_3 and C_4 grasses and C_3 forbs, and these species behave differently due to significant differences in phenology, root morphology, and mycorrhizal dependence. Variability in the fraction of C_3 and C_4 species are correlated with several environmental and management factors [2,3]. Changes in plant growth and plant community composition due to different management practices and disturbances can have profound impacts on both quantity and quality of biomass (hay or forage) production, quantity and quality of organic substrates [4], and ultimately on soil microbes and their activities as soil microbes play a key role in cycling and storage of soil carbon (C) and nitrogen (N) via decomposition of root and litter inputs.

Although previous studies have explored the impacts of management practices and disturbances on responses of tallgrass prairie systems, the findings are inconsistent for different intensities, frequencies, climatic conditions, and timing of disturbances (e.g., grazing and burning) and management practices [5–7]. For examples, burning reduced microbial C during dry years but increased in normal or wet years [8]; application of fertilizers was more effective in wet years than dry years [9]; N fertilization was more effective to increase production on frequently or annually burnt sites compared to infrequently burnt and unburnt sites [10]. In addition, the focus of most past studies has been on a single treatment (e.g., burning or grazing) as the response of vegetation to a treatment might be affected by another treatment. For example, fire favors C_4 growth dominance, while fire and grazing together maximize diversity in tallgrass prairie [11]. In addition, as most studies are short-term (2–3 years), long-term impacts of treatments might change which remain poorly understood. For example, large shifts in plant compositional changes (shift towards C_4 grass dominance) occurred within three years of prairie establishment and became much smaller after three years, most likely due to competition of soil resources [12]. Likewise, the abundance of C_3 grasses decreased as the study progressed in N fertilized prairies [13]. Thus, it is necessary to assess long-term potential consequences of fertilization, grazing or removal of biomass by harvesting, and burning on forage production, microbial biomass and activity, and nutrient cycling in tallgrass prairie.

There have been increasing efforts including the Conservation Reserve Program (CRP) from government and private organizations to restore diverse prairie in agricultural sites with the goal of reestablishment of native plant species [14,15]. Some previous studies have reported that restored prairies little resembled the species richness of native remnants even after several decades of restoration [16]. However, changes in plant species composition and ecosystem properties in restored tallgrass prairies are not well understood [12]. The major goal of this article is to review the independent and interactive impacts of different common management practices and disturbances and tallgrass prairie restoration on plant species composition, plant growth and development, water and nutrient cycles, and microbial activities in tallgrass prairie systems. However, the response of tallgrass prairies to management practices and disturbances may show inconsistent results across space and time due to different functional traits (e.g., different mixture/proportion of C_3 and C_4 species, and root structures) and different geographical distributions (e.g., lowland prairie with deeper and moister soil and upland prairie with shallow and drier soil). By reviewing a wide range of research from different parts of the tallgrass prairie region, this article summarizes key and consistent responses, highlight inconsistent responses of tallgrass prairie to major management practices, and point out knowledge gaps. Having a thorough understanding of the impacts of major management practices and disturbances in different settings of climate, functional traits, and time periods can offer better insights into developing and adopting sustainable management practices and better predicting forage quality and quantity for tallgrass prairie systems.

2. Impacts of Management Practices and Disturbances

2.1. Fertilization

2.1.1. Biomass Production

It is well known that aboveground biomass is increased by fertilization, particularly N [17,18]. For example, aboveground biomass was increased by 24–44% with moderate (84 kg N ha^{-1} yr^{-1}) N fertilization in a prairie site in Iowa [13]. However, previous studies have shown that the response of tallgrass prairie vegetation to N fertilization is influenced by other management practices or disturbance history. For example, N fertilization (100 kg N ha^{-1} yr^{-1}) increased aboveground biomass production by approximately 40% in more than three years older prairies and by >50% in 1–3 year old reconstructed prairies in southern Minnesota [12]. Fertilization (100 kg N ha^{-1} yr^{-1}) of a native tallgrass prairie in the Flint Hills regions of Kansas increased aboveground biomass by 57% in annually burnt sites but only by 15% in unburnt sites [19]. Similarly, N fertilization increased foliage biomass by an average of 9% on unburnt sites for over 15 years, by 68% on annually burnt sites, and by 45% on infrequently burnt sites in Konza Prairie, Kansas [10]. In addition, the response of biomass to N fertilization depends on availability of soil moisture. For example, N and phosphorus (P) fertilizers in a mixed prairie range site in Eastern Montana increased biomass yields by 32% in a dry year, but by up to 218% in a wet year [9]. Root biomass is generally higher in unfertilized prairie [20].

Some greenhouse-based studies showed that cool-season C_3 grasses (highly fibrous root systems) did not respond to mycorrhizae or P fertilization, but warm-season prairie C_4 grasses (coarser root systems) benefited significantly from mycorrhizae or P fertilization [21,22]. However, P responses are rarely observed in tallgrass prairie under natural system because of limitation of N and water [23]. The greenhouse experiments cannot be equated with natural situations. Another study in mixed prairie in eastern Montana also showed that responses to P occurred only when N was not limiting [9].

2.1.2. Plant Diversity

In general, N fertilization reduces plant diversity [18] by favoring plants species that are better competitors for light resource than for soil resources [24]. Species richness of N fertilized tallgrass prairie decreased significantly in Kansas [25]. Forb species showed stronger production response to N enrichment than did grasses [10]. Another study also found that forb species had the strongest response to N fertilization followed by C_3 grasses and C_4 grasses, but growth rates of legumes were not consistently stimulated when grown in monocultures in nutrient-limited soil [17]. As a result, fertilized prairies were characterized by forbs and C_3 grasses, whereas unfertilized prairies were characterized by C_4 grasses and legumes [13], but the abundance of C_3 grasses decreased as the study progressed.

Although the majority of studies have reported that N fertilization reduces prairie species diversity, early spring N application when used with a post-senescence annual harvest increased prairie diversity in August, while unfertilized prairie had higher diversity in June in Iowa [13]. This inconsistency on the influence of N fertilizer on species and functional group diversity was caused by the timing of the N fertilizer application as it was applied as a single dose of N in early spring while most studies apply N either in single or multiple doses during the growing season. Spring N application can stimulate growth of C_3 species and reduce the growth of C_4 grasses, which favors the abundance of forbs later in the growing season due to availability of more light resources [13]. These results highlight the impact of the timing of fertilization on plant diversity.

2.1.3. Soil Chemistry and Microbial Activities

Additions of N can promote aboveground plant biomass, resulting in greater C inputs [26]. Increased C inputs to the soil stimulates the abundance and activity of microbes. Thus, management practices that minimize the addition of C inputs in soils can reduce microbial activities. Consequently, enzyme activity, microbial biomass, and total soil C and N were greater in fertilized prairies than unfertilized prairies [27]. However, in a long-term (1986–1994) burnt tallgrass prairie in Kansas, N fertilization reduced microbial C and N due to increase in acid phosphatase activity (responsible for releasing P) and decrease in urease activity (negative relationship with inorganic N) [28]. Another study reported that short-term N fertilization to a previously unfertilized soil had little or no effect on microbial C and N, but long-term N fertilization increased microbial C and N by >60% in previously unfertilized soil than in previously fertilized soil in a grassland system [29]. Relatively larger fluctuations in microbial biomass N, but not in microbial C was reported in fertilized prairies [27,30], indicating that microbes use added N without simultaneous increase in C.

Seasonality of microbial biomass associated with prairie phenology should also be considered. Microbial biomass showed seasonality in N fertilized tallgrass prairie (higher in early spring and late-summer/early fall, and lower between March and July with the initiation of plant growth) [8]. Decrease in Microbial N coincided with plant N uptake with the initiation of plant growth in tallgrass prairie. As a result, inorganic soil N concentration was lowest in August due to greater plant uptake even though a high amount of N (100 kg N ha^{-1}) was applied after June [28]. However, the concentration of soil N increased from August to October because of mineralized N from plant tissues returned to soil during vegetation senescence. Microbial activity in N fertilized prairie was dominated by bacteria (lower C:N ratio) than fungi [8]. In addition, N enrichment significantly altered bacterial community diversity in Konza prairie, Kansas [31].

2.2. Burning/Fire

Burning of tallgrass prairie is a commonly recognized disturbance regime and it has become a necessary and an important management tool for prairie systems to maintain dominant C_4 grass species, to stimulate plant production, and to improve quality of prairie grasslands. Burning of tallgrass prairie has a profound influence on plant physiological status and growth [32].

2.2.1. Plant Growth and Production

In general, burning of tallgrass prairie increases above- and belowground plant productivity [33]. Enhanced plant growth after burning can be attributed to earlier vegetation green-up due to greater soil heating, greater N availability due to increased N mineralization from soil organic matter (SOM) as a consequence of elevated soil temperature, increased solar radiation at the soil surface due to removal of the litter layer, changes in site biophysical properties, and plant physiological responses [34–36]. Soil warms more rapidly and remains warmer throughout or most of the growing season in burnt areas. Removal of dead vegetation by burning resulted in earlier greening-up and more rapid growth of warm-season C_4 grasses in the early growing season than at unburnt native prairie sites [37]. Burning also increased stem tiller density in tallgrass prairie [38]. Reduction in light levels by litter reduced plant growth in unburnt prairie [39]. Higher forage yields of burnt prairies than ungrazed, unmowed, and unburnt prairies [40], and similar forage yields of mowed or dead vegetation removed plots as compared to burnt plots [41,42] indicate that removal of old or dead vegetation is an important factor for higher yields in tallgrass prairie. Root growth was increased by 25% at the annual burnt site compared to the unburnt (for four years) site at Konza Prairie, Kansas as plants compensate for N limitation in annually burnt sites by increasing allocation of roots [43]. Infrequent fire at certain times of the year tends to increase plant growth and palatability of forages [44].

2.2.2. Species Composition and Nutrient Cycling

Absence of burning in tallgrass prairie systems for a long period may increase invasion of woody species and alter plant species composition [45,46]. Accumulation of detritus in absence of fire reduces available light energy, changes microclimate and plant physiological processes [34], and results in decline in species richness [47]. Warm-season C_4 grasses were favored in burnt tallgrass prairie plots at the expense of woody species and forbs [25].

Nutrients bound in litter are released as ash on the soil surface by burning. Because of the conducive conditions for free-living N fixers under burning and the ability of the dominant perennial C_4 grasses to translocate much of the N from aboveground vegetation to root systems prior to senescence [48], little N is lost by dormant-season fires. However, C and nutrient balance of grasslands are decreased by removing dead aboveground biomass in fires [34], resulting in low C and N inputs into the soil. Thus, frequent spring fires reduced mineralization rates and availability of N in burnt than unburnt prairie [49]. Moreover, plant N demand increases because of greater N uptake by greater plant biomass in burnt sites [28]. This N deficient condition allows a competitive advantage to shift towards C_4 species with high nitrogen use efficiency (NUE), while build-up of a thick surface litter layer in the absence of fire slows down the decomposition of SOM and increases organic C and N, and consequently, favoring C_3 species [38]. Other studies also reported that high light and low N environments caused by burning favored the growth of dominant C_4 grasses in native tallgrass prairie [50,51].

Time of burning also alters species composition. Winter and spring burning increased species richness in ungrazed prairie in Kansas [52], while late spring burning decreased forbs [38]. All burn regimes increased big bluestem, spring burning increased Indian grass, and autumn and spring burning increased forbs. However, total biomass production did not show significant differences among autumn, winter, or spring burns [53]. The dominance of C_4 grasses was increased and the abundance of cool-season species was reduced in Konza prairie, Kansans by frequent spring burning [40], thereby reducing species richness [11]. However, species richness may increase under occasional spring burning [54], likely by opening spaces for seedling establishment. Summer burning increased the abundance of cool-season C_3 species and reduced C_4 grasses, and fall and spring burning increased forb density in a northern mixed prairie site, USA [55].

Frequency of fire is also important for regulating species composition, soil moisture, and nutrient cycling. Infrequent fire increased woody vegetation [56], while frequent fire increased C_4 grasses and decreased C_3 grasses and forbs and overall plant species richness and diversity. Another study also showed that the species richness of forbs was reduced by increased burn frequency in a tallgrass prairie at Konza, Kansas [57]. The frequency of burning affects nutrient (C and N) cycling differently due to differences in input of litter or pyrogenic organic matter (py-OM) [58]. Decomposition of aboveground litter is a major process of SOM accumulation. Decomposition of litter releases C and N in mineralized forms, while the py-OM remained unused and intact by soil decomposers for a long time [58]. Although substitution of litter inputs with more recalcitrant plant inputs such as py-OM by burning might contribute to C and N sequestration in the soil, input of py-OM which is resistant to microbial degradation and lack of fresh litter for microbial decomposition infers N limitation [58]. Consequently, regular burning of tallgrass prairie inhibits C and N cycling in soil. Over the short-term, fire in the tallgrass prairie at the Flint Hills near Manhattan, Kansas enhanced above- and belowground plant biomass, microbial activity, and NUE, but repeated annual burning over a long-term significantly reduced soil organic N, N availability, and microbial biomass, and increased C:N ratios in SOM [59]. Inorganic N increased by 14% after the first-year burning and decreased by 8% after repeated burning in tallgrass prairie [60] due to higher N uptake associated with more plant growth. Unburnt prairie sites had greatest, annually burnt sites had lowest, and infrequent burnt sites had intermediate levels of inorganic soil N and cumulative net N mineralization at Konza Prairie, Kansas [49].

Soil respiration was higher in frequently burnt tallgrass prairie due to greater root respiration and lower N mineralization rates relative to unburnt or infrequently burnt sites at Konza Prairie, Kansas [43]. Similarly, magnitudes and annual estimates of soil C emissions were higher in more productive burnt sites than in unburnt sites at Konza Prairie, Kansas [61]. The NO_3^- concentration in soil solution was higher in unburnt prairie [62]. Consequently, unburnt tallgrass prairie sites had higher denitrification and N_2O fluxes than in burnt sites [63].

2.2.3. Soil Water Availability

Soil water availability is low in burnt sites due to increase in bare soil evaporation and plant transpiration caused by earlier greening up and enhanced vegetation growth [6,41]. As a result, soil water content was significantly reduced by burning at all depths (up to 1.5 m) in bluestem prairie in the Flint Hills, Kansas [64]. In contrast, higher amounts of litter/mulch in unburnt sites intercept much precipitation and increase infiltration of water into soil as well as reduce soil evaporation [39]. Due to differences in soil water availability, the impact of drought on biomass production is higher at burnt sites than unburnt sites [65].

Time of burning also influences soil water availability. Comparison of the effects of time of burning on soil moisture of bluestem prairie in the Flint Hills, Kansas showed that the earliest burning reduced soil moisture the most due to greater runoff and less infiltration, and the reduction was greater in deeper soil layers [6,64]. However, multiple factors associated with burning can alter water use patterns of grasslands [66]. For example, dominance of C_4 grasses due to burning can reduce soil water content in top soil profile as C_4 grasses rely on water from the shallowest soil, but C_3 forbs and shrubs can utilize water from multiple soil layers based on availability of water [67–69].

2.2.4. Microbial Activities

Previous studies have shown that fires can significantly alter enzyme activity, microbial biomass, activity, and community as well. Burning affects soil enzymes, which are essential for SOM decomposition, nutrient cycling, and microbial biomass, which are labile portions of the organic content in soils and serve as source and sink of plant nutrients of tallgrass prairie soils [28]. However, different enzymes respond differently to burning. Long-term burning and N fertilization (1986–1994) of tallgrass prairie in Kansas reduced microbial C and N, but burning alone had little effect on microbial C and N [28]. Long-term burning increased urease and acid phosphatase activities, but decreased deaminase activity (responsible for releasing N from N compounds, contrary to urease enzyme) [28]. Such changes in microbial biomass and the production of enzymes in long-term burning were attributed to the change in rate of organic matter turnover. Long-term annual burning improved active pool of organic N (N availability) due to faster N turnover [70], thereby affecting microbial biomass and the production of enzymes.

Based on a meta-analysis of 42 published papers on microbial responses to fire, a study [71] reported that fire reduced microbial abundance as a whole and fungal abundance by an average of 33.2% and 47.6%, respectively, across sites. Reduction in microbial biomass can be attributed to microbial mortality due to burning of the organic layer and heat-transfer to soil [72], C substrate-limitation [73], and changes in soil moisture and nutrients [74,75]. However, response of microbes to fire varied among biome and fire types. Microbial biomass reduced greatly by wildfires than prescribed fires, and microbial biomass did not decline in grasslands but declined in boreal and temperate forests after fire [71]. This different microbial response among biomes might be related to differences in fire severity. Higher fuel loads can cause more severe fires in forests than in grasslands. In particular, previous studies have reported that high severity fires reduce microbial biomass greatly than low severity fires [76,77]. In addition, no reduction in microbial biomass in grasslands might be related to its adaptation to less severe fire. Despite significant changes in soil environment, plant species composition, and production by burning, there was little response to burning by the soil microbial community [31], indicating that changes in plant communities and soil microbial communities might correlate to each other.

2.3. Grazing or Removal of Biomass by Harvesting

Natural tallgrass prairies are grazed by wild/domestic herbivores. Grazing affects prairie systems in numerous ways as shown below:

2.3.1. Plant Community Composition and Diversity

Grazing by large herbivores can promote plant species diversity and structural heterogeneity in native tallgrass prairies [78,79] due to their preference on dominant C_4 grasses over subdominant C_3 grasses and forbs [50]. The impacts of cattle stocking densities and grazing systems on plant community composition and diversity were assessed for a period of 1992–1997 in tallgrass prairie in the Flint Hills region of Kansas [7]. The study showed that grazing increased diversity and richness of native plant species. However, grazing systems (season-long vs. late-season rest rotation) did not have significant effect on plant diversity. Animal density was a key management to influence plant species diversity and composition as diversity increased along with stocking density. The study reported that the abundance of the dominant perennial tallgrasses decreased and the abundance of the C_4 perennial mid-grasses and short grasses increased with increasing cattle stocking densities, whereas C_3 perennial grasses and perennial forbs changed a little, but annual forbs were more abundant in grazed prairie than in ungrazed prairie. Another study in tallgrass prairie in Oklahoma also reported significantly higher number and cover of annuals on grazed treatments than on ungrazed treatments [11]. The study also reported increase in species richness with increasing disturbance intensity. Some other recent studies have also reported the increased abundance and growth of flowering plant communities [57,80] and forb species [81] in tallgrass prairies due to bison grazing.

Like grazing, removal of biomass from harvesting/mowing also alters aboveground community composition [79]. Clipping of biomass stimulated growth of C_3 species and suppressed growth of C_4 species in tallgrass prairie in Oklahoma [82]. The results indicate that shorter and shaded C_3 grasses can get more light due to more open plant canopy under grazing or clipping. Thus, the overall increase in richness and diversity by grazing, harvesting/mowing, and disturbances can be attributed to increase in light availability for other species due to reduction in aboveground biomass of dominant grasses [11].

2.3.2. Impacts on Soil Properties, Nutrient Cycling, and Microbial Activities

Soil compaction, greater bulk density, and deterioration of soil structure are direct impacts of grazing on soil [83]. Change in shoot-root ratio and resource partitioning [84], and reduction in C and N inputs to soil and substrate limitation to microbes [85] are indirect impacts of grazing/removal of biomass on soils. Heavy grazing of prairies at the Marvin Klemme Range Research Station, Oklahoma reduced the rate of SOM and soil nutrient accumulation due to reduced tallgrasses and litter accumulation [86]. Grazing reduced root growth and induced faster N cycling and increased N availability at Konza Prairie, Kansas [43]. Urination, defecation, and transforming and redistribution of N from recalcitrant plant tissue to more labile forms by grazer animals can also significantly increase soil N availability and cycling in grazed than in ungrazed prairie [43,79]. Enhanced biological activity and decomposition causes higher microbial biomass N in grazed relative to ungrazed prairie [80]. However, grazing increases C limitation as shoots regrow and plants allocate less C to root systems [69], thereby reducing belowground C inputs and microbial growth [87]. A lower soil C:N ratio also induces more rapid N cycling in grazed than in ungrazed prairie [43]. The higher floristic quality and species richness of grazed prairies was correlated to soil fertility traits (total soil N, organic C, and microbial biomass C pool) [80].

Grazing can alter soil moisture and water use dynamics of grasses in several ways. Allocation of more C to regrowth of aboveground biomass after grazing reduces root biomass, thereby limiting shallow water uptake by grasses [43]. Grazing can reduce transpiration due to removal of biomass, resulting in increased soil moisture [88]. In contrast, soil compassion by cattle during grazing can limit infiltration and decrease soil moisture [84].

3. Impacts of Tallgrass Prairie Restorations

3.1. Plant Diversity

Plant functional diversity changed over years in restored tallgrass prairies [12]. Annual and biennials dominated in the first year, perennial native composites dominated in the second year, and there was a significant shift to warm-season C_4 grasses after three years. These findings are consistent with the findings of other studies [16,89] that community composition in restored prairies shifts towards C_4 grass dominance within few years. The cover of C_4 grasses was comparable in a 5-year old restored prairie (81%) relative to an adjacent prairie remnant (79%) [16]. Similarly, the cover of C_4 grasses was 71–92% in a 35-year old restored prairie, while it was 59% in an adjacent remnant prairie [16]. Because of C_4 grass dominance, species richness of restored prairies in Illinois declined by 50% within 15 year of restoration [89], which is consistent with the findings of other studies [25,90] that aboveground plant diversity decreases as restoration continues. Plant-mycorrhizal interactions can also promote C_4 dominance in restored prairies as biomass of C_4 grasses was enhanced by mycorrhizae but not of C_3 grasses or annuals [91]. Another study also showed that arbuscular mycorrhizal inoculum promoted establishment of prairie species in restored tallgrass prairies even though it had no effect on total percent of cover of plants [92].

3.2. Soil Quality

Prairie restorations can improve soil quality over time. Application of 100 kg N ha^{-1} yr^{-1} increased above ground biomass by more than 50% in 1–3 year old restored prairies and approximately by 40% after three years [12]. Belowground productivity, litter mass, and C mineralization rates were increased and N mineralization rate was decreased with the rise of C_4 grasses in restored prairies [12]. Another study also showed that virgin prairie remnant and long-term (21 and 24 years) prairie restorations had significantly greater soil moisture, water holding capacity, organic matter, total C, N, C:N, and microbial biomass, and significantly smaller soil bulk density and smaller levels of poly-β-hydroxybutyrate and phospholipid fatty acid analysis indicators of nutritional stress compared with the agricultural field and recently (7 years) restored prairie [93]. However, some studies have shown that C and N mass did not show significant change over a decade in CRP and restored grasslands [94,95]. Soil C and N on previously cultivated restored prairies was 30–40% lower than that in native prairies even after 30–50 years [86]. The results indicate that restoration prairie sites had intermediate soil quality indicators and microbial community structures compared with the virgin prairie and the agricultural sites even after multiple decades.

3.3. Microbial Community

Cessation of tillage-based agriculture increased the abundance of fungi relative to bacteria most likely due to reduced soil disturbance, and the ratio of fungi to bacteria declined through subsequent succession when tilling ceases [96]. Note that increasing the proportion of fungi relative to bacteria favors C accumulation as fungi are more C efficient and are composed of more recalcitrant C compounds. However, the ratio of fungi to bacteria decreased with soil organic carbon (SOC) in prairie soils [96], which is opposite of what has been reported in agricultural soils.

Comparison of soil microbial communities under recently established (2 years) intensive (located in research stations) and older (at least 10 years) extensive (located in working farms or reserves) corn, switchgrass, and prairie sites shows that soil type was more important than plant community at recently established intensive sites and plant community was more important than soil type at older extensive sites in determining microbial communities [97]. Higher bacterial and fungal biomass under perennial grasses than corn indicated a greater microbial processing potential for retention of carbon, water, and nutrients in perennial grasslands [97].

4. Discussion

4.1. Fertilization

Limitations of N and water are common for tallgrass prairies. Due to these limitations, P responses are rarely observed in tallgrass prairie [23]. Although N fertilization generally increases aboveground biomass and decreases root biomass in tallgrass prairies, the range of stimulation due to fertilization varies greatly across prairie sites, depending on several factors such as nutritional status of the sites, composition of dominant plant species, soil moisture availability, frequency of burning, and other management practices. For example, N fertilization was more effective in wet years than dry years [9]. Similarly, N fertilization was more effective to increase production on frequently or annually burnt sites (N limiting) compared to infrequently burnt and unburnt sites (light limiting) [10]. Overall, results illustrated that N limited sites such as annually burnt sites or newly constructed sites benefitted more by addition of N. Results also suggest that unfertilized prairies should not be burnt frequently (N limiting) to maintain higher forage production.

Although N fertilization increases forage production, it can affect forage quality by altering plant species composition. Application of N fertilizers can bolster the competitive advantage to C_3 grasses and forbs. Consequently, forbs and C_3 grasses dominated fertilized prairies, and C_4 grasses and legumes dominated unfertilized prairies [13]. However, dominance of C_3 grass species declined over time, indicating the necessity of monitoring compositional changes for long-term in N fertilized prairies. In addition, the impact of the timing of fertilization on plant species composition should be considered. Differential production responses of plant species to N fertilization [10] further suggest that long-term production response of each species to N fertilization should be determined rather than examining the response of overall forage production. Because of temporal differences in plant species diversity [13], plant diversity should be monitored throughout the growing season. In addition, interaction effects of other disturbances or management practices and climatic variability on species diversity should be considered.

Inconsistent effects (increase or decrease) of N fertilization on microbial N were observed in tallgrass prairie [8,28], most likely due to differences in composition of microbial population and nutritional status of the sites. Microbial activity shows seasonal fluctuations, and seasonal dynamics varied in fertilized and unfertilized prairies [27]. These results indicate the necessity of investigating seasonality of biotic (i.e., plant growth, plant community composition) and abiotic (i.e., temperature, precipitation) factors that drive seasonal fluctuations in microbial activity to more fully understand the responses of microbial community to N fertilization.

4.2. Burning/Fire

In general, burning increases above- and belowground biomass production, decreases invasion of woody species, increases dominance of C_4 grasses, and reduces soil moisture, soil C, and nutrient balance in tallgrass prairie. However, the effect of burning in tallgrass prairie can vary across sites due to differences in soil types, burning time, duration, frequency, weather conditions (i.e., dry and normal rainfall years), dominant plant species, and several other factors. For example, drought impact was higher at burnt prairie sites than unburnt prairie sites [65] due to reduction in soil water availability at burnt sites. Burning reduced microbial C during dry years, but increased in normal and wet years in tallgrass prairie in Kansas [8]. These results highlight the interacting effects of disturbances/management practices and climatic variability on the responses of tallgrass prairie to burning.

Most of the previous studies have generally investigated a single cause of the effect of burning and compared results from burnt and unburnt sites. A study of the interaction of disturbances in tallgrass prairie in Oklahoma showed that burning reduced species diversity on ungrazed treatments and increased on grazed treatments [11], indicating the important interaction effects of burning and grazing on plant community structure in tallgrass prairie. Similarly, infrequent burning and grazing maximized plant species diversity, while frequently burnt ungrazed sites had lowest diversity [51]. As a result, grass cover was highest in infrequently burnt ungrazed sites and lowest in frequently burnt grazed sites, and abundance of C_3 forbs was maximum in infrequently burnt grazed sites [51]. Annual burning without grazing lowered N mineralization and induced severe N limitation, but combination of annul burning and grazing enhanced faster N cycling through deposition of labile N in urine and dung [98]. Mowing reduced root biomass and root C storage and resulted in shallower root distribution in annually burnt plots but root biomass in unburnt plots was not affected at Konza Prairie, Kansas [99]. In that study, soil N and C concentrations were not significantly influenced by burning alone, but burning and mowing reduced soil C and N concentrations by 20% and 17%, respectively. Similarly, the interaction effect of burning and N fertilization on microbial C and N was highly significant [28]. These results suggest that the interacting effects of other management practices or disturbances (e.g., interacting effect of burning and grazing) should be examined to better understand the response of tallgrass prairie systems to burning as well as to recommend any particular timing and frequency of burning. In addition, the response of burning should be evaluated for long-terms since some responses might not evident in short periods. For example, long-term (>40 years) annual burning reduced microbial C and N, but they were not affected by 1–2 years of burning in tallgrass prairie [100].

4.3. Grazing or Removal of Biomass by Harvesting

In general, grazing or removal of biomass by harvesting decreases the abundance of dominant tallgrasses and increases species richness and diversity due to more light availability for shorter and annual species as shorter and shaded species can get more light under more open plant canopy. Consequently, species diversity increased along with stocking density [7] or disturbance intensity [11]. Removal of C inputs to soil by grazing or removal of biomass greatly influences soil characteristics, nutrient cycling, and microbial activities in prairie systems. Due to changes in light energy and nutrient availability, microclimate, and plant physiological status and growth, frequent burning and absence of burning reduce species richness and diversity of prairie, while infrequent burning and grazing maximize species richness. Grazing can modulate the effect of frequent fire on plant diversity by reducing C_4 grass dominance and increasing species diversity and richness [51]. Different fire and grazing regimes can cause heterogeneous landscape to maximize plant diversity [57]. Thus, occurrence of fire and grazing together can create highly productive, diverse, and heterogeneous ecosystems [50,51]. However, timing of burning and grazing should be considered since they influence species diversity and richness.

4.4. Tallgrass Prairie Restorations

Due to a significant community composition shifts towards C_4 grass dominance within 3–5 years of prairie restorations [12], studies have shown lower levels of species richness in restored prairies compared to native remnants [16,90]. A potential mechanism for the shift towards C_4 grass dominance can be competition of soil resources as restored prairie sites are limited by N and have been degraded by agricultural practices over a century [12]. The C_4 grass species can tolerate lower levels of N compared to early successional species [101]. In addition, reduction in light availability for shorter C_3 species with the dominance of taller C_4 grasses can be another possible reason for lower levels of species richness in restored prairies. However, it is not fully understood why C_4 grass dominance is higher in restored prairies than in remnant parries. In our opinion, still poor soil quality indicators and microbial community structures in restored prairies relative to virgin prairie should have been favored the dominance of C_4 grasses.

Studies have shown that restorations of prairie improve soil quality. As a result, biomass production had greater response to N in newly reconstructed prairies (<3 years old) compared to more than three years old prairies [12]. Although prairie restorations increase active pools of C and N rapidly [12], lack of a significant change in total C and N over several decades indicates that restored prairies may require additional inputs to restore SOM, C, and N at the level of the virgin prairie.

5. Summary and Future Directions

Although N fertilization increases aboveground biomass production in tallgrass prairie, the range of stimulation due to N fertilization varies greatly across tallgrass prairie sites due to differences in nutritional status of the sites, composition of dominant plant species, soil moisture availability, frequency of burning, and other management practices. Since N fertilization reduces C_4 grass dominance by bolstering the competitive advantage to other species, judicious use of N fertilization is important to minimize undesirable and invasive species and to maintain healthy native prairie stands.

Higher forage yields at burnt prairies than unburnt prairies, and similar forage yields of mowed or dead vegetation removed sites as compared to burnt sites indicate that removal of old and dead vegetation by burning or mowing is important for higher yields in tallgrass prairie. However, prairie sites should be burnt infrequently to avoid N limitation to maintain higher forage production. Meanwhile, attention should be paid on the invasion of woody species since infrequent fire can increase woody encroachments. Fire-induced changes in nutrient cycling (i.e., N deficient condition under frequent burning) can influence plant species composition by shifting competitive advantage towards C_4 species with high NUE.

Fire and grazing together can create highly productive, diverse, and heterogeneous tallgrass prairie systems due to modulating the effect of frequent fire by grazing on plant diversity by reducing C_4 grass dominance and increasing species diversity and richness. As infrequent fire and grazing often maximize species richness in prairies and many restorations do not include infrequent burning and regular grazing, infrequent burning and regular grazing can be practiced to maximize species richness and diversity in restored prairies for biodiversity conservation, aesthetics, and support of wildlife.

Although prairie restoration improves soil quality over decades, additional inputs are required to restore soil quality indicators and microbial community structures at the level of virgin prairie. The succession process in restored prairies can be enhanced by increasing the amount of mycorrhizae because of the competitive advantage that mycorrhizae bestow on prairie species.

Since microbial biomass and soil enzymes respond immediately to management practices and tend to stabilize later, their temporal changes should be evaluated while evaluating the effect of management practices. In addition, temporal changes can vary between fertilized (more dynamic) and unfertilized prairies.

The responses of C_3/C_4 mixed tallgrass prairie systems to managements are complicated as the responses may vary as the growing season progresses due to shifts in plant species composition. Thus, the effect of management practices on biomass, plant community composition, soil properties, nutrient cycling, and microbial activities should be evaluated over various times (i.e., winter, spring, summer, fall) of the year.

Since interacting effects of multiple management practices and interannual climatic variability can modify the response of tallgrass prairie systems, a thorough understanding of the main and interactive effects of multiple management practices under diverse plant community (different proportion of C_3 and C_4 species), climatic conditions (different geographical distributions), and time periods (short- vs. long-term) is needed for a holistic assessment.

Author Contributions: P.W. wrote the manuscript and P.H.G. revised and contributed for intellectual content.

Funding: This study was partly supported by a research grant (Project No. 2013-69002) through the USDA-NIFA's Agriculture and Food Research Initiative (AFRI). The article processing charge was paid by vouchers issued for P. Wagle for reviewing MDPI journal articles.

Conflicts of Interest: The authors declare no conflicts of interest for this study.

Disclaimer: The U.S. Department of Agriculture (USDA) prohibits discrimination in all its programs and activities on the basis of race, color, national origin, age, disability, and where applicable, sex, marital status, familial status, parental status, religion, sexual orientation, genetic information, political beliefs, reprisal, or because all or part of an individual's income is derived from any public assistance program. (Not all prohibited bases apply to all programs.) Persons with disabilities who require alternative means for communication of program information (Braille, large print, audiotape, etc.) should contact USDA's TARGET Center at (202) 720-2600 (voice and TDD). To file a complaint of discrimination, write to USDA, Director, Office of Civil Rights, 1400 Independence Avenue, S.W., Washington, D.C. 20250-9410, or call (800) 795-3272 (voice) or (202) 720-6382 (TDD). USDA is an equal opportunity provider and employer.

References

1. Samson, F.; Knopf, F. Prairie conservation in north america. *BioScience* **1994**, *44*, 418–421. [CrossRef]
2. Goodin, D.G.; Henebry, G.M. A technique for monitoring ecological disturbance in tallgrass prairie using seasonal ndvi trajectories and a discriminant function mixture model. *Remote. Sens. Environ.* **1997**, *61*, 270–278. [CrossRef]

3. Nelson, D.M.; Hu, F.S.; Tian, J.; Stefanova, I.; Brown, T.A. Response of c3 and c4 plants to middle-holocene climatic variation near the prairie-forest ecotone of minnesota. *Proc. Natl. Acad. Sci. USA* **2003**, *101*, 562–567. [CrossRef] [PubMed]
4. Saleska, S.R.; Shaw, M.R.; Fischer, M.L.; Dunne, J.A.; Still, C.J.; Holman, M.L.; Harte, J. Plant community composition mediates both large transient decline and predicted long-term recovery of soil carbon under climate warming. *Glob. Biogeochem. Cycles* **2002**, *16*, 3:1–3:18. [CrossRef]
5. Hulbert, L.C. Fire effects on tallgrass prairie. In *Proceedings of the Ninth North American Prairie Conference*; Clambey, G.K., Pemble, R.H., Eds.; Tri-College University Center for Environmental Studies: Fargo, ND, USA, 1986; pp. 138–142.
6. Anderson, K.L. Time of burning as it affects soil moisture in an ordinary upland bluestem prairie in the Flint Hills. *J. Range Manag.* **1965**, *18*, 311–316. [CrossRef]
7. Hickman, K.R.; Hartnett, D.C.; Cochran, R.C.; Owensby, C.E. Grazing management effects on plant species diversity in tallgrass prairie. *J. Range Manag.* **2004**, *57*, 58–65. [CrossRef]
8. Garcia, F.O.; Rice, C.W. Microbial biomass dynamics in tallgrass prairie. *Soil Sci. Soc. Am. J.* **1994**, *58*, 816–823. [CrossRef]
9. Wight, J.R.; Black, A. Range fertilization: Plant response and water use. *J. Range Manag.* **1979**, *32*, 345–349. [CrossRef]
10. Seastedt, T.; Briggs, J.; Gibson, D. Controls of nitrogen limitation in tallgrass prairie. *Oecologia* **1991**, *87*, 72–79. [CrossRef]
11. Collins, S.L. Interaction of disturbances in tallgrass prairie: A field experiment. *Ecology* **1987**, *68*, 1243–1250. [CrossRef]
12. Camill, P.; McKone, M.J.; Sturges, S.T.; Severud, W.J.; Ellis, E.; Limmer, J.; Martin, C.B.; Navratil, R.T.; Purdie, A.J.; Sandel, B.S. Community-and ecosystem-level changes in a species-rich tallgrass prairie restoration. *Ecol. Appl.* **2004**, *14*, 1680–1694. [CrossRef]
13. Jarchow, M.E.; Liebman, M. Nitrogen fertilization increases diversity and productivity of prairie communities used for bioenergy. *GCB Bioenergy* **2012**, *5*, 281–289. [CrossRef]
14. Mlot, C. Restoring the prairie. *BioScience* **1990**, *40*, 804–809. [CrossRef]
15. Skold, M.D. Cropland retirement policies and their effects on land use in the Great Plains. *J. Prod. Agric.* **1989**, *2*, 197–201. [CrossRef]
16. Kindscher, K.; Tieszen, L.L. Floristic and soil organic matter changes after five and thirty-five years of native tallgrass prairie restoration. *Restor. Ecol.* **1998**, *6*, 181–196. [CrossRef]
17. Reich, P.B.; Buschena, C.; Tjoelker, M.G.; Wrage, K.; Knops, J.; Tilman, D.; Machado, J.L. Variation in growth rate and ecophysiology among 34 grassland and savanna species under contrasting n supply: A test of functional group differences. *New Phytol.* **2003**, *157*, 617–631. [CrossRef]
18. Suding, K.N.; Collins, S.L.; Gough, L.; Clark, C.; Cleland, E.E.; Gross, K.L.; Milchunas, D.G.; Pennings, S. Functional-and abundance-based mechanisms explain diversity loss due to N fertilization. *Proc. Natl. Acad. Sci. USA* **2005**, *102*, 4387–4392. [CrossRef] [PubMed]
19. Turner, C.L.; Blair, J.M.; Schartz, R.J.; Neel, J.C. Soil n and plant responses to fire, topography, and supplemental n in tallgrass prairie. *Ecology* **1997**, *78*, 1832–1843. [CrossRef]
20. Dietzel, R. A Comparison of Carbon Storage Potential in Corn-and Prairie-Based Agroecosystems. Ph.D. Thesis, Iowa State University, Ames, IA, USA, 2014.
21. Hetrick, B.D.; Wilson, G.; Todd, T. Differential responses of C3 and C4 grasses to mycorrhizal symbiosis, phosphorus fertilization, and soil microorganisms. *Can. J. Bot.* **1990**, *68*, 461–467. [CrossRef]
22. Hetrick, B.A.D.; Kitt, D.G.; Wilson, G.T. The influence of phosphorus fertilization, drought, fungal species, and nonsterile soil on mycorrhizal growth response in tall grass prairie plants. *Can. J. Bot.* **1986**, *64*, 1199–1203. [CrossRef]
23. Owensby, C.E.; Hyde, R.M.; Anderson, K.L. Effects of clipping and supplemental nitrogen and water on loamy upland bluestem range. *J. Range Manag.* **1970**, *23*, 341–346. [CrossRef]
24. Tilman, D. *Plant Strategies and the Dynamics and Structure of Plant Communities*; Princeton University Press: Princeton, NJ, USA, 1988.
25. Gibson, D.J.; Seastedt, T.; Briggs, J.M. Management Practices in Tallgrass Prairie: Large-and Small-Scale. In *Ecosystem Management: Selected Readings*; Springer Science & Business Media: Berlin/Heidelberg, Germany, 1996; p. 106.

26. Bruce, J.P.; Frome, M.; Haites, E.; Janzen, H.; Lal, R.; Paustian, K. Carbon sequestration in soils. *J. Soil Water Conserv.* **1999**, *54*, 382–389.
27. Bach, E.M.; Hofmockel, K.S. Coupled carbon and nitrogen inputs increase microbial biomass and activity in prairie bioenergy systems. *Ecosystems* **2015**, *18*, 417–427. [CrossRef]
28. Ajwa, H.A.; Dell, C.J.; Rice, C.W. Changes in enzyme activities and microbial biomass of tallgrass prairie soil as related to burning and nitrogen fertilization. *Soil Biol. Biochem.* **1999**, *31*, 769–777. [CrossRef]
29. Lovell, R.D.; Jarvis, S.C.; Bardgett, R.D. Soil microbial biomass and activity in long-term grassland: Effects of management changes. *Soil Biol. Biochem.* **1995**, *27*, 969–975. [CrossRef]
30. Baer, S.G.; Blair, J.M. Grassland establishment under varying resource availability: A test of positive and negative feedback. *Ecology* **2008**, *89*, 1859–1871. [CrossRef] [PubMed]
31. Coolon, J.D.; Jones, K.L.; Todd, T.C.; Blair, J.M.; Herman, M.A. Long-term nitrogen amendment alters the diversity and assemblage of soil bacterial communities in tallgrass prairie. *PLoS ONE* **2013**, *8*, e67884. [CrossRef] [PubMed]
32. Knapp, A.K. Effect of fire and drought on the ecophysiology of andropogon gerardii and panicum virgatum in a tallgrass prairie. *Ecology* **1985**, *66*, 1309–1320. [CrossRef]
33. Briggs, J.M.; Knapp, A.K. Interannual variability in primary production in tallgrass prairie: Climate, soil moisture, topographic position, and fire as determinants of aboveground biomass. *Am. J. Bot.* **1995**, *82*, 1024–1030. [CrossRef]
34. Knapp, A.K.; Seastedt, T.R. Detritus accumulation limits productivity of tallgrass prairie. *BioScience* **1986**, *36*, 662–668. [CrossRef]
35. Knapp, A.K. Post-burn differences in solar radiation, leaf temperature and water stress influencing production in a lowland tallgrass prairie. *Am. J. Bot.* **1984**, *71*, 220–227. [CrossRef]
36. Sharrow, S.H.; Wright, H.A. Effects of fire, ash, and litter on soil nitrate, temperature, moisture and tobosagrass production in the rolling plains. *J. Range Manag.* **1977**, *30*, 266–270. [CrossRef]
37. Ehrenreich, J.H.; Aikman, J.M. An ecological study of the effect of certain management practices on native prairie in iowa. *Ecol. Monogr.* **1963**, *33*, 113–130. [CrossRef]
38. Towne, G.; Owensby, C. Long-term effects of annual burning at different dates in ungrazed kansas tallgrass prairie. *J. Range Manag.* **1984**, *37*, 392–397. [CrossRef]
39. Weaver, J.E.; Rowland, N.W. Effects of excessive natural mulch on development, yield, and structure of native grassland. *Bot. Gaz.* **1952**, *114*, 1–19. [CrossRef]
40. Hulbert, L.C. Causes of fire effects in tallgrass prairie. *Ecology* **1988**, *69*, 46–58. [CrossRef]
41. Hulbert, L.C. Fire and litter effects in undisturbed bluestem prairie in kansas. *Ecology* **1969**, *50*, 874–877. [CrossRef]
42. Old, S.M. Microclimate, fire, and plant production in an illinois prairie. *Ecol. Monogr.* **1969**, *39*, 355–384. [CrossRef]
43. Johnson, L.C.; Matchett, J.R. Fire and grazing regulate belowground processes in tallgrass prairie. *Ecology* **2001**, *82*, 3377–3389. [CrossRef]
44. Aldous, A.E. *Effect of Burning on Kansas Bluestem Pastures*; Technical Bulletin 88; Agricultural Experiment Station, Kansas State Agricultural College: Manhattan, KS, USA, 1934; Volume 38, pp. 1–65.
45. Hadley, E.B.; Kieckhefer, B.J. Productivity of two prairie grasses in relation to fire frequency. *Ecology* **1963**, *44*, 389–395. [CrossRef]
46. Bragg, T.B.; Hulbert, L.C. Woody plant invasion of unburned Kansas bluestem prairie. *J. Range Manag.* **1976**, *29*, 19–24. [CrossRef]
47. Leach, M.K.; Givnish, T.J. Ecological determinants of species loss in remnant prairies. *Science* **1996**, *273*, 1555–1558. [CrossRef]
48. Adams, D.E.; Wallace, L.L. Nutrient and biomass allocation in five grass species in an oklahoma tallgrass prairie. *Am. Midl. Nat.* **1985**, *113*, 170–181. [CrossRef]
49. Blair, J.M. Fire, n availability, and plant response in grasslands: A test of the transient maxima hypothesis. *Ecology* **1997**, *78*, 2359–2368. [CrossRef]
50. Veen, G.F.; Blair, J.M.; Smith, M.D.; Collins, S.L. Influence of grazing and fire frequency on small-scale plant community structure and resource variability in native tallgrass prairie. *Oikos* **2008**, *117*, 859–866. [CrossRef]
51. Collins, S.L.; Calabrese, L.B. Effects of fire, grazing and topographic variation on vegetation structure in tallgrass prairie. *J. Veg. Sci.* **2011**, *23*, 563–575. [CrossRef]

52. Towne, E.G.; Kemp, K.E. Vegetation dynamics from annually burning tallgrass prairie in different seasons. *J. Range Manag.* **2003**, *56*, 185–192. [CrossRef]
53. Towne, E.G.; Craine, J.M. Ecological consequences of shifting the timing of burning tallgrass prairie. *PLoS ONE* **2014**, *9*, e103423. [CrossRef]
54. Collins, S.L.; Gibson, D.J. *Effects of Fire on Community Structure in Tallgrass and Mixed-Grass Prairie, Fire in North American Tallgrass Prairies*; University of Oklahoma Press: Norman, OK, USA, 1990; pp. 81–98.
55. Biondini, M.E.; Steuter, A.A.; Grygiel, C.E. Seasonal fire effects on the diversity patterns, spatial distribution and community structure of forbs in the northern mixed prairie, USA. *Vegetatio* **1989**, *85*, 21–31. [CrossRef]
56. Collins, S.L.; Adams, D.E. Succession in grasslands: Thirty-two years of change in a central oklahoma tallgrass prairie. *Vegetatio* **1983**, *51*, 181–190. [CrossRef]
57. Welti, E.A.; Joern, A. Fire and grazing modulate the structure and resistance of plant–floral visitor networks in a tallgrass prairie. *Oecologia* **2018**, *186*, 517–528. [CrossRef] [PubMed]
58. Soong, J.L.; Cotrufo, M.F. Annual burning of a tallgrass prairie inhibits c and n cycling in soil, increasing recalcitrant pyrogenic organic matter storage while reducing n availability. *Glob. Chang. Biol.* **2015**, *21*, 2321–2333. [CrossRef] [PubMed]
59. Ojima, D.S.; Schimel, D.S.; Parton, W.J.; Owensby, C.E. Long- and short-term effects of fire on nitrogen cycling in tallgrass prairie. *Biogeochemistry* **1994**, *24*, 67–84. [CrossRef]
60. Ojima, D.S.; Parton, W.; Schimel, D.; Owensby, C. Simulated impacts of annual burning on prairie ecosystems. In *Fire in North American Tallgrass Prairies*; University of Oklahoma Press: Norman, OK, USA, 1990; pp. 118–132.
61. Knapp, A.K.; Conard, S.L.; Blair, J.M. Determinants of soil co 2 flux from a sub-humid grassland: Effect of fire and fire history. *Ecol. Appl.* **1998**, *8*, 760–770. [CrossRef]
62. Groffman, P.M.; Rice, C.W.; Tiedje, J.M. Denitrification in a tallgrass prairie landscape. *Ecology* **1993**, *74*, 855–862. [CrossRef]
63. Groffman, P.M.; Turner, C.L. Plant productivity and nitrogen gas fluxes in a tallgrass prairie landscape. *Landsc. Ecol.* **1995**, *10*, 255–266. [CrossRef]
64. Anderson, K.L.; Smith, E.F.; Owensby, C.E. Burning bluestem range. *J. Range Manag.* **1970**, *23*, 81–92. [CrossRef]
65. Weaver, J.E. *North American Prairie*; Johnsen Publishing: Lincoln, NE, USA, 1954.
66. O'Keefe, K.; Nippert, J.B. Grazing by bison is a stronger driver of plant ecohydrology in tallgrass prairie than fire history. *Plant Soil* **2016**, *411*, 423–436. [CrossRef]
67. Nippert, J.B.; Knapp, A.K. Linking water uptake with rooting patterns in grassland species. *Oecologia* **2007**, *153*, 261–272. [CrossRef]
68. Kulmatiski, A.; Beard, K.H. Root niche partitioning among grasses, saplings, and trees measured using a tracer technique. *Oecologia* **2012**, *171*, 25–37. [CrossRef]
69. Nippert, J.B.; Wieme, R.A.; Ocheltree, T.W.; Craine, J.M. Root characteristics of c4 grasses limit reliance on deep soil water in tallgrass prairie. *Plant Soil* **2012**, *355*, 385–394. [CrossRef]
70. Hunt, H.; Trlica, M.; Redente, E.; Moore, J.; Detling, J.; Kittel, T.; Walter, D.; Fowler, M.; Klein, D.; Elliott, E. Simulation model for the effects of climate change on temperate grassland ecosystems. *Ecol. Model.* **1991**, *53*, 205–246. [CrossRef]
71. Dooley, S.R.; Treseder, K.K. The effect of fire on microbial biomass: A meta-analysis of field studies. *Biogeochemistry* **2012**, *109*, 49–61. [CrossRef]
72. Cairney, J.W.G.; Bastias, B.A. Influences of fire on forest soil fungal communitiesthis article is one of a selection of papers published in the special forum on towards sustainable forestry—The living soil: Soil biodiversity and ecosystem function. *Can. J. For. Res.* **2007**, *37*, 207–215. [CrossRef]
73. Choromanska, U.; DeLuca, T.H. Prescribed fire alters the impact of wildfire on soil biochemical properties in a ponderosa pine forest. *Soil Sci. Soc. Am. J.* **2001**, *65*, 232–238. [CrossRef]
74. Smith, N.R.; Kishchuk, B.E.; Mohn, W.W. Effects of wildfire and harvest disturbances on forest soil bacterial communities. *Appl. Environ. Microbiol.* **2007**, *74*, 216–224. [CrossRef] [PubMed]
75. Capogna, F.; Persiani, A.M.; Maggi, O.; Dowgiallo, G.; Puppi, G.; Manes, F. Effects of different fire intensities on chemical and biological soil components and related feedbacks on a mediterranean shrub (*phillyrea angustifolia* L.). *Plant Ecol.* **2009**, *204*, 155–171. [CrossRef]

76. Hamman, S.T.; Burke, I.C.; Stromberger, M.E. Relationships between microbial community structure and soil environmental conditions in a recently burned system. *Soil Biol. Biochem.* **2007**, *39*, 1703–1711. [CrossRef]
77. Palese, A.M.; Giovannini, G.; Lucchesi, S.; Dumontet, S.; Perucci, P. Effect of fire on soil C, N and microbial biomass. *Agronomie* **2004**, *24*, 47–53. [CrossRef]
78. Collins, S.L.; Knapp, A.K.; Briggs, J.M. Modulation of diversity by grazing and mowing in native tallgrass prairie. *Science* **1998**, *280*, 745–747. [CrossRef]
79. Knapp, A.K.; Fay, P.A.; Blair, J.M.; Collins, S.L.; Smith, M.D.; Carlisle, J.D.; Harper, C.W.; Danner, B.T.; Lett, M.S.; McCarron, J.K. Rainfall variability, carbon cycling, and plant species diversity in a mesic grassland. *Science* **2002**, *298*, 2202–2205. [CrossRef]
80. Manning, G.C.; Baer, S.G.; Blair, J.M. Effects of grazing and fire frequency on floristic quality and its relationship to indicators of soil quality in tallgrass prairie. *Environ. Manag.* **2017**, *60*, 1062–1075. [CrossRef]
81. Elson, A.; Hartnett, D.C. Bison Increase the Growth and Reproduction of Forbs in Tallgrass Prairie. *Am. Midl. Nat.* **2017**, *178*, 245–259. [CrossRef]
82. Niu, S.; Sherry, R.; Zhou, X.; Wan, S.; Luo, Y. Nitrogen regulation of the climate-carbon feedback: Evidence from a long-term global change experiment. *Ecology* **2010**, *91*, 3261–3273. [CrossRef] [PubMed]
83. Greenwood, K.; McKenzie, B. Grazing effects on soil physical properties and the consequences for pastures: A review. *Aust. J. Exp. Agric.* **2001**, *41*, 1231–1250. [CrossRef]
84. Bardgett, R.D.; Wardle, D.A. *Aboveground-Belowground Linkages: Biotic Interactions, Ecosystem Processes, and Global Change*; Oxford University Press: Oxford, UK, 2010.
85. Wan, S.; Luo, Y. Substrate regulation of soil respiration in a tallgrass prairie: Results of a clipping and shading experiment. *Glob. Biogeochem. Cycles* **2003**, *17*. [CrossRef]
86. Fuhlendorf, S.D.; Zhang, H.; Tunnell, T.R.; Engle, D.M.; Cross, A.F. Effects of grazing on restoration of southern mixed prairie soils. *Restor. Ecol.* **2002**, *10*, 401–407. [CrossRef]
87. Rains, J.R.; Owensby, C.E.; Kemp, K.E. Effects of nitrogen fertilization, burning, and grazing on reserve constituents of big bluestem. *J. Range Manag.* **1975**, *28*, 358–362. [CrossRef]
88. Archer, S.; Detling, J.K. Evaluation of potential herbivore mediation of plant water status in a north american mixed-grass prairie. *Oikos* **1986**, *47*, 287–291. [CrossRef]
89. Sluis, W.J. Patterns of species richness and composition in re-created grassland. *Restor. Ecol.* **2002**, *10*, 677–684. [CrossRef]
90. Baer, S.G.; Blair, J.M.; Collins, S.L.; Knapp, A.K. Plant community responses to resource availability and heterogeneity during restoration. *Oecologia* **2004**, *139*, 617–629. [CrossRef] [PubMed]
91. Wilson, G.W.T.; Hartnett, D.C. Interspecific variation in plant responses to mycorrhizal colonization in tallgrass prairie. *Am. J. Bot.* **1998**, *85*, 1732–1738. [CrossRef] [PubMed]
92. Smith, M.R.; Charvat, I.; Jacobson, R.L. Arbuscular mycorrhizae promote establishment of prairie species in a tallgrass prairie restoration. *Can. J. Bot.* **1998**, *76*, 1947–1954.
93. McKinley, V.L.; Peacock, A.D.; White, D.C. Microbial community plfa and phb responses to ecosystem restoration in tallgrass prairie soils. *Soil Biol. Biochem.* **2005**, *37*, 1946–1958. [CrossRef]
94. Barker, J.R.; Baumgardner, G.A.; Turner, D.P.; Lee, J.J. Potential carbon benefits of the conservation reserve program in the united states. *J. Biogeogr.* **1995**, *22*, 743–751. [CrossRef]
95. Baer, S.G.; Kitchen, D.J.; Blair, J.M.; Rice, C.W. Changes in ecosystem structure and function along a chronosequence of restored grasslands. *Ecol. Appl.* **2002**, *12*, 1688–1701. [CrossRef]
96. Allison, V.J.; Miller, R.M.; Jastrow, J.D.; Matamala, R.; Zak, D.R. Changes in soil microbial community structure in a tallgrass prairie chronosequence. *Soil Sci. Soc. Am. J.* **2005**, *69*, 1412–1421. [CrossRef]
97. Jesus, E.d.C.; Liang, C.; Quensen, J.F.; Susilawati, E.; Jackson, R.D.; Balser, T.C.; Tiedje, J.M. Influence of corn, switchgrass, and prairie cropping systems on soil microbial communities in the upper Midwest of the United States. *GCB Bioenergy* **2016**, *8*, 481–494. [CrossRef]
98. McNaughton, S.; Ruess, R.; Seagle, S. Large mammals and process dynamics in African ecosystems. *BioScience* **1988**, *38*, 794–800. [CrossRef]
99. Kitchen, D.J.; Blair, J.M.; Callaham, M.A. Annual fire and mowing alter biomass, depth distribution, and c and n content of roots and soil in tallgrass prairie. *Plant Soil* **2009**, *323*, 235–247. [CrossRef]

100. Ojima, D.S. The Short-Term and Long-Term Effects of Burning on Tallgrass Ecosystem Properties and Dynamics. Ph.D. Thesis, Colorado State University, Fort Collins, CO, USA, 1987.
101. Tilman, D. *Resource Competition and Community Structure*; Princeton University Press: Princeton, NJ, USA, 1982.

© 2018 by the authors. Licensee MDPI, Basel, Switzerland. This article is an open access article distributed under the terms and conditions of the Creative Commons Attribution (CC BY) license (http://creativecommons.org/licenses/by/4.0/).

MDPI
St. Alban-Anlage 66
4052 Basel
Switzerland
Tel. +41 61 683 77 34
Fax +41 61 302 89 18
www.mdpi.com

Agronomy Editorial Office
E-mail: agronomy@mdpi.com
www.mdpi.com/journal/agronomy

www.ingramcontent.com/pod-product-compliance
Lightning Source LLC
LaVergne TN
LVHW071942080526
838202LV00064B/6653